Advances in Fischer-Tropsch Synthesis, Catalysts, and Catalysis

CHEMICAL INDUSTRIES

A Series of Reference Books and Textbooks

Founding Editor

HEINZ HEINEMANN
Berkeley, California

Series Editor

JAMES G. SPEIGHT
CD & W, Inc.
Laramie, Wyoming

Advances in Fischer-Tropsch Synthesis, Catalysts, and Catalysis

Burtron H. Davis
Center for Applied Energy Research
Lexington, Kentucky, U.S.A.

Mario L. Occelli
MLO Consulting
Atlanta, Georgia, U.S.A.

CRC Press
Taylor & Francis Group
Boca Raton   London   New York

CRC Press is an imprint of the
Taylor & Francis Group, an **informa** business

CRC Press
Taylor & Francis Group
6000 Broken Sound Parkway NW, Suite 300
Boca Raton, FL 33487-2742

© 2010 by Taylor and Francis Group, LLC
CRC Press is an imprint of Taylor & Francis Group, an Informa business

No claim to original U.S. Government works

Printed in the United States of America on acid-free paper
10 9 8 7 6 5 4 3 2 1

International Standard Book Number: 978-1-4200-6256-4 (Hardback)

Library of Congress Cataloging-in-Publication Data

Advances in Fischer-Tropsch synthesis, catalysts, and catalysis / editors, B.H. Davis, Mario L. Occelli.
p. cm. -- (Chemical industries ; 127)
Includes bibliographical references and index.
ISBN 978-1-4200-6256-4 (hardcover : alk. paper)
1. Fischer-Tropsch process. 2. Catalysts. I. Davis, Burtron H., 1934- II. Occelli, Mario L., 1942- III. Title. IV. Series.

TP352.A38 2009
660'.2995--dc22   2009024442

Visit the Taylor & Francis Web site at
http://www.taylorandfrancis.com

and the CRC Press Web site at
http://www.crcpress.com

Contents

Preface

Interest in the Fischer-Tropsch synthesis (FTS) has undergone periods of significant development and periods of near neglect since its discovery by Fischer and Tropsch in 1924. It was developed in Germany during the 1930s in competition with direct liquefaction. Up to the end of the war in 1945, direct liquefaction was the dominant process for converting coal to transportation fuels; in Germany the FTS was used mainly for the production of chemicals and substitutes for natural products, including butter. After 1945, the commercial application of FTS in Germany rapidly declined, although Professor Kölbel, among others, continued work on the scientific and engineering aspects of the reaction.

Following WWII, efforts to develop commercial operations were initiated in the United States, led by Dobie Keith of Hydrocarbon Research, Inc. (HRI), and in South Africa. The U.S. effort was an industrial undertaking and, with the potential oil production in the Middle East and the rising cost of natural gas feedstock, was terminated in the mid-1950s. The effort in South Africa had the support of the government and continued to be developed. The South African government and Sasol Technology agreed upon a base price for the FTS product; when the price of oil was above that of the base price, Sasol would pay the government the difference, and when a lower price prevailed, the government would pay Sasol the difference. The South African payments were similar to the agreements that were in place in Germany prior to 1945. In both instances, the government had to make payments to the developers during the initial years of operation, but eventually the payments to the government began and, in Germany, greatly exceeded what was paid out initially by the government. Sasol succeeded in developing various processes for FTS and eventually became an independent company without government support. Sasol continues to grow and has recently opened a 35,000 bbl/day plant based on natural gas in Qatar. Shell, with some support from the Malaysian government, has developed a commercial plant based on natural gas with a capacity of about 15,000 bbl/day and is now constructing a plant in Qatar with a capacity of 70,000 bbl/day. PetroSA (formerly Mossgas) has developed, with government support, a commercial operation, based on Sasol technology, in South Africa that is based on natural gas.

Because of the immense initial investment, FTS has only been commercialized when some form of government support was available. In each instance, the developer generated a profitable technology and more than repaid the government subsidy.

The current high price for oil has again developed significant interest in FTS. The present volume is based on a symposium held during the 236th meeting of the American Chemical Society in Philadelphia in August 2008. The renewed interest in FTS was evidenced by the excellent attendance and lively discussions following each presentation.

The majority of the presentations can be grouped into three subject areas: catalyst preparation and activation, reaction mechanism, and process-related topics.

The presentations demonstrated that, while FTS has advanced in maturity, many issues remain concerning the preparation of increasingly active catalysts and the method of activation to attain the maximum catalytic activity and catalyst life. To assist in an understanding of the structural features and their changes, increasingly sophisticated characterization techniques are being applied, and some of them are represented in this volume. As usual, the reaction mechanism is never completely understood, and several of the presentations made efforts to advance our understanding of the reaction mechanism, both of the FTS reaction, and the water-gas shift reaction, which is used to adjust the H_2/CO ratio upward when low H_2/CO ratio syngas is available. Finally, commercialization requires advances of a more practical nature. For obvious reasons, industrial organizations are reluctant to provide details of advances in this area. However, several of the presentations provide glimpses into the processes and the reaction rates under realistic commercial process conditions.

The Editors

Dr. Burtron H. Davis received his PhD in physical chemistry from the University of Florida and was a research associate under Professor Paul H. Emmett at the Johns Hopkins University. Dr. Davis is responsible for catalysis, Fischer-Tropsch synthesis, and direct coal liquefaction research at the Center for Applied Energy Research, University of Kentucky. He developed a program that involved both academic research and cooperative research with industry. He has developed a laboratory with extensive capability in the use of radioactive and stable isotopes in reaction mechanism studies and materials characterization and developed research programs in Fischer-Tropsch synthesis, surface science studies, heterogeneous catalysis, materials science, organic analysis, one-fourth ton per day direct coal liquefaction pilot plant operation, liquefaction mechanistic studies, clean gasoline reforming with superacid catalysts, and upgrading naphthas. Dr. Davis has held various offices and memberships in several professional societies, including the American Chemical Society, the Catalysis Society, and the Materials Research Society. He is the recipient of the H. H. Storch Award for 2002, and is the author of more than six hundred technical publications.

Dr. Mario L. Occelli came to the United States in 1963 on a Fulbright scholarship to study chemical engineering at Iowa State University (ISU). After receiving a BS degree in chemical engineering in 1967 and a PhD degree in physical chemistry in 1972, he has spent his entire career researching the synthesis and characterization of microporous materials and their application in the preparation of fluid cracking catalysts (FCCs) and hydrocracking catalysts for the petrochemical and petroleum refining industry. While working at Gulf, UNOCAL, and the Georgia Tech Research Institute, his dual background in physical chemistry and chemical engineering facilitated his participation and contribution in multidisciplinary research projects involving chemists, surface scientists, physicists, material scientists, and chemical engineers working in national and industrial laboratories and in academia. As a result of these activities, he has published extensively, received thirty U.S. patents, and presented papers and lectures at national and international meetings. Currently, he works as an independent consultant.

Contributors

Zuzana Ballova
Dipartimento di Energia
Politecnico di Milano
Milan, Italy

Peter T. Bishop
Johnson Matthey Technology Centre
Reading, United Kingdom

Christopher J. Brooks
Honda Research Incorporated USA
Columbus, Ohio

John L. Casci
Johnson Matthey Technology Centre
Billingham, United Kingdom

Neil J. Coville
School of Chemistry
University of the Witwatersrand
Johannesburg, South Africa

Donald Cronauer
Chemical Sciences and Engineering
    Division
Argonne National Laboratory
Argonne, Illinois

Burtron H. Davis
Center for Applied Energy Research
University of Kentucky
Lexington, Kentucky

Arno de Klerk
Sasol Technology R&D Ltd.
Sasolburg, South Africa
and
Department of Chemical and
    Materials Engineering
University of Alberta
Edmonton, Canada

Peter R. Ellis
Johnson Matthey Technology Centre
Reading, United Kingdom

Pio Forzatti
Dipartimento di Energia
Politecnico di Milano
Milan, Italy

Johann Gaube
Ernst Berl-Institut für Technische and
    Makromolekulare Chemie
Technische Universität Darmstadt
Darmstadt, Germany

David Glasser
School of Chemical and Metallurgical
    Engineering
University of the Witwatersrand
Johannesburg, South Africa

Muthu K. Gnanamani
Center for Applied Energy Research
University of Kentucky
Lexington, Kentucky

Uschi M. Graham
Center for Applied Energy Research
University of Kentucky
Lexington, Kentucky

Thelma Grobler
Sasol Technology R&D Ltd.
Sasolburg, South Africa

Yasuhiko Hayasaka
Department of Applied Chemistry
Graduate School of Engineering
Tohoku University
Aramaki, Aoba-ku, Sendai, Japan

Diane Hildebrandt
School of Chemical and Metallurgical
  Engineering
University of the Witwatersrand
Johannesburg, South Africa

Daichi Hongo
Department of Applied Chemistry
Graduate School of Engineering
Tohoku University
Aramaki, Aoba-ku, Sendai, Japan

Johan Huyser
Sasol Technology R&D Ltd.
Sasolburg, South Africa

Yukiya Ibi
Department of Applied Chemistry
Graduate School of Engineering
Tohoku University
Aramaki, Aoba-ku, Sendai, Japan

Megumu Inaba
National Institute of Advanced
  Industrial Science & Technology
Tsukuba, Ibaraki, Japan

Gary Jacobs
Center for Applied Energy Research
University of Kentucky
Lexington, Kentucky

David James
Johnson Matthey Technology Centre
Reading, United Kingdom

Matthys Janse van Vuuren
Sasol Technology R&D Ltd.
Sasolburg, South Africa

Andreas Jess
Department of Chemical Engineering
University Bayreuth
Bayreuth, Germany

Yaying Ji
Center for Applied Energy Research
University of Kentucky
Lexington, Kentucky

Anke Jung
Department of Chemical Engineering
University Bayreuth
Bayreuth, Germany

Gordon J. Kelly
Johnson Matthey Catalysts
Billingham, United Kingdom

C. Kern
Department of Chemical Engineering
University Bayreuth
Bayreuth, Germany

Syed Khalid
National Synchrotron Light Source
Brookhaven National Laboratory
Upton, New York

Hans-Friedrich Klein
Eduard Zintl-Institut für Anorganische
  und Physikalische Chemie
Technische Universität Darmstadt
Darmstadt, Germany

Naoto Koizumi
Department of Applied Chemistry
Graduate School of Engineering
Tohoku University
Aramaki, Aoba-ku, Sendai, Japan

A. Jeremy Kropf
Chemical Sciences and Engineering
  Division
Argonne National Laboratory
Argonne, Illinois

Godfrey Kupi
Sasol Technology R&D Ltd.
Sasolburg, South Africa

Luca Lietti
Dipartimento di Energia
Politecnico di Milano
Milan, Italy

C. Martin Lok
Johnson Matthey Technology Centre
Billingham, United Kingdom

Wenping Ma
Center for Applied Energy Research
University of Kentucky
Lexington, Kentucky

Christopher L. Marshall
Chemical Sciences and Engineering
    Division
Argonne National Laboratory
Argonne, Illinois

Denzil J. Moodley
Eindhoven University of Technology
Eindhoven, The Netherlands

Peter Mukoma
School of Chemical and Metallurgical
    Engineering
University of the Witwatersrand
Johannesburg, South Africa

Kazuhisa Murata
National Institute of Advanced
    Industrial Science & Technology
Tsukuba, Ibaraki, Japan

James K. Neathery
Center for Applied Energy Research
University of Kentucky
Lexington, Kentucky

Hans J. W. Niemantsverdriet
Eindhoven University of Technology
Eindhoven, The Netherlands

Mohammad Nurunnabi
National Institute of Advanced
    Industrial Science & Technology
Tsukuba, Ibaraki, Japan

Kiyomi Okabe
National Institute of Advanced
    Industrial Science & Technology
Tsukuba, Ibaraki, Japan

Bilal Patel
Centre of Material and Process
    Synthesis (COMPS)
University of the Witwatersrand
Johannesburg, South Africa

John M. Pigos
Honda Research Incorporated USA
Columbus, Ohio

Abdool M. Saib
Sasol Technology R&D Ltd.
Sasolburg, South Africa

Amitava Sarkar
Imperial Oil Resources
Calgary, Alberta, Canada

Hans Schulz
University of Karlsruhe
Engler-Bunte-Institute
Karlsruhe, Germany

Robert L. Spicer
Center for Applied Energy Research
University of Kentucky
Lexington, Kentucky

Isao Takahara
National Institute of Advanced
    Industrial Science & Technology
Tsukuba, Ibaraki, Japan

Enrico Tronconi
Dipartimento di Energia
Politecnico di Milano
Milan, Italy

Jan van de Loosdrecht
Sasol Technology R&D Ltd.
Sasolburg, South Africa

Carlo Giorgio Visconti
Dipartimento di Energia
Politecnico di Milano
Milan, Italy

Muneyoshi Yamada
Department of Applied Chemistry
Graduate School of Engineering
Tohoku University
Aramaki, Aoba-ku, Sendai, Japan

Roberto Zennaro
Exploration & Production Division
Eni S.p.A.
San Donato Milanese, Italy

1 Synthesis of High Surface Area Cobalt-on-Alumina Catalysts by Modification with Organic Compounds

Peter R. Ellis, David James, Peter T. Bishop, John L. Casci, C. Martin Lok, and Gordon J. Kelly

CONTENTS

Cobalt-on-alumina catalysts with increased dispersion and catalytic activity are prepared by addition of mannitol to the cobalt nitrate solution prior to impregnation. Thermogravimetric analysis (TGA) and *in situ* visible microscopy of the impregnation solution show that the organic compound reacts with cobalt nitrate, forming a foam. The foam forms because significant amounts of gas are released through a viscous liquid. The structure of the foam is retained in the final calcined product. It is this effect that is responsible for the increased dispersion.

1.1   INTRODUCTION

Cobalt catalysts are preferred for the Fischer-Tropsch (FT) reaction in many applications because they give greater yields of straight-chain alkanes than iron catalysts [1,2]. These catalysts are typically prepared by cobalt nitrate impregnation, followed by a low-temperature drying step and calcination at a higher temperature to decompose cobalt nitrate. The product is a supported cobalt oxide, which is then reduced to give the active cobalt metal phase. The final cobalt particle size is critical in determining the activity of the catalyst. Bezemer and coworkers [3] showed that the optimum cobalt particle size for a Co/carbon nanofiber catalyst is 6 to 8 nm. Below 6 nm, a rapid decrease in catalytic activity and an increase in methane selectivity were observed with decreasing particle size. A typical cobalt/ alumina catalyst prepared via nitrate impregnation might have an average cobalt particle size nearer to 11 nm, depending on a number of factors, such as loading and reduction conditions [4]. Despite this, impregnation with cobalt nitrate is a preferred route for the preparation of cobalt FT catalysts. Hence, methods that decrease cobalt particle size are needed for the preparation of improved FT catalysts prepared by cobalt nitrate impregnation.

There are a number of ways in which the dispersion of a cobalt-based FT catalyst can be increased. One of these is to change support. The cobalt dispersion obtained on a transition alumina-supported catalyst has been shown to be greater than that obtained using silica, titania [5,6], or alpha alumina [7]. These references also show that the cobalt loading used in the catalyst has an effect on the dispersion produced. Another method of increasing the dispersion is to add a promoter. Many promoters for cobalt FT catalysts are described in the literature, and a diverse range of those, including rhodium [8], manganese [9], zirconium [10], zinc [11], and platinum [12], have been shown to increase the dispersion of supported cobalt catalysts compared with their unpromoted equivalents. The catalyst preparation method also has an impact on the catalyst dispersion. Cobalt nitrate impregnation is generally regarded as giving catalysts with low dispersions, while other technologies, such as Johnson Matthey's Highly Displaced Cobalt (HDC) method, give high dispersions even at high cobalt loadings [13–15]. Many attempts have been made within the materials science community to synthesize small crystallites of cobalt [16] or cobalt oxide [17,18], but few of these have been investigated as catalysts for the Fischer-Tropsch reaction.

One way in which cobalt dispersion can be increased is the addition of an organic compound to the cobalt nitrate prior to calcination. Previous work in this area is summarized in Table 1.1. The data are complex, but there are a number of factors that affect the nature of the catalyst prepared. One of these is the cobalt loading. Preparation of catalysts containing low levels of cobalt tends to lead to high concentrations of cobalt-support compounds. For example, Mochizuki et al. [37] used x-ray photoelectron spectroscopy (XPS) and temperature-programmed reduction (TPR) to identify cobalt silicate-like species in their 5% Co/SiO$_2$ catalysts modified with nitrilotriacetic acid (NTA). The nature of the support also has

TABLE 1.1

Summary of Relevant Literature on Organic Modification of Cobalt Catalysts

Organic Compound	Catalyst	Reference	Notes
Simple carboxylic acids (e.g., acetic acid)	10% Co/SiO$_2$	[19]	Organic added to support before cobalt impregnation
	10% Co/SiO$_2$	[20]	Washing of catalyst after impregnation
	10% Co/SiO$_2$	[21, 22]	
Simple alcohols (e.g., ethanol)	10% Co/SiO$_2$	[19]	
	9% Co/SiO$_2$	[23]	
	20% Co/SiO$_2$	[24]	
	6% Co/SiO$_2$	[25]	
	10% Co/SiO$_2$	[22]	
	3% Co/C		
	10% Co/C	[26]	
Sugars (e.g., sucrose)	3% Co/ZrO$_2$	[27]	Violent reaction observed between saccharose and cobalt nitrate that damaged the catalyst particle
	7% Co/TiO$_2$	[28]	
	Re-17% Co/TiO$_2$	[28]	Decreased methane selectivity observed on sucrose addition
	Re-17% Co/SiO$_2$		
	Re-7% Co/SiO$_2$	[29]	
	Co/SiO$_2$	[30]	
	Re-7% Co/SiO$_2$	[29]	
DMSO	9% Co/SiO$_2$	[23]	Complex formed
DMF	9% Co/SiO$_2$	[23]	Complex formed
Polyols (e.g., sorbitol)	7% Co/TiO$_2$	[31]	
	0.8% Re-18% Co/SiO$_2$	[31]	
Polyacids (e.g., citric acid)	10% Co/Al$_2$O$_3$	[32]	Reaction tested was methane combustion
	7% Co/TiO$_2$		
	Re-12%Co/TiO$_2$		
	17% Co/Al$_2$O$_3$		
	17% Co/SiO$_2$		
	Re-17% Co/TiO$_2$	[33]	Addition of Re reduction promoter increased activity markedly
NTA, CyTA	1% Co/Al$_2$O$_3$		
	5% Co/Al$_2$O$_3$	[34]	Reaction tested was reduction of NO to N$_2$ by propene

TABLE 1.1

Summary of Relevant Literature on Organic Modification of Cobalt Catalysts (continued)

Organic Compound	Catalyst	Reference	Notes
	1–21% Co/Al$_2$O$_3$	[35]	Reaction tested was complete oxidation of benzene to CO$_2$
	5% Co/SiO$_2$	[36, 37]	Stepwise and co-impregnation
Ethylenediamine	12% Co/Al$_2$O$_3$	[38]	Complex formed
	3% Co/SiO$_2$	[39]	Complex formed
Mixed cobalt nitrate-acetate	10% Co/SiO$_2$	[40]	

The structures of the organic compounds are shown in Figure 1.1.

an impact. Interestingly, the majority of the work in this area has been performed using a silica support, although the increased dispersion has also been observed for alumina-, titania-, and zirconia-supported catalysts (Table 1.1). The nature of the organic modifier has a significant influence on the catalyst prepared. For instance, some compounds, such as ethylenediamine [38,39] or NTA [29,35–37], are shown to complex to cobalt. In contrast, Mauldin [33] reports that no complex is formed between cobalt and either glutamic acid or citric acid. Also, the decomposition temperature of the organic can vary. For instance, acetic acid [21] is a liquid at room temperature and boils at 117.9°C, while citric acid [33] is solid at room temperature and melts at 153°C [41]. The decomposition of citric acid is complex and many decomposition products are observed, depending on the environment and temperature [42]. In practical terms, this means that the organic may decompose during the drying or calcination steps of catalyst preparation. Mauldin [31] reported that exothermic decomposition of organics during drying was disadvantageous to the properties of the final catalyst. A number of authors report that the organic-modified catalysts are difficult to reduce. For instance, van Steen et al. [23] observed lower degrees of reduction by TPR for 9% Co/SiO$_2$ catalysts modified using alcohols. This is countered in some cases by the addition of reduction promoters to make the catalyst easier to reduce [33]. Preparation of a catalyst using mixed cobalt nitrate and cobalt acetate is also relevant here [40] and yields increased dispersion over catalysts prepared using either precursor singly. Cobalt acetate alone tends to yield catalysts with high levels of cobalt-support compounds [21]. The nature of the support is important in determining the levels of cobalt-support compound observed [43]. Interestingly, the effect of organic modification has also been observed outside of supported FT catalysts. Soled et

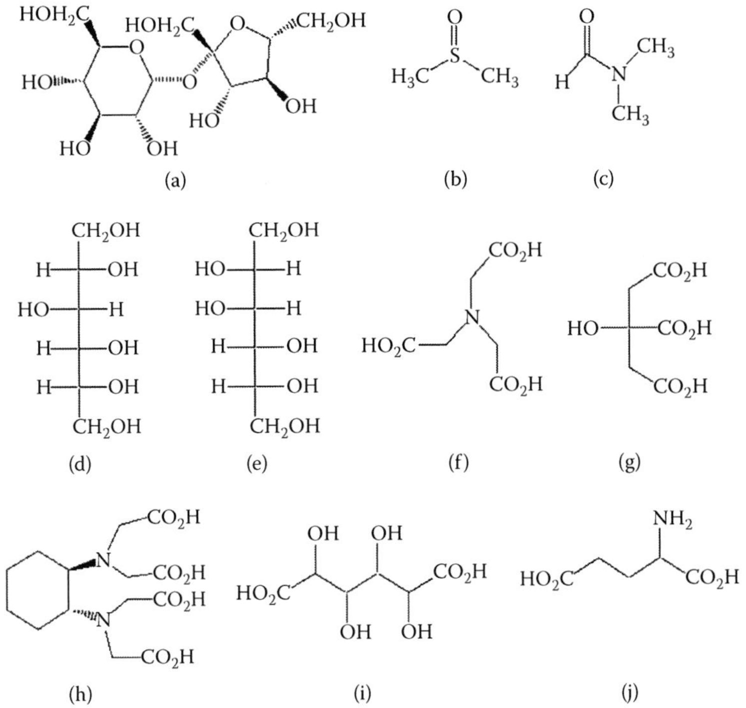

FIGURE 1.1 Structures of organic compounds referred to in the text: (a) sucrose (also known as saccharose), (b) dimethyl sulfoxide (DMSO), (c) dimethylformamide (DMF), (d) sorbitol, (e) mannitol, (f) nitrilotriacetic acid (NTA), (g) citric acid, (h) N,N,N′,N′-trans-1,2-diaminocyclohexane-tetraacetic acid (CyTA), (i) saccharic acid, (j) glutamic acid.

al. [44] prepared CoMn spinels with increased surface area using citric acid as an additive.

A number of mechanisms for the increase in dispersion have been proposed. Many authors talk about modification of the interaction between cobalt and the support. Van Steen et al. [23] and Zhang et al. [19] describe organic solvents modifying the silanol OH group, which changes the interaction between Co^{2+} and the support. It is clear that some organic compounds will complex with Co^{2+}, especially those with nitrogen donor atoms in their structure [29,36–39]. Nitrogen donor complexes of cobalt are well known [45], not least in biological systems such as vitamin B_{12}. Other authors report that the addition of acetic acid to the catalyst can form cobalt acetate *in situ* [21], giving an increased dispersion as for the mixed cobalt acetate–cobalt nitrate system [40]. Culross and Mauldin [28,33] propose a mechanism where the organic compound forms a "blanket" over the cobalt particles and prevents sintering. Girardon et al. [29] added sucrose to their promoted 7% Co/SiO$_2$ catalysts and assigned the increased dispersion to changes in the decomposition and nucleation mechanisms. They also noted that the oxidation of sucrose to saccharic acid by nitric ions is known. Saccharic acid may polymerize on the catalyst surface, and thus template the metal particles formed. They also used TGA to observe that the decomposition of nitrate became exothermic in the presence of sucrose, while it was endothermic without.

Outside of catalyst preparation, reaction of sucrose with metal nitrates has been used to prepare nanocomposite mixed oxide materials. Wu et al. [46] reported the synthesis of $MgO-Al_2O_3$ and $Y_2O_3-ZrO_2$ mixed oxides by reaction of nitrate precursors with sucrose. The resulting powders had smaller particles than those prepared without sucrose. Das [47] used a similar method in the presence of polyvinylalcohol to produce nanocrystalline lead zirconium titanate and metal ferrierites (MFe_2O_4, M = Co, Ni, or Zn). The materials prepared using sucrose had smaller crystallites than those made without. Both authors observed an exothermic decomposition of the precursors during calcination.

As described above, understanding the mechanism of the dispersion increase is a difficult task. In this work we compare a catalyst prepared by cobalt nitrate impregnation onto alumina with one modified by the addition of mannitol, and use TGA and *in situ* microscopy to investigate the increased dispersion. Mannitol is a sugar alcohol that is structurally similar to sorbitol [31], as shown in Figure 1.1.

1.2   EXPERIMENTAL

1.2.1   MATERIALS

Cobalt nitrate hexahydrate (97%) and D-mannitol (99%) were obtained from Alfa Aesar and used as received. Alumina HP14-150 is a gamma alumina with surface area around 150 $m^2\ g^{-1}$, which was supplied by Sasol and used as received.

1.2.2   CATALYST PREPARATION

The unmodified catalyst was prepared by dissolving cobalt nitrate hexahydrate (19.76g, 68 mmol) in water (9 ml) and impregnating the solution into alumina (16.0 g). The catalyst was dried at 105°C in static air for 3 h and calcined at 400°C for 1 h. The mannitol-containing catalyst was prepared as described above, except that mannitol (2 g, 11 mmol) was dissolved in the cobalt nitrate solution prior to impregnation. The Co/mannitol ratio was 6.2.

1.2.3   CATALYST CHARACTERIZATION

The catalysts were characterized by inductively coupled plasma emission spectroscopy (ICP-ES; Perkin Elmer Optima 3300RL) to determine cobalt content, x-ray diffraction (XRD; Bruker A-500) with crystallite size determination using the Rietveld method, and temperature-programmed reduction (Zeton Altamira AMI-200) using 30 ml/min 10% H_2/Ar and a ramp rate of 10°C/min. Surface area measurements were by hydrogen chemisorption (Micromeretics ASAP 2010), where the catalysts were reduced at 425°C for 7 h prior to evacuation and analysis at 150°C. X-ray photoelectron spectroscopy (VG Escalab 250) was recorded using samples of the powder pressed into pellets. Spectra were referenced to adventitious carbon at 286.4 eV. Samples were prepared for transmission electron microscopy (TEM; Tecnai F20) by setting in resin, sectioning, and drying onto a holey carbon-coated copper TEM grid.

1.2.4 CATALYST TESTING

Catalysts were tested for activity in the Fischer-Tropsch reaction using a fixed-bed reactor. The catalyst (0.4 g) was reduced *in situ* in flowing hydrogen at 425°C for 7 h prior to testing. The test was performed under 2/1 H_2/CO at 20 bar total pressure. The initial flow was 64 ml/min, but this was reduced after 24 h to increase the conversion. A final reading of activity and selectivity was taken after 100 h on stream.

1.2.5 *IN SITU* MICROSCOPY AND THERMOGRAVIMETRIC ANALYSIS

The microscopy and thermogravimetric analysis were performed using solutions of cobalt nitrate hexahydrate (2 g) in water (1 g) with or without mannitol (0.1 g). When used, the Co/mannitol molar ratio was 12.5. TGA (TA Instruments SDT2960) was performed by heating to 300°C at 1°C/min in flowing air. Microscopy was performed on a single drop of the solution (approximately 2 mm diameter) placed into a quartz sample holder. A Linkam heating stage (THMS 600) was used to heat the sample at 1°C/min to 300°C, with digital images recorded during the temperature ramp. Visible spectroscopy (Unicam) was performed from 400 to 600 nm wavelengths. The concentration of cobalt nitrate in the solutions was approximately 0.03 M. Fourier transform infrared spectra (FTIR; Perkin Elmer Spectrum One) were obtained in the attenuated total reflectance mode using solid-state samples.

1.3 RESULTS AND DISCUSSION

1.3.1 CATALYST PREPARATION, CHARACTERIZATION, AND TESTING

Catalysts were prepared with or without addition of mannitol, as described in the experimental section. The presence of cobalt nitrate means that catalyst preparation takes place at low pH, making it a complementary method of producing well-dispersed cobalt catalysts to Johnson Matthey's high-pH HDC technology [13–15]. Both catalysts were characterized by XRD and hydrogen chemisorption, and had cobalt content measured by ICP (Table 1.2). XRD only detected two phases—the gamma alumina support and cobalt oxide, Co_3O_4. The absence of Co and CoO as crystalline phases suggests that there is no reduction of cobalt oxide by the mannitol; instead, some cobalt is oxidized by nitrate or air from Co^{II} to

TABLE 1.2
Characterization of Catalysts

Co/Mannitol	Co Content/%	Co$_3$O$_4$ Crystallite Size (XRD)/nm	Cobalt Surface Area	
			m^2 g$_{cat}^{-1}$	m^2 g$_{Co}^{-1}$
0	19.3	12.4	8.8	42.3
6.2	17.8	4.0	16.9	84.5

Co^{III} to form Co$_3$O$_4$. The catalyst containing mannitol was found to have smaller Co$_3$O$_4$ crystallites before reduction and a higher cobalt surface area (greater Co dispersion) afterwards (Table 1.2), in agreement with literature precedent. This was also evident from dark-field TEM images (Figure 1.2). TPR demonstrated that the mannitol-modified catalyst was significantly harder to reduce than the unmodified comparator (Figure 1.3). The peak temperature of the higher temperature (Co^{2+} → Co^0 [4]) peak is shifted from 500°C to 665°C by the mannitol treatment. There is also a significant increase in the amount of cobalt-support compound, which is reduced above 800°C. Again, this has been reported previously for catalysts modified by organic compounds [48]. XPS analysis of the two catalysts found little difference in the levels of carbon on the catalyst surface. This suggests that the mannitol had been removed during the calcination step.

The mannitol-modified catalyst showed significantly increased activity in the Fischer-Tropsch reaction (Table 1.3). After 20 h on stream, the mannitol-modified catalyst is 286% as active as the unmodified catalyst, and 262% as active after 100 h. The selectivity values of the catalysts are similar. Hence, the characteristics of a mannitol-modified catalyst are that it has a higher activity than but the same selectivity as an unmodified catalyst. Increased activity for the FT reaction

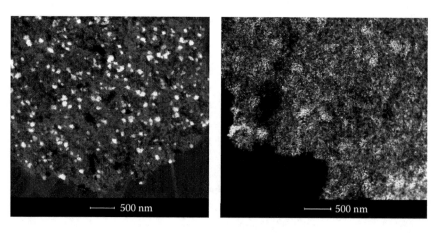

FIGURE 1.2 Dark-field TEM images of (a) unmodified and (b) mannitol-modified catalysts.

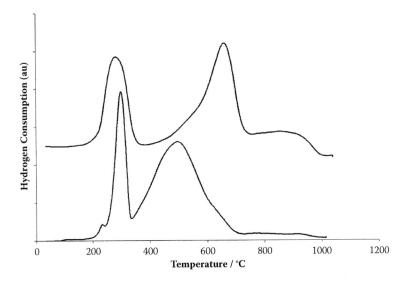

FIGURE 1.3   TPR analysis of unmodified (bottom) and mannitol-modified (top) catalysts.

TABLE 1.3
Fischer-Tropsch Test Data

Co/ Mannitol	GHSV/l kg_{cat}^{-1} h^{-1}	CO Conversion/%	Selectivity/%			
			CO_2	CH_4	C_2–C_4	C_{5+}
		Data after 20 h On Stream				
0	9600	6.1	0.57	11.0	8.40	80.0
6.2	9600	18.0	0.10	9.63	8.98	81.3
		Data after 100 h On Stream				
0	4800	11.6	0.50	11.3	7.92	80.3
6.2	4800	30.5	0.13	9.63	10.3	80.0

has been reported by many authors, often in line with increased dispersion. The effect on selectivity is less well understood. Culross and Mauldin [28(a)] report a decrease in methane selectivity on sucrose modification of rhenium-promoted 16% Co/TiO_2 catalysts. For rhenium-promoted 18% Co/SiO_2 catalysts, they observed a decrease in methane selectivity up to a concentration of 5.71% sucrose in the impregnating solution, above which level it increased back to the value of the unmodified catalyst. Mauldin [31] observed the same effect for sorbitol modification of 0.8% Re-16% Co/SiO_2 catalysts, the maximum being at 6.5% sorbitol in solution. Liu et al [21] also observed a decrease in methane selectivity on acetic acid addition to Co/SiO_2. However, Zhang et al. [24] found decreased methane selectivity on modification with ethanol, acetone, DMF, or THF, but sharply increased methane selectivity on modification with cyclohexanol. The addition of propanol or methanol did not affect the methane selectivity. Increased methane selectivity has also been reported for catalysts modified with NTA [36,37].

1.3.2 ANALYSIS OF COBALT NITRATE SOLUTIONS

Cobalt nitrate solution was analyzed by TGA and found to decompose [49,50] in several steps (Figure 1.4). The first step is the loss of solution water, giving the hydrated salt. This also loses water on heating, until anhydrous cobalt nitrate is present. At this point, decomposition of the nitrate anion begins immediately. Differential thermal analysis (DTA) shows that all these steps are endothermic, and that the loss of the last two water molecules is a complex multistage process. By 300°C, the decomposition is complete. The mass remaining is that of cobalt oxide, Co_3O_4. FTIR confirms the presence of Co_3O_4 with bands at 660 and 557 cm^{-1} [51]. The addition of mannitol changes the TGA profile significantly. The most striking feature is a period of noise around 80°C, which is assigned to the production of a large amount of gas. In catalyst preparation terms, this would occur during the drying stage. Boot et al. [27] observed a violent reaction between cobalt nitrate and saccharose, which caused damage to the catalyst particle. The mass loss during this event is approximately 16 wt%. As the solution contains around 3 wt% mannitol, the gas released is likely to contain nitrogen oxides and vaporized water that was coordinated to cobalt, as well as the decomposition products of mannitol oxidation. The decomposition of nitrate occurs at a slightly lower temperature than in the unmodified case. The decomposition is more rapid in the presence of mannitol, but it is still endothermic, in contrast to the results of Girardon et al. [29], who observed an exothermic decomposition, albeit with a different system (promoted 7% Co/SiO$_2$ modified with sucrose).

The two solutions were also analysed by *in situ* microscopy. The solution that did not contain mannitol showed a number of different stages of decomposition, as shown in Figure 1.5. The initial solution is a pale red color. The color deepens on heating, and the droplet appears to solidify (Figure 1.5b), followed by a crystallization (Figure 1.5c). Above 100°C, loss of water can be observed as a bubble of water vapor breaks through the surface of the droplet. The final product was a black pellet, confirmed as Co_3O_4 by its FTIR spectrum.

The solution containing mannitol behaves quite differently (Figure 1.6). The initial solution is very similar to the solution without mannitol. However, in this case crystallization is not observed, but the production of a large amount of gas is (Figure 1.6b and c). This is presumably largely CO$_2$ and a small amount of NO$_x$ from the reaction of nitrate with mannitol. The evolution of the gas causes a foam to form (Figure 1.6c). This all happens below 100°C; in terms of catalyst preparation, it would occur in the drying step. The structure of the foam is retained to a large extent during the subsequent drying and calcination processes.

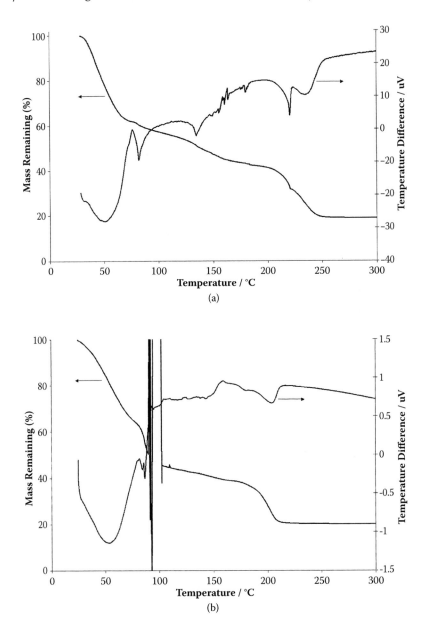

FIGURE 1.4 Decomposition of (a) cobalt nitrate solution and (b) cobalt nitrate–mannitol solution, measured by TGA and DTA.

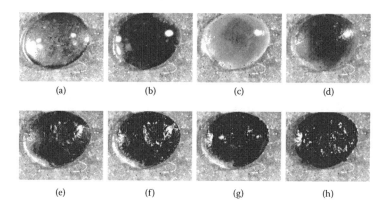

FIGURE 1.5 (**See color insert following page 12.**) Cobalt nitrate solution decomposition at (a) 21°C, (b) 59°C, (c) 67°C, (d) 110°C, (e) 126°C, (f) 142°C, (g) 152°C, and (h) 200°C.

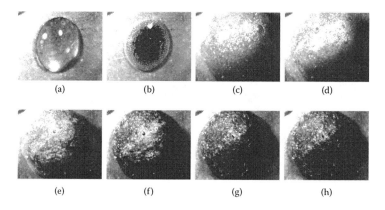

FIGURE 1.6 (**See color insert following page 12.**) Cobalt nitrate–mannitol solution decomposition at (a) 21°C, (b) 61°C, (c) 75°C, (d) 123°C, (e) 159°C, (f) 185°C, (g) 220°C, and (h) 250°C.

The final product is a fine powder that is shown to be Co_3O_4 by FTIR. The paler color observed at the top of the particle in Figure 1.6g arises because the sample is heated from below.

It is the formation of the foam and its associated release of gas that is responsible for the large noise event in the TGA. The two events occur at the same temperature. It is also responsible for the increase in dispersion observed in the related catalysts.

Visible spectroscopy (Figure 1.7) shows that mannitol does not form a complex with cobalt nitrate hexahydrate in water, even when heated to reflux. This

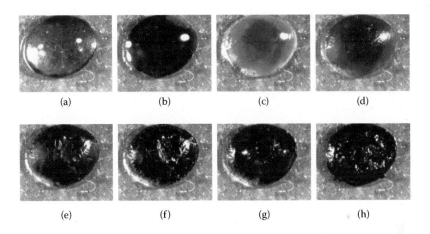

FIGURE 1.5 Cobalt nitrate solution decomposition at (a) 21°C, (b) 59°C, (c) 67°C, (d) 110°C, (e) 126°C, (f) 142°C, (g) 152°C, and (h) 200°C.

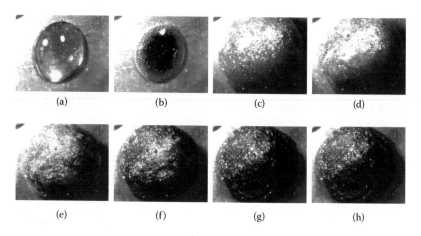

FIGURE 1.6 Cobalt nitrate–mannitol solution decomposition at (a) 21°C, (b) 61°C, (c) 75°C, (d) 123°C, (e) 159°C, (f) 185°C, (g) 220°C, and (h) 250°C.

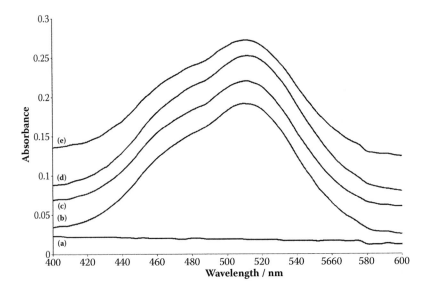

FIGURE 1.7   Visible spectroscopy of (a) mannitol solution; (b) cobalt nitrate solution; (c) cobalt nitrate with six equivalents of mannitol; (d) cobalt nitrate with six equivalents of mannitol after heating to boiling; and (e) cobalt nitrate solution with excess mannitol.

is believed to be a general trend for oxygen donor ligands, which are unable to displace the six water molecules coordinated to the Co^{2+} ion in aqueous solution. The same is not true of nitrogen donor ligands, for instance, where complexes are well known [45].

1.4   CONCLUSION

The addition of mannitol to the catalyst increases the dispersion and hence the activity of the catalyst. The reason for this is the formation of a foam when large amounts of CO_2 and NO_x gas are released following reaction of mannitol with the nitrate anion at temperatures around 80°C. We believe that the formation of this foam in the alumina pores is responsible for the increased dispersion observed in the mannitol-modified catalyst.

ACKNOWLEDGMENTS

The authors thank Matt Gregory and Ian Briggs for ICP analysis, Rob Fletcher and Sandra Riley for hydrogen chemisorption, James McNaught for XRD, Gregory Goodlet for TEM, and Richard Smith for XPS. We also thank Johnson Matthey for permission to publish this work.

REFERENCES

1. Khodakov A.Y., Chu W., and Fongarland P. 2007. Advances in the development of novel cobalt Fischer-Tropsch catalysts for synthesis of long-chain hydrocarbons and clean fuels. *Chem. Rev.* 107:1692–744.
2. Dry M.E. 2004. Present and future applications of the Fischer-Tropsch process. *Appl. Catal. A Gen.* 276:1–3.
3. Bezemer G.L., Bitter J.H., Kuipers H.P.C.E., Oosterbeek H., Holewijn J.E., Xu X., Kapteijn F., van Dillon A.J., and de Jong K.P. 2006. Cobalt particle size effects in the Fischer-Tropsch reaction studied with carbon nanofibre supported catalysts. *J. Am. Chem. Soc.* 128:3956–64.
4. Jacobs G., Ji Y., Davis B.H., Cronauer D., Kropf J., and Marshall C.L. 2007. Fischer-Tropsch synthesis: Temperature programmed EXAFS/XANES investigation of the influence of support type, cobalt loading and noble metal promoter addition to the reduction behaviour of cobalt oxide particles. *Appl. Catal. A Gen.* 333:179–91.
5. Jacobs G., Das T.K., Zhang Y., Li J., Racoillet G., Davis B.H. 2002. Fischer-Tropsch synthesis: Support, loading and promoter effects on the reducibility of cobalt catalysts. *Appl. Catal. A Gen.* 233:263–81.
6. Storsaeter S., Totdal B., Walmsley J.C., Tanem B.S., and Holmen A. 2005. Characterisation of alumina-, silica- and titania-supported cobalt Fischer-Tropsch catalysts. *J. Catal.* 236:139–52.
7. Eri S., Kinnari K.J., Dag S., and Hilmen A.M. 2008. Fischer Tropsch catalyst, preparation, and use thereof. U.S. Patent 7351679B2.
8. Van't Blik H.F.J. and Prins R. 1986. Characterisation of supported cobalt and cobalt-rhodium catalysts. 1. Temperature-programmed reduction (TPR) and oxidation (TPO) of Co-Rh/Al$_2$O$_3$. *J. Catal.* 97:188–99.
9. Morales F., de Smit E., de Groot F.M.F., Visser T., and Weckhuysen B.M. 2007. Effects of manganese oxide promoter on the CO and H$_2$ adsorption properties of titania-supported cobalt Fischer-Tropsch catalysts. *J. Catal.* 246:91–99.
10. Xiong H., Zhang Y., Liew K., and Li J. 2005. Catalytic performance of zirconium modified Co/Al$_2$O$_3$ for Fischer-Tropsch synthesis. *J. Mol. Catal. A Chem.* 231:145–51.
11. Madikizela-Mnqanqeni N.N. and Coville N.J. 2004. Surface and reactor study of zinc on titania-supported Fischer-Tropsch cobalt catalysts. *Appl. Catal. A Gen.* 272:339–46.
12. Schanke D., Vada S., Blekkan E.A., Hilmen A.M., Hoff A., and Holmen A. 1995. Study of Pt-promoted cobalt CO hydrogenation catalysts. *J. Catal.* 156:85–95.
13. Lok C.M., Bailey S., and Gray G. 2003. Method for the production of cobalt catalysts supported on silicon dioxide and their use. U.S. Patent 6534436B2.
14. Bonne R.L. and Lok C.M. 1999. Cobalt on alumina catalysts. U.S. Patent 5874381.
15. Lok C.M., Kelly G.J., and Gray G. 2005. Catalysts with high surface area. U.S. Patent 6927190.
16. Lisiecki I. and Pileni M.P. 2003. Synthesis of well-defined and low size distribution cobalt nanocrystals: The limited influence of reverse micelles. *Langmuir* 19:9486–89.
17. Sugimoto T. and Matuevic E. 1979. Colloidal cobalt hydrous oxides. Preparation and properties of monodispersed Co$_3$O$_4$. *J. Inorg. Nucl. Chem.* 41:165–72.
18. Xu R. and Zeng H.C. 2003. Mechanistic investigation on salt-mediated formation of free-standing Co$_3$O$_4$ nanocubes at 95°C. *J. Phys. Chem. B* 107:926–30.

19. Zhang Y., Hanayama K., and Tsubaki N. 2006. The surface modification effects of silica support by organic solvents for Fischer-Tropsch synthesis catalysts. *Catal. Commun.* 7:251–54.
20. Tsubaki N., Sakota H., and Takahashi S. 2004. Production method of Fischer-Tropsch synthesis catalyst and hydrocarbon. Japanese Patent Application JP2004237254.
21. Liu Y., Zhang Y., and Tsubaki N. 2007. The effect of acetic acid pretreatment for cobalt catalysts prepared from cobalt nitrate. *Catal. Commun.* 8:773–76.
22. Zhang Y., Hanayama K., and Tsubaki N. 2006. The surface modification effects of silica support by organic solvents for Fischer-Tropsch synthesis catalysts. *Catal. Commun.* 7:251–54.
23. van Steen E., Sewell G.S., Makhothe R.A., Micklethwaite C., Manstein H., de Lange M., and O'Connor C.T. 1996. TPR study on the preparation of impregnated Co/SiO$_2$ catalysts. *J. Catal.* 162:220–29.
24. Zhang Y., Liu Y., Yang G., Sun S., and Tsubaki N. 2007. Effect of impregnation solvent on Co/SiO$_2$ catalyst with bimodal sized cobalt particles. *Appl. Catal. A Gen.* 321:79–85.
25. Ho S.-W. and Su Y.-S. 1997. Effects of ethanol impregnation on the properties of silica-supported cobalt catalysts. *J. Catal.* 168:51–59.
26. Reuel R.C. and Bartholomew C.H. 1984. The stoichiometries of H$_2$ and CO adsorptions on cobalt: Effects of support and preparation. *J. Catal.* 85:63–77.
27. Boot L.A., Kerkhoffs M.H.J.V., van der Linden B.Th., van Dillen A.J., Geus J.W., and van Buren F.R. 1996. Preparation, characterisation and catalytic testing of cobalt oxide and manganese oxide catalysts supported on zirconia. *Appl. Catal. A Gen.* 137:69–86.
28. (a) Culross C.C. and Mauldin C.H. 1999. Preparation of high activity catalysts; the catalysts and their use. U.S. Patent 5856261. (b) Culross C.C. and Mauldin C.H. 2000. Preparation of high activity catalysts; the catalysts and their use. U.S. Patent 6136868. (c) Culross C.C. and Mauldin C.H. 2002. Preparation of high activity catalysts and their use. European Patent EP 977632. (d) Culross C.C. and Mauldin C.H. 1998. Preparation of high activity catalysts; the catalysts and their use. International Patent WO98/47620.
29. Girardon J.-S., Quinet E., Constant-Griboval A., Chernavskii P.A., Gengembre L., and Khodakov A.Y. 2007. Cobalt dispersion, reducibility and surface sites in promoted silica-supported Fischer-Tropsch catalysts. *J. Catal.* 248:143–57.
30. Girardon J.-S., Constant-Griboval A., Gengembre L., Chernavskii P.A., and Khodakov A.Y. 2005. Optimisation of the pretreatment procedure in the design of cobalt silica supported Fischer Tropsch catalysts. Paper presented at the Gas-Fuel 05 Conference, Bruges, Belgium.
31. Mauldin C.H. 1999. Preparation of high activity catalysts; the catalysts and their use. U.S. Patent 5856260.
32. Zavyalova U., Scholz P., and Ondruschka B. 2007. Influence of cobalt precursor and fuels on the performance of combustion-synthesised Co$_3$O$_4$/γ-Al$_2$O$_3$ catalysts for total oxidation of methane. *Appl. Catal. A Gen.* 323:226–33.
33. (a) Mauldin C.H. 1998. Preparation of high activity catalysts, the catalysts and their use. International Patent WO 98/47618. (b) Mauldin C.H. 1999. Preparation of high activity catalysts, the catalysts and their use. U.S. Patent 5863856.
34. Sarellas A., Niakolas D., Bourikas K., Vakros J., and Kordulis C. 2006. The influence of the preparation method and the Co loading on the structure and activity of cobalt oxide/γ-alumina catalysts for NO reduction by propene. *J. Colloid. Interf. Sci.* 295:165–72.

35. Ataloglou T., Vakros J., Bourikas K., Fountzoula C., Kordulis C., and Lycourghiotis A. 2005. Influence of the preparation method on the structure-activity of cobalt oxide catalysts supported on alumina for complete benzene oxidation. *Appl. Catal. B Environ.* 57:299–312.

36. Koizumi N., Mochizuki T., and Yamada M. 2009. Preparation of highly active catalysts for ultra-clean fuels. *Catal. Today* 141:34–42.

37. Mochizuki T., Hara T., Koizumi N., and Yamada M. 2007. Surface structure and Fischer-Tropsch synthesis activity of highly active Co/SiO_2 catalysts prepared from the impregnating solution modified with some chelating agents. *Appl. Catal. A Gen.* 317:97–104.

38. Dumond F., Marceau E., and Che M. 2007. A study of cobalt speciation in Co/Al_2O_3 catalysts prepared from solutions of cobalt-ethylenediamine complexes. *J. Phys. Chem. C* 111:4780–89.

39. Trujillano R., Villain F., Louis C., and Lambert J.-F. 2007. Chemistry of silica-supported cobalt catalysts prepared by cation adsorption. 1. Initial localised adsorption of cobalt precursors. *J. Phys. Chem. C* 111:7152–64.

40. (a) Sun S., Fujimoto K., Yoneyama Y., and Tsubaki N. 2002. Fischer-Tropsch synthesis using Co/SiO_2 catalysts prepared from mixed precursors and addition effect of noble metals. *Fuel* 81:1583–91. (b) Sun S., Tsubaki N., and Fujimoto K. 2000. The reaction performances and characterisation of Fischer-Tropsch synthesis Co/SiO_2 catalysts prepared from mixed cobalt salts. *Appl. Catal. A Gen.* 202:121–31.

41. Lide D.R. 1990. *CRC Handbook of Chemistry and Physics.* 71st ed. Boca Raton, FL: CRC Press.

42. Carlsson M., Habenicht C., Kam L.C., Antal M.J. Jr., Bian N., Cunningham R.J., and Jones M. Jr. 1994. Study of the sequential conversion of citric to itaconic to methacrylic acid in near-critical and supercritical water. *Ind. Eng. Chem. Res.* 33:1989–96.

43. Ellis P.R. and Bishop P.T. 2006. Supported cobalt catalysts for the Fischer-Tropsch synthesis. International Patent Application WO2006/136863.

44. Soled S.L., Iglesia E., and Fiato R.A. 1992. Copper-promoted cobalt manganese spinel catalyst and method for making the catalyst for Fischer-Tropsch synthesis. U.S. Patent 5162284.

45. Greenwood N.N. and Earnshaw A. 1984. *Chemistry of the elements*, 1300–20. Oxford, UK: Pergamon Press.

46. Wu Y., Bandyopadhyay A., and Bose S. 2004. Processing of alumina and zirconia nano-powders and compacts. *Mater. Sci. Eng.* 380:349–55.

47. Das R.N. 2001. Nanocrystalline ceramics from sucrose process. *Mater. Lett.* 47:344–50.

48. van de Loosdrecht J., van der Haar M., van der Kraan A.M., van Dillen A.J., and Geus J.W. 1997. Preparation and properties of supported cobalt catalysts for Fischer-Tropsch synthesis. *Appl. Catal. A Gen.* 150:365–76.

49. Malecki A., Gajerski R., Labus S., Prochowska-Klisch B., and Wojciechowski K.T. 2000. Mechanism of thermal decomposition of d-metals nitrates hydrates. *J. Therm. Anal. Calorim.* 60:17–23.

50. Parkyns N.D. 1973. Adsorption sites on oxides. Infra-red studies of adsorption of oxides of nitrogen. *Proc. 5th Int. Congress Catal.* 1: 255–64.

51. Christokova St. G., Stoyanova M., and Georgieva M. 2001. Low-temperature iron-modified cobalt oxide system. Part 1. Preparation and characterisation. *Appl. Catal. A Gen.* 208:235–42.

2 Carbon Nanomaterials as Supports for Fischer-Tropsch Catalysts

Anke Jung, C. Kern, and Andreas Jess

CONTENTS

The potential of carbon nanomaterials for the Fischer-Tropsch synthesis was investigated by employing three different nanomaterials as catalyst supports. Herringbone (HB) and platelet (PL) type nanofibers as well as multiwalled (MW) nanotubes were examined in terms of stability, activity, and selectivity for Fischer-Tropsch synthesis (FTS).

For this purpose, all three catalyst supports were initially synthesized by a chemical vapor deposition (CVD) process and thereafter, using a wet impregnation method, loaded with cobalt as the active component for FTS. The as-synthesized Co/nanocatalysts were then characterized by applying electron microscopic analysis as well as temperature-programmed desorption, chemi- and physisorption measurements, thermogravimetric analysis, and inductively coupled plasma

(ICP) measurements. The catalytic performance of the nanocatalysts was finally tested in the Fischer-Tropsch synthesis carried out in a fixed bed reactor. The obtained results were compared with literature data of commercially used Fischer-Tropsch catalysts.

Regarding the Fischer-Tropsch reaction, stable activity was obtained for all three Co/nanomaterial catalysts within the synthesis period (approximately 50 h). The Co/nanotube catalyst exhibited the highest (Co mass-related) intrinsic activity, followed by the Co/platelet nanofibers. Although the Co/herringbone material showed the lowest activity for FTS, a comparably high chain growth probability (α-value) of 0.83 was reached at a given temperature (513 K). For the Co/platelet and Co/tube catalysts, in contrast, α-values of only around 0.73 were achieved. For all nanomaterial catalysts a comparably low CO_2 and a rather high CH_4 selectivity was obtained.

2.1 INTRODUCTION

Recently, the Fischer-Tropsch synthesis regained much attention mainly due to the (political) desire for cleaner fuels and the potential shortage of crude oil. Therefore, research activity is focusing on the development of improved reactor concepts as well as on novel and promising catalysts for an economic production of clean fuels via FTS.

Since the discovery of carbon nanomaterials by Iijima[1] in 1991, a lot of research has been done on these new materials. There are a number of applications for new carbon nanomaterials in many different fields, such as in sensor technology,[2] in biomedicine as drug delivery systems,[3] or in polymer processing.[4] Reaction engineering may also benefit from the unique properties of carbon nanotubes and nanofibers. In heterogeneous catalysis much research is focused on implementing carbon nanomaterials as catalyst,[5] catalyst promoter,[6] or a catalyst support.[7-16] Owing to their high mechanical stability and heat conductivity, carbon nanomaterials are said to be promising support materials for heterogeneous catalyst preparation.[17]

Concerning the Fischer-Tropsch synthesis, carbon nanomaterials have already been successfully employed as catalyst support media on a laboratory scale. The main attention in literature has been paid so far to subjects such as the comparison of functionalization techniques,[9-11] the influence of promoters on the catalytic performance,[1,12] and the investigations of metal particle size effects[7,8] as well as of metal-support interactions.[14,15] However, research was focused on one nanomaterial type only in each of these studies. Yu et al.[16] compared the performance of two different kinds of nanofibers (herringbones and platelets) in the Fischer-Tropsch synthesis. A direct comparison between nanotubes and nanofibers as catalyst support media has not yet been an issue of discussion in Fischer-Tropsch investigations. In addition, a comparison with commercially used FT catalysts has up to now not been published.

Therefore, carbon nanofibers (CNFs) as well as carbon nanotubes (CNTs) were synthesized,[18,19] functionalized (with the catalytic active metal Co), and finally tested in a reactor conducting the Fischer-Tropsch synthesis. In this publication, the potential of carbon nanomaterial-based Fischer-Tropsch catalysts will be demonstrated by examples of applied experiments.

2.2 EXPERIMENTAL

2.2.1 CARBON NANOMATERIAL SYNTHESIS

The multiwalled nanotubes as well as the herringbone type carbon nanofibers were synthesized in-house in a quartz glass fluidized bed reactor via chemical vapor deposition (CVD). The method is described in detail elsewhere.[19] The platelet nanofibers, in contrast, were purchased from the company FutureCarbon GmbH (Bayreuth, Germany).

2.2.2 FUNCTIONALIZATION OF CARBON NANOMATERIALS

Prior to functionalization the carbon nanomaterials were washed in concentrated nitric acid (65%; Fisher Scientific) for 8 h using a Soxhlet device in order to remove catalyst residues of the nanomaterial synthesis as well as to create anchor sites (surface oxides) for the Co on the surface of the nanomaterials. After acid treatment the feedstock was treated overnight with a sodium hydrogen carbonate solution (Gruessing) for neutralization reasons. For the functionalization of the support media with cobalt particles, a wet impregnation technique was applied. For this purpose 10 g of the respective nanomaterial and 10 g of cobalt(II)-nitrate hexahydrate ($Co(NO_3)_2 \cdot 6\ H_2O$, Fluka) were suspended in ethanol (1 l) and stirred for 24 h. Thereafter, the suspension was filtered via a water jet pump and finally entirely dried using a high-vacuum pump (5 mbar).

2.2.3 CATALYST CHARACTERIZATION

2.2.3.1 Electron Microscopy

The materials were examined by scanning electron microscopy (SEM; Zeiss LEO 1530) in order to determine the morphology of the catalyst surface. By SEM analysis using a special backscattering technique as well as by transmission electron microscopy (TEM; Zeiss 922 Omega), information about the particle size distribution and about the metal anchorage on the support surface was gained. The sample specimens for TEM analysis were prepared by ultrasonic dispersion of the nanomaterial catalysts in acetone, and by dropping the suspension onto a carbon-coated copper grid.

2.2.3.2 ICP Measurements

The cobalt content of the catalyst was analyzed by ICP measurements (Perkin Elmer, Plasma 400) at the Institute of Chemical Engineering of the University of Erlangen-Nuremberg.

2.2.3.3   Nitrogen Physisorption

The pore size distribution based on BJH (Barrett-Joyner-Halenda) calculations, the micropore fraction (t-plot analysis), and the BET (Brunauer-Emmett-Teller) surface area of the catalysts were acquired by physisorption measurements of nitrogen at 77 K (Micrometrics Gemini 2360). Prior to BET analysis the samples were evacuated at 373 K for at least 12 h.

2.2.3.4   Carbon Monoxide Chemisorption

Chemisorption measurements (Quantachrome Instruments, ChemBET 3000) were conducted in order to determine the metal (Co) dispersion. Therefore, the nanomaterial catalysts were reduced under a hydrogen flow (10% H_2 in Ar) at 633 K for 3 h. The samples were then flushed with helium for another hour at the same temperature in order to remove the weakly adsorbed hydrogen. Chemisorption was carried out by applying a pulse-titration method with carbon monoxide as adsorbing agent at 77 K. The calculation of the dispersion is based on a molar adsorption stoichiometry of CO to Co of 1.

2.2.3.5   Thermogravimetric Analysis

The catalyst purity and stability with respect to oxidation was investigated by thermogravimetric/differential thermal analysis (TG/DTA) in a Seiko EXSTAR 6000. During the measurement an oxygen-containing gas mixture (2 vol.% O_2 in N_2) was applied to the sample at a constant heating rate of 2 K/min. The mass change (decrease) at a given temperature characterizes the beginning of the oxidation of the carbonaceous support. The appearance of more than one oxidation peak indicates the presence of impurities such as amorphous carbon.

2.2.3.6   Temperature-Programmed Reduction

Temperature-programmed reduction (TPR) gives information on the reduction behavior of the Co catalysts. The spectra were recorded by the instrument ChemBET 3000 (Quantachrome Instruments) equipped with a thermal conductivity detector. Before analysis the samples were dried overnight (at least 12 h) at 373 K. The reduction was carried out in a hydrogen mixture of 10% H_2 in Ar with a heating rate of 10 K/min.

2.3   RESULTS AND DISCUSSION

2.3.1   CATALYST CHARACTERIZATION

As can be seen from Figure 2.1, cobalt was deposited on the carbon nanomaterials quite homogeneously. Hence, the cobalt particle sizes of the three catalyst types vary only little. The Co/nanofiber materials exhibit cobalt particle diameters of roughly 10 nm. In case of the nanotubes, particle sizes ranging from 5 to 7 nm were observed.

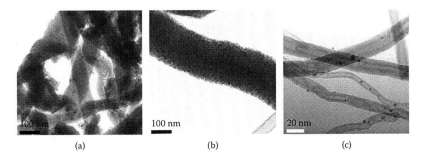

(a)   (b)   (c)

FIGURE 2.1   Cobalt functionalized (a) platelet type carbon nanofibers, (b) herringbone type carbon nanofibers, and (c) carbon multiwalled nanotubes.

To some extent the cobalt particles in Figure 2.1(c) seem to be distributed within the tubular structure of the multiwalled nanotubes. TEM analysis could not fully clarify if this is an artifact or if the particles are truly situated inside the hollow space of the tubes. However, Tavasoli et al.[14] observed Co particles captured inside the tubes after incipient wetness impregnation. Thus, it can be assumed that this is the case here as well.

Other important results of the catalyst characterization that will be discussed in detail below are summarized in Table 2.1.

Initially, for all three support materials a Co content of 20 wt.% was supposed to be deposited. Table 2.1 shows that functionalization of the herringbone nanofibers (HB-CNF) appeared to be most efficient since nearly all of the applied metal was adsorbed by the nanomaterial surface with impregnation (over 95%).

TABLE 2.1

Results of Carbon Nanomaterial Catalyst Characterization: ICP, Chemisorption, Physisorption, Thermogravimetric Analysis (TG)

Catalyst Type	Co Content[a] wt.%	Dispersion[b] %	BET Surface Area m²/g	Micropore Fraction %	TG[c] K
Co/HB-CNF	19.4	10	120	14	673
Co/PL-CNF	16.0	56	102	15	523
Co/MW-CNT	11.0	75	287	8	503

[a] Obtained from ICP measurement.

[b] Obtained from chem-BET pulse-titration with carbon monoxide at 77 K.

[c] Thermogravimetric oxidation (2 vol.% O_2 in N_2, heating rate 2 K/min): The given temperature characterizes the onset of a detectable oxidation of the carbonaceous support.

However, the platelet nanofibers (PL-CNF) carry only about two-thirds of the loaded cobalt, and the nanotubes (MW-CNT) showed the most inefficient functionalization performance, exhibiting barely 50% of metal adsorption.

Also, the dispersion degree (measured by physisorption of CO) differs strongly (see Table 2.1), ranging from only 10% for herringbones to up to 75% for multi-walled nanotubes with the finest metal distribution. The deviation between the values for the nanotubes and the platelet fibers was confirmed by microscopic analysis, where we found larger particles on the fiber surface (see Figure 2.1), leading to a lower dispersion degree. However, it is impossible to explain the extremely small dispersion degree of the herringbone nanofibers by TEM analysis since we measured identical Co particle sizes for both fiber catalysts (platelets and herringbones). Most likely, there exist much bigger Co agglomerates beyond the limited image section of the microscope.

Physisorption measurements showed that carbon nanomaterials exhibit rather meso- and macroporous structures (maximum micropore fraction, 15%; see Table 2.1). The lowest specific surface area was measured with the platelet fiber catalyst exhibiting slightly more than 100 m^2/g. The Co/HB material offers 120 m^2/g of surface area, and the highest BET value was determined with the Co/MW catalyst featuring nearly 290 m^2/g. Carbon nanomaterials, though, are not really porous, as the space between the graphene layers is too small for nitrogen molecules to enter. The only location of adsorption is the external surface of the nanomaterials and the inner surface of the nanotubes.

According to Table 2.1, the highest stability concerning oxidation was reached with the Co/HB material, where combustion started first at 673 K. In the case of the Co/PL and Co/MW catalysts, weight loss initiates at 523 and 503 K, respectively. None of the catalysts showed major impurities (more than one oxidation peak) within TG analysis.

The reduction behavior of the Co-based nanomaterial catalysts was studied by temperature-programmed reduction. The respective TPR profiles are plotted in Figure 2.2. Basically, there are three main peaks present in all curves, whereas the herringbones additionally show a shoulder right at the beginning of the graph, which can be ascribed to the reduction or decomposition of organic impurities resulting from catalyst production. In all TPR profiles, the first two main peaks are caused by the stepwise reduction of cobalt from Co_3O_4 via CoO to Co. The peaks observed at temperatures above 700 K are attributed to methane formation.[9,11,13] This was confirmed by gas chromatographic analysis of the TPR outlet gas. Thus, the supported metal not only catalyzes the Fischer-Tropsch reaction but also assists the formation of methane via the reaction of hydrogen with the carbon support at high temperatures (above 700 K). The comparison of the TPR profiles of the different nanomaterial catalysts shows that in the case of the nanotubes the last reduction peak (CoO to Co) is shifted to lower temperatures than the peaks obtained with the other two catalysts. This indicates lower metal-support interactions in the Co/MW material. Tavasoli et al.[15] even assume that in the case of Co/nanotube catalysts the degree of interaction between active metal and support is

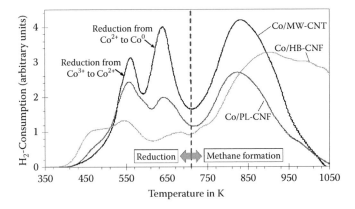

FIGURE 2.2 TPR profiles of carbon nanomaterial Fischer-Tropsch catalysts (gas mixture: 10% H_2 in Ar; heating rate: 10 K/min).

zero. To the contrary, the broad peaks in the profile of the Co/HB catalyst are a sign for strong metal-support interactions.[13]

2.3.2 PERFORMANCE OF THE NANOMATERIALS IN FISCHER-TROPSCH SYNTHESIS

The catalytic performance of the nanofiber/nanotube-supported catalysts was tested in a continuous fixed bed reactor consisting of a tube with a diameter of 14 mm and a length of 0.3 m. Gaseous products were analyzed by gas chromatography (GC; Caldos 17 and Uras 14 from ABB; analysis of H_2, CO, and CO_2) and by an online GC (column: CP-PoraPLOT Q; analysis of C_1–C_3). Differently tempered cold traps (294, 273, and 193 K) were applied to separate liquid products and higher hydrocarbons (waxes), which were weighed after synthesis for mass balance calculations. In order to obtain the carbon content and the product distribution of the organic liquid products, GC measurements (Varian CP 3800) as well as elemental analysis (EuroVector Euro EA 3000) were performed.

The FTS was conducted at varying temperatures (from 483 to 513 K) over approximately 50 h of reaction time in order to investigate the reaction kinetics achieved with the respective catalysts. A typical conversion curve using the Co/HB catalyst as an example is shown in Figure 2.3. After a short settling phase (caused by the pore filling of liquid Fischer-Tropsch products) of only about 4 h, steady-state conditions were reached. In the observed synthesis period of 50 h no deactivation of the catalysts was detected. However, industrially relevant experiments over several weeks are still outstanding.

Assuming a reaction order of one concerning hydrogen and a reaction order of zero regarding carbon monoxide (according to Post et al.[20]), the activation energy E_A and the collision factor k_0 can be derived via the Arrhenius relation:

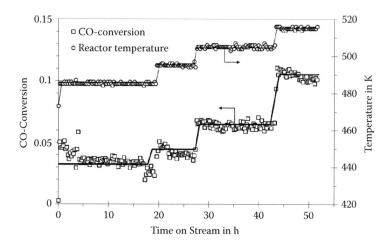

FIGURE 2.3   Typical conversion curve depending on the reaction temperature (reaction conditions: $p = 3$ MPa, $CO/H_2 = \frac{1}{2}$, $\dot{V}_{ges} = 18.5$ l/h (NTP)).

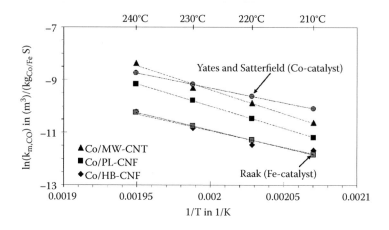

FIGURE 2.4   Arrhenius plots of the tested carbon nanomaterial catalysts and commercially used Fischer-Tropsch catalysts (reaction conditions: $p = 3$ MPa, $CO/H_2 = \frac{1}{2}$, $\dot{V}_{ges} = 18.5$ l/h (NTP)).

$$k_m = k_0 \exp\left(-\frac{E_A}{RT}\right) \tag{2.1}$$

with k_m as the catalyst mass (Co)-related rate constant.

Taking the logarithm of Equation 2.1 allows plotting $\ln(k_{m,CO})$ against the reciprocal temperature (Figure 2.4). In Figure 2.4 the experimental results are compared to data of a commercial Co catalyst[21] and a precipitated iron material.[22]

TABLE 2.2
Activation Energy, E_A, and Collision Factor, k_m, of Carbon Nanomaterial-Supported Co Catalysts and Commercially Used Fischer-Tropsch Catalysts

Catalyst Type	E_A kJ/mol	$k_{m,CO,0}$ $m^3/(kg_{Co}\ s)$
Co/HB-CNF	103	$1.1 \cdot 10^6$
Co/PL-CNF	140	$2.0 \cdot 10^{10}$
Co/MW-CNT	151	$5.7 \cdot 10^{11}$
Co catalyst (Yates and Satterfield[21])	93	$4.7 \cdot 10^5$
Fe catalyst (Raak[22])	112	$9.7 \cdot 10^6$

TABLE 2.3
Activities of Carbon Nanomaterial-Supported Co Catalysts and of a Commercially Used Fischer-Tropsch Catalyst[23] at a Reaction Temperature of 493 K

	Co/HB-CNF	Co/PL-CNF	Co/MW-CNT	Commercial Co Catalyst
Activity in $mol_{CO}/(g_{Co}\ s)$	$6.9 \cdot 10^{-6}$	$1.9 \cdot 10^{-5}$	$3.2 \cdot 10^{-5}$	$5.8 \cdot 10^{-5}$
Relative activity to commercial Co catalyst in %	12	33	55	100

The Arrhenius plot (Figure 2.4) shows that literature data are in good agreement with values obtained from the nanomaterial catalysts.

The resulting activation energies E_A as well as the collision factors $k_{m,CO,0}$ are displayed in Table 2.2. The most active material among the nanomaterials is the Co/MW catalyst, with the highest values for both kinetic parameters (E_A and $k_{m,CO,0}$). The lowest activation energy and collision factor, in contrast, is seen with the herringbone material.

A direct comparison of the productivities of the Co/nanomaterials and a typical Co catalyst[23] (promoted Co/Ru-alumina catalyst) is presented in Table 2.3. Bearing in mind that the nanocatalysts are unpromoted systems and that only a simple wetness impregnation technique was employed for catalyst production, the obtained activities are quite promising, especially in the case of the Co/MW catalyst.

When determining the product selectivities, all compounds of equal carbon numbers (paraffines, olefins, isomers, and oxygen compounds) were summarized to one product fraction. The chain growth probability was determined by the Anderson-Schulz-Flory (ASF) distribution:

$$x_i = (1-a)a^{i-1} \tag{2.2}$$

with the molar fraction x_i. Figure 2.5 displays the α-values (chain growth probabilities) obtained from the wax fraction from C_{30} to C_{40}.

In the diagram, the chain growth probabilities of the Co/PL and the Co/MW materials are roughly the same (see also Table 2.4). The Co/HB catalyst exhibits the highest selectivity toward higher hydrocarbons, with an α-value of 0.83. According to Jager and Espinoza,[24] Shell specifies chain growth probabilities for "classical" cobalt-based catalysts between 0.78 and 0.82, while for "new" cobalt-based catalysts the α-value ranges between 0.8 and about 0.93 to 0.94. These values apply to typical FT reaction temperatures of a maximum of 493 K. However, the chain growth probability decreases significantly with rising synthesis temperatures,[25] but only a few references exist where Co catalysts are employed at

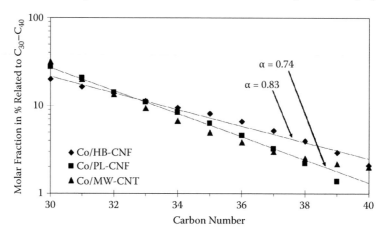

FIGURE 2.5 Determination of the growth probability (α-value; reaction conditions: T = 513 K, p = 3 MPa, CO/H$_2$ = ½, $\dot{V}_{ges}$.= 18.5 l/h (NTP)).

TABLE 2.4

Comparison of the C_2/C_3 Ratio Calculated by the ASF Distribution with Measured Data by the Online GC (Reaction Conditions: T = 513 K, p = 3 MPa, CO/H$_2$ = ½, $\dot{V}_{ges}$.= 18.5 l/h (NTP))

Catalyst Type	α-Value (Measured)	Ratio C_2/C_3 (Calculated, Based on α)	Ratio C_2/C_3 (Measured)
Co/HB-CNF	0.83	0.8	0.7
Co/PL-CNF	0.74	0.9	1
Co/MW-CNT	0.73	0.9	1.3

TABLE 2.5
CO Conversion, Selectivities, and α-Values Reached with Carbon Nanomaterial Co Catalysts (Reaction Conditions: T = 513 K, p = 3 MPa, CO/H$_2$ = ½, $\dot{V}_{ges}$ = 18.5 l/h (NTP))

Catalyst Type	CO Conversion %	S_{CO_2} %	S_{CO_4} %	$S_{C_2\text{-}C_3}$ %	$S_{C_{4+}}$ %	α-Value —
Co/HB-CNF	9.6	1.1	19.2	8.9	70.7	0.83
Co/PL-CNF	26.6	0.2	13.4	19.9	66.5	0.74
Co/MW-CNT	49.2	1.6	15.3	20.0	63.1	0.73

temperatures above 493 K. Huang et al.,[26] for example, observed a drop of the α-value from 0.86 to 0.71 within a temperature range of 483 to 523 K (p = 4.5 MPa, CO/H$_2$ = ½) for an alumina-supported cobalt catalyst from United Catalyst. At the specific reaction temperature of 513 K used in this work, a growth probability of 0.77 was reached. O'Shea et al.[27] measured an α-value of 0.7 at 513 K when using a silica-supported bimetallic Co/Fe catalyst (Co:Fe = 10:1) at 2 MPa (CO/H$_2$ = ½). Thus, the examined nanocatalysts show comparable or, in the case of the Co/HB catalyst, even improved selectivities toward higher hydrocarbons (high α-value) at a given temperature (513 K).

The molar C$_2$/C$_3$ ratios calculated from the obtained α-values in Figure 2.5 were also compared with the C$_2$/C$_3$ ratios received from online GC measurements. According to Table 2.4 a good agreement of the measured (online GC) and calculated data (α-values) was reached, which confirms the correctness of the results shown in Figure 2.5, even in the range of short-chain hydrocarbons.

The methane and the carbon dioxide selectivities as well as the selectivities concerning the product fractions C$_2$ to C$_3$ and C$_{4+}$ with the respective conversion degrees and α-values at 513 K are shown in Table 2.5.

As usual, employing Co catalysts leads to only little CO$_2$ formation. However, the methane selectivity is quite high compared to modern Co catalysts, where a CH$_4$ selectivity of less than 10% is reached at 493 K.[23] At this temperature the nanomaterial catalysts, except for the Co/PL catalyst (9% selectivity), exceed this limit already with 11 and 16% for the Co/HB and Co/MW materials, respectively.

2.4   CONCLUSIONS AND PERSPECTIVES

In initial experiments carbon nanomaterial-supported catalysts showed acceptable activities and comparatively high selectivities toward higher hydrocarbons. Nevertheless, the applicability of these new materials in large-scale fixed bed reactors is limited due to their powdery appearance. Concerning this challenge research has already started, and hopefully carbon nanomaterial pellets will be available soon. Macro-shaped carbon nanomaterial catalysts might have the advantage of diminished pore diffusion limitations compared to conventional catalysts due to their meso-/macroporous structure. Another benefit might be the extraordinary heat removal capacity, especially of carbon nanotubes, which render complex cooling devices unnecessary or at least simplify them. Concerning fiber/tube functionalization, improvements regarding uniform decomposition of the metal are still essential.

Although research on carbon nanomaterials has by far not yet been finished and the price of carbon nanomaterials remains high, their unique properties justify continuing research on these remarkable materials.

REFERENCES

1. Iijima, S. 1991. Helical microtubules of graphitic carbon. *Nature* 354:56–58.
2. Trojanowicz, M. 2006. Analytical applications of carbon nanotubes: A review. *Trends in Analytical Chemistry* 25:480–89.
3. Bianco, A., Kostarelos, K., Partidos, C. D., and Prato, M. 2005. Biomedical applications of functionalised carbon nanotubes. *Chemical Communications* 5:571–77.
4. Coleman, J. N., Khan, U., Blau, W. J., and Gunko, Y. K. 2006. Small but strong: A review of the mechanical properties of carbon nanotube-polymer composites. *Carbon* 44:1624–52.
5. Su, D. S., Maksimova, N., Delgado, J. J., Keller, N., Mestl, G., Ledoux, M. J., and Schlogl, R. 2005. Nanocarbons in selective oxidative dehydrogenation reaction. *Catalysis Today* 102:110–14.
6. Zhang, H., Dong, X., Lin, G., Liang, X. L., and Li, H. Y. 2005. Carbon nanotube-promoted Co-Cu catalyst for highly efficient synthesis of higher alcohols from syngas. *Chemical Communications* 40:5094–96.
7. Bezemer, G. L., Bitter, J. H., Kuipers, H. P. C. E., Oosterbeek, H., Holewijn, J. E., Xu, X. D., Kapteijn, F., van Dillen, A. J., and de Jong, K. P. 2006. Cobalt particle size effects in the Fischer-Tropsch reaction studied with carbon nanofiber supported catalysts. *Journal of the American Chemical Society* 128:3956–64.
8. Bezemer, G. L., van Laak, A., van Dillen, A. J., and de Jong, K. P. 2004. Cobalt supported on carbon nanofibers—A promising novel Fischer-Tropsch catalyst. *Natural Gas Conversion* 147:259–64.
9. Van Steen, E., and Prinsloo, F. F. 2002. Comparison of preparation methods for carbon nanotubes supported iron Fischer-Tropsch catalysts. *Catalysis Today* 71:327–34.
10. Guczi, L., Stefler, G., Geszti, O., Koppány, Zs., Molnár, E., Urbán, M., and Kiricsi, I. 2006. CO hydrogenation over cobalt and iron catalysts supported over multiwall carbon nanotubes: Effect of preparation. *Journal of Catalysis* 244:24–32.
11. Bezemer, G. L., Radstake, P.B., Koot, V., van Dillen, A. J., Geus, J. W., and de Jong, K. P. 2006. Preparation of Fischer-Tropsch cobalt catalysts supported on carbon nanofibers and silica using homogeneous deposition-precipitation. *Journal of Catalysis* 237:291–302.
12. Bahome, M., Jewell, L., Hildebrandt, D., Glasser, D., and Coville, N. J. 2005. Fischer-Tropsch synthesis over iron catalysts supported on carbon nanotubes. *Applied Catalysis A: General* 287:60–67.

13. Bezemer, G. L., Radstake, P. B., Falke, U., Oosterbeek, H., Kuipers, H. P. C. E., van Dillen, A., and de Jong, K. P. 2006. Investigation of promoter effects of manganese oxide on carbon nanofiber-supported cobalt catalysts for Fischer-Tropsch synthesis. *Journal of Catalysis* 237:152–61.

14. Bahome, M. C., Jewell, K., Padayachy, L. L., Padayachy, K., Hildebrandt, D., Glasser, D., Datye, A. K., and Coville, N. J. 2007. Fe-Ru small particle bimetallic catalysts supported on carbon nanotubes for use in Fischer-Tropsch synthesis. *Applied Catalysis A: General* 328:243–51.

15. Tavasoli, A., Abbaslou, R. M. M., Trepanier, M., and Dalai, A. K. 2008. Fischer-Tropsch synthesis over cobalt catalyst supported on carbon nanotubes in a slurry reactor. *Applied Catalysis A: General* 345:134–42.

16. Yu, Z., Borg, Ø., Chen, D., Enger, B. C., Frøseth, V., Rytter, E., Wigum, H., and Holmen, A. 2006. Carbon nanofiber supported cobalt catalysts for Fischer-Tropsch synthesis with high activity and selectivity. *Catalysis Letters* 109:43–47.

17. Serp, P., Corrias, M., and Kalck, P. 2003. Carbon nanotubes and nanofibers in catalysis. *Applied Catalysis A: General* 253:337–358.

18. Jess, A., Kern, C., Schrögel, K., Jung, A., and Schütz, W. 2006. Manufacturing of carbon nanotubes and -fibers through gas phase separation. *Chemie Ingenieur Technik* 78:94–100.

19. Schrögel, K. 2007. Reaktionskinetische und verfahrenstechnische Aspekte der Synthese von Kohlenstoff-Nanotubes und -Nanofasern. PhD thesis, Faculty of Chemical Engineering, University Bayreuth, Germany.

20. Post, M. F. M., Vanthoog, A. C., Minderhoud, J. K., and Sie, S. T. 1989. Diffusion limitations in Fischer-Tropsch catalysts. *AIChE Journal* 35:1107–14.

21. Yates, I. C., and Satterfield, C. N. 1991. Intrinsic kinetics of the Fischer-Tropsch synthesis on a cobalt catalyst. *Energy & Fuels* 5:168–73.

22. Raak, H. 1995. Reaktionskinetische Untersuchungen in der Anfangsphase der Fischer-Tropsch-Synthese an einem technischen Eisenfällungskatalysator. PhD thesis, Faculty of Chemical Engineering, University Karlsruhe, Germany.

23. Dathe, H., Finger, K.-F., Haas, A., Kolb, P., and Sundermann, A. 2008. High throughput catalyst optimization program for the gas-to-liquid (GTL) technologies methanol-to-gasoline (MTG), higher alcohol synthesis (HAS) and Fischer-Tropsch synthesis (FTS). Paper presented at Future Feedstocks for Fuels and Chemicals (DGMK), Berlin.

24. Jager, B., and Espinoza, R. 1995. Advances in low-temperature Fischer-Tropsch synthesis. *Catalysis Today* 23:17–28.

25. Claeys, M. 1997. Selektivität, Elementarschritte und kinetische Modellierung bei der Fischer-Tropsch-Synthese. PhD thesis, Faculty of Chemical Engineering, University Karlsruhe, Germany.

26. Huang, X. W., Elbashir N. O., and Roberts, C. B. 2004. Supercritical solvent effects on hydrocarbon product distributions from Fischer-Tropsch synthesis over an alumina-supported cobalt catalyst. *Industrial & Engineering Chemistry Research* 43:6369–81.

27. O'Shea, V. A. D., Alvarez-Galvan, M. C., Campos-Martin, J. M., and Fierro, J. L. G. 2007. Fischer-Tropsch synthesis on mono- and bimetallic Co and Fe catalysts in fixed-bed and slurry reactors. *Applied Catalysis A: General* 326:65–73.

3 Effect of a Novel Nitric Oxide Calcination on the Catalytic Behavior of Silica-Supported Cobalt Catalysts during Fischer-Tropsch Synthesis, and Impact on Performance Parameters

Wenping Ma, Gary Jacobs, Muthu K. Gnanamani, Uschi M. Graham, and Burtron H. Davis

CONTENTS

Fischer-Tropsch synthesis (FTS) kinetic parameters, including a water effect term, were measured over air calcined 15% Co/SiO_2 and 25% Co/SiO_2 catalysts. Moreover, the sensitivity of Co cluster size to the FTS reaction was studied using reduced 15% Co/SiO_2 and 25% Co/SiO_2 catalysts pretreated by two different calcination methods: the traditional air calcination and a novel calcination with nitric oxide. To assess kinetic parameters of FTS over air calcined 15% and 25% Co/SiO_2 (PQ) catalysts, a 1 L continuously stirred tank reactor (CSTR) was utilized. This was accomplished by varying hydrogen partial pressure at a constant CO partial pressure (0.51 MPa), and then conversely by varying CO partial pressure at a constant H_2 partial (0.81 MPa) pressure in order to vary the H_2/CO ratio from 1.0 to 2.5 in the temperature range of 205 to 220°C. Two Langmuir Hinshelwood CO consumption models and three empirical CO consumption models (including or excluding water inhibition) were employed to fit the experimental data. The model that provided the best fit was obtained with the expression $r_{CO} = kP_{CO}{}^{a}$ $P_{H_2}{}^{b}/(1 + m P_{H_2O}/P_{H_2})$, resulting in the parameters $k = 0.0187$ mol/g-cat/h/$MPa^{0.38}$ (220°C), $E_a = 85.9$ kJ/mol, $a = -0.22$, $b = 0.6$, and $m = -0.33$ for the reduced air calcined 15% Co/SiO_2 catalyst and $k = 0.0381$ mol/g-cat/h/$MPa^{0.32}$ (220°C), $E_a = 93.7$ kJ/mol, $a = -0.19$, $b = 0.51$, and $m = -1.11$ for the reduced air calcined 25% Co/SiO_2 catalyst. Thus, the rate exhibits a positive water effect, and the effect is greater for the catalyst with particles small enough to reside in the pore, consistent with the proposal that the kinetic effect of water is to displace heavy hydrocarbons residing in the pore and, thereby, remove intraparticle transport restrictions. NO calcination not only significantly improved catalyst activity of SiO_2-supported Co catalysts but also increased the formation rate of heavier hydrocarbons by suppressing the formation of CH_4 and CO_2. NO calcination led to a smaller average Co cluster size for the Co/SiO_2 catalyst, but the catalysts exhibited a greater sensitivity to deactivation phenomena during FTS, which may be influenced by water. Despite a lower extent of reduction of Co oxide species, the NO calcination generated smaller Co clusters and increased the active site densities of surface Co^0 atoms. The catalysts displayed improvements in initial hydrocarbon productivity rates and product selectivities.

3.1 INTRODUCTION

It is well known that water, possessing as high as about 50% of total FTS products, can significantly affect the performance parameters (e.g., CO conversion rate, product selectivities, and catalyst deactivation rate) of supported cobalt catalysts. Water co-feeding studies indicate that the support type influences the size distribution of Co clusters, as well as their extent of reduction, as determined by the strength of the support-cobalt oxide interaction. This, in turn, in large part governs the way in which H_2O affects catalytic behavior. The Davis group [1–6] and others [7–9] conducted water co-feeding studies and identified that at low volume percentages of co-fed H_2O, where inert balancing gas was replaced by H_2O to maintain constant reactant partial pressures and space velocity, H_2O exerts a reversible negative effect on CO consumption rate for Co/Al_2O_3, and this has been suggested to be due to adsorption inhibition. At higher volume percentages of co-fed H_2O, however, extended x-ray absorption fine structure (EXAFS)/x-ray absorption near-edge spectroscopy (XANES) results suggest that a fraction of tiny Co crystallites, those influenced by the support, may oxidize to either CoO or irreducible cobalt aluminate-like species, depending on the crystallite size (i.e., as determined by the Co loading). An empirical kinetic model has proven to be reliable in defining the reversible kinetic H_2O effect [10]:

$$-r_{CO} = kP_{CO}{}^a P_{H_2}{}^b/(1 + m P_{H_2O}/P_{H_2}) \qquad (3.1)$$

In the case of Co/Al_2O_3 catalysts, regression results showed a negative water effect on the catalyst, in good agreement with the results from the water co-feeding studies.

There is less agreement regarding the impact of water during FTS over SiO_2-supported cobalt catalysts. Davis et al. [2,12] and Krishnamoorthy et al. [14] performed water co-feeding experiments over 12.4 to 12.7% Co/SiO_2 catalysts and found that the addition of less than 25% by volume water increased the CO consumption rate, while Minderhoud et al. [15] have reported a negligible impact on CO conversion, but rather an increase in the selectivity of heavier hydrocarbons in the range 0 to 33% of water over 20% Co/SiO_2 catalysts. At least two competing theories exist to explain the positive water effect: (1) water may be adsorbed at reaction conditions to an extent that is sufficient to alter the pore-filling by heavier hydrocarbons, thereby removing intraparticle transport restrictions and thus enhancing FTS, as proposed by Dalai et al. [5], or (2) water may increase the amount of active surface carbon present predominantly as monomeric species, as proposed by Bertole et al. [16]. To confirm the water effect resulting by changing the support from Al_2O_3 to SiO_2, a kinetic study using the same kinetic model described above was conducted to define the water effect on the catalyst [12]. Preliminarily kinetic results in our group confirmed the positive water effect over a SiO_2-supported catalyst. Thus, taking into account the consistency in the kinetic model in describing the results from kinetic and water co-feeding studies over both (though different) Co/SiO_2 and Co/Al_2O_3 catalysts [10,17], the effectiveness of the kinetic method for defining the water effect term over different Co/support systems is demonstrated. In this contribution, we continue to investigate the catalyst performance parameters of Co/SiO_2 catalysts with different inherent properties resulting from differences in pretreatment methods and Co loading. In the latter case, the kinetic approach was utilized.

There is mounting evidence to suggest that water effects during FTS may also be linked to Co particle size, as determined by the support interaction [18,19], even though the turnover frequencies of CO over the surface Co metal atoms on supported Co catalysts have been reported to be quite constant at a given temperature [20]. Due to the rather weak interaction of SiO_2 with Co oxide species, the traditional impregnation plus air calcination usually led to larger metal particles over Co/SiO_2 catalysts. Recently, a novel calcination procedure using nitric oxide gas was reported to result in smaller cobalt crystallites over silica [19]. With the objective of investigating the sensitivity of Co particle size to water formed during FTS, we employed the traditional air calcination procedure and the new nitric oxide gas procedure to prepare both moderately and more heavily loaded supported cobalt catalysts, 15% Co/SiO_2 and 25% Co/SiO_2. The catalytic performance characteristics of the two catalysts have been examined by testing in a 1 L stirred tank slurry reactor. Thus, in this chapter, the kinetics of the two air calcined catalysts and the sensitivity of the catalysts pretreated by air and nitric oxide are reported. Also, the catalysts studied in this chapter have been systematically characterized by hydrogen chemisorption/pulse reoxidation and EXAFS/XANES methods. Catalytic reaction results are explained quite well in the context of the characterization results, which are reported separately in this book.

3.2   EXPERIMENTAL

3.2.1   Catalyst Preparation

Details of procedures for the preparation of 15% Co/SiO_2 and 25% Co/SiO_2 catalysts are reported in Chapter 8 [18]. Briefly, a slurry phase method involving sequential impregnations (due to the solubility limit of cobalt nitrate) was used to load the desired amount of cobalt nitrate to the support. Following cobalt addition, the catalyst was dried at 80 and 100°C in a rotary evaporator. Finally, catalysts were calcined in either flowing air or flowing 5% nitric oxide in nitrogen at a rate of 1 L/min for 4 h at 350°C.

3.2.2   Catalyst Characterization

The 15% Co/SiO_2 and 25% Co/SiO_2 prepared by either air calcination or nitric oxide calcination have been extensively characterized using BET measurements, temperature-programmed reduction, hydrogen chemisorption with pulse reoxidation, and EXAFS/XANES, and a detailed description of the characterization techniques is reported in Chapter 8 [18]. Reduced catalysts prepared by the nitric oxide calcination method displayed higher Co active site densities, and Co cluster size decreased by $\Delta 13$ and $\Delta 19$ nm by employing the nitric oxide calcination over the air calcination procedure for the 25% Co/SiO_2 and 15% Co/SiO_2 catalysts, respectively.

3.2.3   Catalyst Pretreatment

The calcined (~10 g) 15% Co or 25% Co supported on SiO_2 catalyst was ground and sieved to 80 to 325 mesh before loading into a fixed-bed reactor for *ex situ* reduction at 350°C and atmospheric pressure for 10 h using a gas mixture of $H_2/$

He with the molar ratio of 1:3. The reduced catalyst was then transferred to a 1 L CSTR under the protection of N_2 inert gas, which was previously charged with 315 g of melted Polywax 3000. The transferred catalyst was further reduced *in situ* at 230°C at atmospheric pressure using pure hydrogen for another 10 h before starting the FTS reaction.

3.2.4 SLURRY PHASE CATALYST TESTING

The catalyst was tested in a 1 L CSTR. The kinetic experiment was conducted on the air calcined 15% Co/SiO_2 and 25% Co/SiO_2 catalysts through varying H_2 partial pressure between 1.27 and 0.51 MPa while maintaining a constant CO partial pressure of 0.51 MPa, and, conversely, by varying CO partial pressure between 0.32 and 0.81 MPa while holding a constant H_2 partial pressure of 0.81 MPa in the temperature range of 205 to 220°C. Throughout the test, total reaction pressure was maintained at 2.03 MPa by adjusting the inert gas (N_2) flow to give a range of 12.5 to 50 vol.%. In total, four ratios of H_2/CO (2.5, 2.0, 1.5, and 1.0) and four space velocities (20, 10, 5.5, and 3.3 Nl/g-cat/h) were used in the experiment. Space velocity was usually changed in a decreasing order for each ratio of $H_2/$ CO. After testing at the four different space velocities for each ratio of H_2/CO was completed, the reaction conditions were returned to a set of reference conditions of 220°C, 2.03 MPa atm, H_2/CO = 2.5, and N_2% =12.5, 10 Nl/g-cat/h in order to measure the extent of deactivation of the catalyst. In general, the reaction period for each space velocity was 24 h to ensure that the data point was achieved at close to steady state. Total mass closure during the kinetics experiment was 100 ± 3%. For testing the sensitivity of the reduced 25% Co/SiO_2 and 15% Co/SiO_2 catalysts prepared by air and nitric oxide calcination methods to deactivation phenomena, time on stream data were collected.

3.2.5 PRODUCT ANALYSIS

Inlet and outlet gases were analyzed online by a Micro GC equipped with four packed columns. The liquid organic and aqueous products were analyzed using an HP 5890 GC with capillary column DB-5 and an HP 5790 GC with Porapak Q packed column, respectively. The reactor wax withdrawn periodically was analyzed by a high-temperature HP5890 GC employing an alumina-clad column.

3.3 RESULTS AND DISCUSSION

3.3.1 KINETICS ON AIR CALCINED 15% Co/ SiO_2 AND 25% Co/SiO_2 CATALYSTS

The kinetic methodology used by The University of Kentucky Center for Applied Energy Research (CAER) for analyzing the kinetic data has been reported elsewhere [10,12,17]. Using the same approach, we obtained the kinetic parameters for the reduced air calcined 15% Co/SiO_2 and 25% Co/SiO_2 catalysts (see Table 3.1).

TABLE 3.1

Summary of Kinetic Results on the 15% Co/SiO$_2$ and 25% Co/SiO$_2$ Catalysts[a]

Catalyst	Reaction Order		Activation Energy Ea, kJ/mol	Water Effect Constant m	Reaction Rate Constant k, mol/g-cat/h/MPa$^{(a+b)}$
	a	b			
15% Co/SiO$_2$	–0.22	0.6	85.9	–0.33	0.0187
25% Co/SiO$_2$	–0.19	0.51	93.7	–1.11	0.0381

[a] Kinetic model: $-r_{CO} = kP_{CO}{}^a P_{H_2}{}^b/(1 + mP_{H_2O}/P_{H_2})$.

From Table 3.1, the orders of a and b for partial pressure of CO and H$_2$ in the current study (15% Co/SiO$_2$) are –0.22 and 0.6, respectively, which are in agreement with the values of –0.19 and 0.51 on the 25% Co/SiO$_2$ as well as with other values reported in literature [21]. This illustrates that the a and b values in this study are reasonable, and were not measurably impacted in the Co loading range of 15 to 25%. The results are consistent with the assumption that the orders of a and b reflect the extent of H$_2$ and CO adsorption based on the nature of the supported Co catalyst [5]. As discussed, the reaction kinetic rate constant is a function of the catalyst and can be used to effectively compare catalyst activity. The k value at 220°C for the 15% Co/SiO$_2$ catalyst is 0.0187 mol/g-cat/h/MPa$^{0.38}$, which is less than half the value of 0.0381 obtained for the 25% Co/SiO$_2$ catalyst. This is also consistent with the fact that higher per gram catalyst activities are observed with the more highly loaded Co catalysts. Characterization results [18] indicate lower degrees of reduction (~68 vs. ~80%) and smaller average Co cluster sizes (~27 vs. ~38 nm) for the reduced air calcined 25% Co/SiO$_2$ catalyst relative to the reduced air calcined 15% Co/SiO$_2$ catalyst. This suggests that the higher activity of the 25% Co/SiO$_2$ catalyst is due to a particle size effect (i.e., a higher active site density of surface Co^0).

The activation energies of the 15% Co/SiO$_2$ and 25% Co/SiO$_2$ catalysts were calculated to be 85.9 and 93.7 kJ/mol, respectively. These values are in agreement with the value of 86 kJ/mol in a study on Co/Zr/SiO$_2$ by Chang et al. [22] and are consistent with several studies of Co-based catalysts [23,24].

For the water effect constant, m, negative values are obtained on the two catalysts (Table 3.1), indicating a positive water effect for the Co/SiO$_2$ catalysts in terms of the empirical kinetic model. However, the m value of –0.33 for 15% Co/SiO$_2$ differed significantly from the value of –1.11 obtained for the 25% Co/SiO$_2$ catalyst. This is clearly an indication of a stronger positive water effect on FTS for 25% Co/SiO$_2$. Considering the two proposed explanations for the water effect, the increased sensitivity of the 25% Co/SiO$_2$ is better explained in terms of the ability of water to remove intraparticle transport restrictions by displacing heavy hydrocarbons residing in the pores. In that case, Co particles residing in the pores should be particularly sensitive to this effect. Comparing the two catalysts, despite the higher loading, the reduced air calcined 25% Co/SiO$_2$ had an average Co cluster size of ~27 nm, making it able to fit well into the pores of the SiO_2, which possesses an average single-point pore diameter of ~28 nm. On the other hand, the reduced air calcined 15% Co/SiO_2 catalyst was observed to have an average Co cluster size of ~38 nm and would, therefore, possess a fraction of very large clusters that must reside externally to the pore interior, and perhaps even block the pores of SiO_2. Therefore, the results of the reduced air calcined 25% Co/SiO_2 (with the smaller Co particles relative to the 15% Co/SiO_2 catalyst) exhibiting a greater positive kinetic effect of water are consistent with the proposal that water can remove transport restrictions caused by pore filling from heavy liquid hydrocarbons.

To examine the fitness of the kinetic model CAER used for study of the water effect, we conducted model discrimination using the kinetic data of 15% Co/SiO_2 by comparing values of the standard function of mean absolute relative residual (MARR), which is simply defined as

$$MARR,\% = 100 \times \frac{\sum_{i=1} \left| (\overline{x_{CO}} - x_{CO})/x_{CO} \right|}{N_{exp}} \tag{3.2}$$

where i represents experiment, $\overline{x_{CO}}$ and x_{CO} represent experimental and calculated CO conversion values, respectively, and N_{exp} is the number of experiments.

In this study, four other kinetic models with or without water inhibition from references [11,13,20,25] plus the CAER model are used to fit the experimental results of the 15% Co/SiO_2 catalyst. The kinetic parameter values obtained based on the same analysis method along with MARR values are listed in Table 3.2.

Clearly, kinetic parameter values varied greatly with the kinetic models. The value of the water effect constant, m, generated by the CAER model is −0.33, but increased to −0.02 to −0.2 if using kinetic models e–f (Table 3.2). This shows that choosing the kinetic model is very important for analyzing the kinetic effect of water over supported cobalt catalysts. Among the kinetic models listed in Table 3.2, c–e are empirical power law equations, while f and g are Langmuir-Hinshelwood-Hougen-Watson (LHHW) equations and are derived from carbide and enol mechanisms, respectively. Overall, empirical kinetic models are better fitted to the experimental data than LHHW models for the cobalt catalyst. However, the best fit was obtained with the CAER model since the MARR value, 17.6, is the smallest among the five numbers.

3.3.2 Impact of Varying Calcination Procedures on FT Performance Parameters over 15% Co/SiO₂

A comparison of catalyst activity and selectivities to hydrocarbon and CO_2 over the reduced air calcined and nitric oxide calcined 15% Co/SiO_2 catalysts is shown in Figures 3.1 through 3.4 and Table 3.3. Initial CO conversion at an SV of 10 Nl/g-cat/h over the reduced air calcined sample was 33%, but was significantly

TABLE 3.2

Summary of Kinetic Parameter Values from Different Kinetic Models[a]

	Kinetic Parameter					MARR[b]
Kinetic Model	a	b	Ea, kJ/mol	m	k, mol/g-cat/h/ MPa$^{(a+b)}$	%
$r_{CO} = -kP_{CO}{}^aP_{H_2}{}^b/(1 + m\ P_{H_2O}/P_{H_2})^{(c)}$	−0.22	0.6	85.9	−0.33	0.0187	17.6
$r_{CO} = -kP_{CO}{}^aP_{H_2}{}^{b(d)}$	−0.22	0.6	78.1	N/A	0.0207	20.8
$r_{CO} = -kP_{CO}{}^aP_{H_2}{}^b/(1 + mP_{CO})^{(e)}$	−0.22	0.6	78.2	−0.02	0.0192	18.0
$r_{CO} = -kP_{H_2}/(1 + m\ P_{H_2O}/P_{CO}/P_{H_2})^{(f)}$	0	1	70.1	−0.19	0.0228	20.9
$r_{CO} = -kP_{H2}/(1 + m\ P_{H_2O}/P_{CO})^{(g)}$	0	1	72.1	−0.20	0.0229	21.2

[a] Catalyst: 15%Co/SiO2.
[b] Mean absolute relative residual.
[c] CAER kinetic model.
[d-g] From references [20], [25], [11], and [13], respectively.

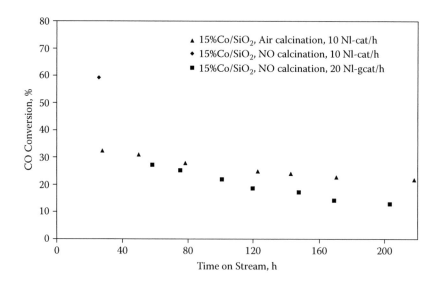

FIGURE 3.1 Change of CO conversion with time on stream on 15% Co/SiO$_2$ (220°C, 2.03 MPa, H$_2$/CO = 2.5, and N$_2$ = 12.5%).

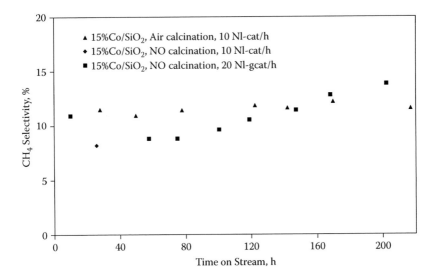

FIGURE 3.2   Change of CH_4 selectivity with time on stream on 15% Co/SiO_2 (220°C, 2.03 MPa, H_2/CO = 2.5, and N_2 = 12.5%).

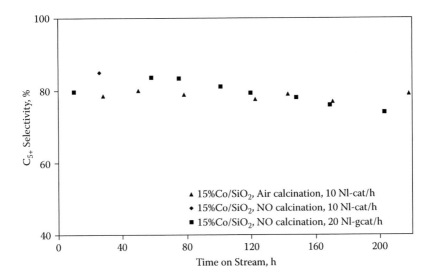

FIGURE 3.3   Change of C_{5+} selectivity with time on stream on 15% Co/SiO_2 (220°C, 2.03 MPa, H_2/CO = 2.5, and N_2 = 12.5%).

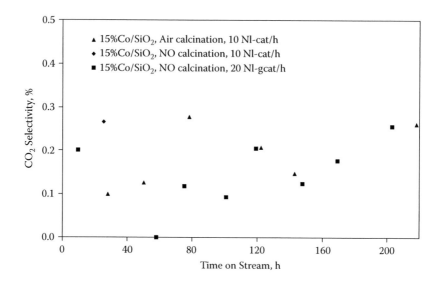

FIGURE 3.4   Change of CO_2 selectivity with time on stream on 15% Co/SiO$_2$ (220°C, 2.03 MPa, H_2/CO = 2.5, and N_2 = 12.5%).

TABLE 3.3

Effect of Calcination on Activity and Selectivity to Hydrocarbon and CO_2 of FTS over 15% Co/SiO$_2$

	CO Consumption Rate, mol/g-cat/h		CH$_4$ Selectivity, mol%	C$_{5+}$ Selectivity, mol%	CO$_2$ %
Calcination	0–75 h	0–220 h	0–220 h	0–220 h	0–220 h
Air	0.034	0.0297	11.7	78.9	0.29
Nitric oxide	0.0606	0.044	10.9	79.7	0.19

higher, 60%, for the reduced nitric oxide calcined sample. In order to decrease the high deactivation rate observed at high CO conversion levels, we increased SV to 20 for the NO calcined sample after 28 h on stream. It can be seen that the SV used for the nitric oxide calcined sample is twice that of the air calcined sample, and the two catalysts showed similar CO conversion rates before 80 h of testing, indicating much higher activity on a per gram catalyst basis for the nitric oxide calcined 15% Co/SiO$_2$ catalyst. A summary of CO consumption rates on a per gram catalyst basis is shown in Table 3.3 and indicates that in the first 75 h, the nitric oxide calcined sample was more active on a per gram catalyst basis than the air calcined sample by 77%, but decreased to 48% during 220 h of testing. This result, along with the curves in Figure 3.1, reflects faster deactivation for the nitric oxide calcined sample (2.12%/day vs. 1.35%/day), even though twice the space velocity was used as the catalyst. Characterization results (see Chapter 8) [18] show that the nitric oxide calcination method resulted in, upon reduction, a higher Co active site density (versus the reduced air calcined catalyst) in spite of a decrease in the extent of reduction of Co oxide species, due to a smaller average cobalt crystallite size obtained (i.e., higher dispersion of metal clusters). This is consistent with the higher activity observed for the reduced nitric oxide treated sample on a per gram catalyst basis.

The application of the novel NO calcination method in lieu of the air calcination method not only benefited CO conversion, but also improved selectivities to heavier hydrocarbons (Figures 3.2 and 3.3 and Table 3.3). In the first 100 h, the nitric oxide method decreased methane selectivity and increased C_{5+} hydrocarbon selectivity by ~4.5%. Because of the faster deactivation of the nitric oxide calcined sample due to loss of Co active sites, CH_4 increased and C_{5+} selectivities decreased after 100 h. Yet, measuring average values of the two parameters during 220 h of testing still showed lower selectivity of CH_4 and higher selectivity of C_{5+} for the nitric oxide treated sample (Table 3.3). This might be due to a Co cluster size effect (i.e., small Co particles may enhance the selectivity toward heavy hydrocarbons [26]). Both catalysts showed an increasing trend of CH_4 selectivity with time on stream, but a more rapid increase was observed for the nitric oxide treated sample. This may be related to one of a number of factors: (1) cobalt oxide formation for the smallest crystallites, (2) the formation of cobalt silicate complexes, (3) carbon formation, or (4) sintering during FTS. The fourth factor may be a consequence of higher surface free energies of particles supported on SiO_2 versus Al_2O_3, since it is well known that even air calcination results in small Co particles with the more strongly interacting Co/Al_2O_3 catalyst.

Figure 3.4 and Table 3.3 show that CO_2 selectivity over the 15% Co/SiO_2 catalyst is quite low, less than 0.4% regardless of the calcination procedure used, indicating a small extent of the water-gas shift (WGS) reaction over the catalysts. However, as shown in Table 3.3, a slightly lower average CO_2 selectivity was observed over the NO calcined 15% Co/SiO_2 catalyst compared to the air calcined one (0.19 vs. 0.29%), another indication that the NO calcination benefited FTS performance.

3.3.3 IMPACT OF VARYING CALCINATION PROCEDURES ON FT PERFORMANCE PARAMETERS OVER 25% Co/SiO$_2$

The reaction results for the reduced air and nitric oxide calcined 25% Co/SiO_2 catalysts are given in Figures 3.5 to 3.8 and Table 3.4. In agreement with the results obtained with 15% Co/SiO_2, the nitric oxide calcination significantly increases catalyst activity on a per gram catalyst basis, as well as promotes heavier hydrocarbon selectivity. From Table 3.4, using the nitric oxide calcination over the air calcination procedure led to an increase in the average CO consumption rate by

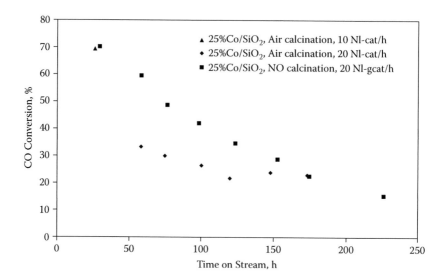

FIGURE 3.5 Change of CO conversion with time on stream on 25% Co/SiO$_2$ (220°C, 2.03 MPa, H$_2$/CO = 2.5, and N$_2$ = 12.5%).

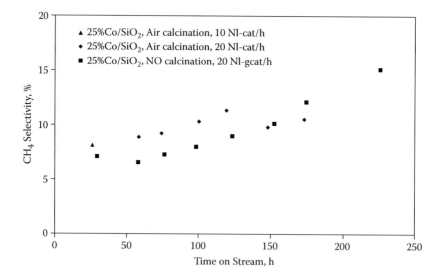

FIGURE 3.6 Change of CH$_4$ selectivity with time on stream on 25% Co/SiO$_2$ (220°C, 2.03 MPa, H$_2$/CO = 2.5, and N$_2$ = 12.5%).

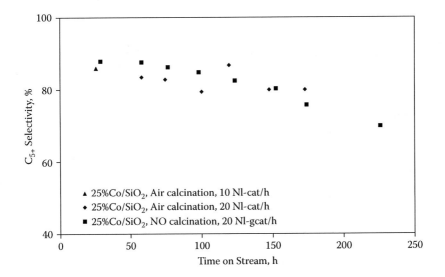

FIGURE 3.7 Change of C_{5+} selectivity with time on stream on 25% Co/SiO$_2$ (220°C, 2.03 MPa, H$_2$/CO = 2.5, and N$_2$ = 12.5%).

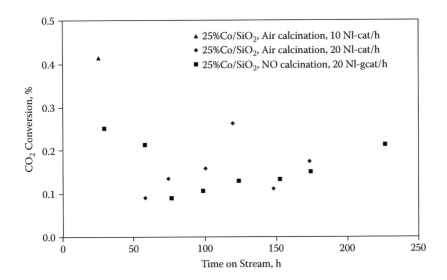

FIGURE 3.8 Change of CO$_2$ selectivity with time on stream on 25% Co/SiO$_2$ (220°C, 2.03 MPa, H$_2$/CO = 2.5, and N$_2$ = 12.5%).

TABLE 3.4

Effect of Calcination on Activity and Selectivity to Hydrocarbon and CO$_2$ of FTS over 25% Co/SiO$_2$

Calcination	CO Consumption Rate, mol/g-cat/h		CH$_4$ Selectivity, mol%	C$_{5+}$ Selectivity, mol%	CO$_2$ %
	0–75 h	0–175 h	0–175 h	0–175 h	0–175 h
Air	0.073	0.0619	9.81	82.6	0.19
Nitric oxide	0.133	0.0976	8.6	83.6	0.15

82.2% in the first 75 h for the 25% Co/SiO$_2$ catalyst, but the increase was reduced to 58% by 175 h.

Examining the data for catalysts at SV 20 in Figure 3.5 and beginning near 30% conversion, a much faster deactivation with the reduced nitric oxide calcined 25% Co/SiO$_2$ was observed (~5.7%/day in Figure 3.5) in comparison with the reduced 25% Co/SiO$_2$ air calcined catalyst (~1.7%/day). Also, a faster deactivation was observed with the reduced 25% Co/SiO$_2$ nitric oxide calcined catalyst (again, ~5.7%/day) relative to the reduced 15% Co/SiO$_2$ nitric oxide calcined catalyst (~2.1%/day). These results could signify a higher sensitivity of water for smaller cobalt particles. Recall that smaller Co particles were measured with the reduced 25% Co/SiO$_2$ nitric oxide calcined catalyst (~5.7%/day, ~14 nm) relative to the reduced 25% Co/SiO$_2$ air calcined catalyst (~1.7%/day, ~27 nm); also, smaller Co particles were formed over the reduced 25% Co/SiO$_2$ nitric oxide calcined catalyst (again, ~5.7%/day, ~14 nm) relative to the reduced nitric oxide calcined 15% Co/SiO$_2$ catalyst (~2.1%/day, ~19 nm).

Comparing the effect of cobalt loading on activity and selectivity of FTS leads to other meaningful results. From Figures 3.1 through 3.8 and Tables 3.3 and 3.4, the higher Co loading catalysts produce less CH$_4$ and CO$_2$ and higher selectivity to C$_{5+}$ hydrocarbons. The methane selectivity on the 25% Co catalyst was 8.6 to 9.6%, which increased to 11 to 11.7% on the 15% Co sample, while C$_{5+}$ decreased from 83% to 80%, regardless of the calcination procedures employed. For the CO$_2$ selectivity, we obtained 0.15 to 0.19% over the 25% Co catalyst, but it increased to 0.19 to 0.28% on the 15% Co catalyst. As discussed, Co particle size may affect product selectivity, with smaller Co particles inhibiting methane and CO$_2$ formation. This is evidenced by our correlations of CH$_4$, CO$_2$, and C$_{5+}$ selectivities with Co cluster size, as shown in Figure 3.9a–c. There is an outlier point for selectivity of CH$_4$ and C$_{5+}$ at Co cluster size of 19 nm in Figure 3.9a and c, which is correlated with the NO calcined 15% Co/SiO$_2$ catalyst. The outlier may be a result of the fast deactivation of the catalyst. Nevertheless, the general trends reported are retained.

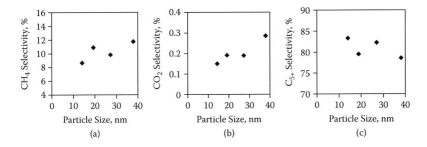

FIGURE 3.9 Effect of Co particle size on selectivities to (a) CH_4, (b) CO_2, and (c) C_{5+} on SiO_2-supported Co catalysts (220°C, 2.03 MPa, H_2/CO = 2.5, 10–20 Nl/g-cat/h, and N_2 = 12.5%).

3.4 CONCLUSIONS

The regression of kinetic parameters was conducted on experimental data obtained using a 1 L slurry reactor over reduced air calcined 15% Co/SiO_2 and 25% Co/SiO_2 catalysts. The model that provided the best fit was $r_{CO} = kP_{CO}{}^a P_{H_2}{}^b/(1 + mP_{H_2O}/P_{H_2})$ with parameter values of k = 0.0187 mol/g-cat/h/MPa$^{0.38}$ (220°C), E_a = 85.9 kJ/mol, a = −0.22, b = 0.6, and m = −0.33 on the 15% Co/SiO_2, and of k = 0.0381 mol/g-cat/h/MPa$^{0.32}$ (220°C), E_a = 93.7 kJ/mol, a = −0.19, b = 0.51, and m = −1.11 on the 25% Co/SiO_2. Thus, the rate exhibits a positive water effect. Comparison of the kinetic results between the two air calcined Co catalysts suggests that the smaller Co particles (~27 nm) residing inside the SiO_2 pores (~28 nm diameter) on the reduced air calcined 25% Co/SiO_2 catalyst are more sensitive to the kinetic effect of water in the SiO_2-supported catalyst system than the larger Co particles (~38 nm) found on the reduced air calcined 15% Co/SiO_2 catalyst, which possess a greater fraction external to the SiO_2 pores. Consistent with these findings, the water effect may be explained by the ability of water to displace heavy hydrocarbons from the pore, thereby removing intraparticle transport restrictions.

The nitric oxide calcination significantly improved catalyst activity on a per gram catalyst basis, as well as heavier hydrocarbon selectivity. This is ascribed to the ability of the nitric oxide treatment to (1) prevent the rapid decomposition of the cobalt nitrate precursor and (2) act as an oxygen scavenger, thus limiting agglomeration and resulting in a smaller average Co particle size upon reduction. Catalysts with smaller Co particles, and especially those produced by the nitric oxide calcination method, were found to be more sensitive to deactivation phenomena during FTS, and water may influence this behavior. Nevertheless, the nitric oxide treatment is a very useful procedure to increase the productivity toward heavier hydrocarbons during FTS over Co/SiO_2. Further studies should focus on improving the stability of these catalysts.

ACKNOWLEDGMENTS

This work was supported by NASA Contract NNX07AB93A and the Commonwealth of Kentucky.

REFERENCES

1. Jacobs, G., Chaney, J.A., Patterson, P.M., Das, T.K., Maillot, J. C., and Davis, B.H. 2004. Fischer-Tropsch synthesis: Study of the promotion of Pt on the reduction property of Co/Al$_2$O$_3$ catalysts by in situ EXAFS of Co K and Pt LIII edges and XPS. *J. Synch. Rad.* 11:414–22.

2. Li, J.L., Jacobs, G., Das, T.K., Zhang, Y.Q., and Davis, B.H. 2002. Fischer-Tropsch synthesis: Effect of water on the catalytic properties of a Co/SiO2 catalyst. *Appl. Catal.* 236:67–76.

3. Li, J.L., Jacobs, G., Das, T.K., and Davis, B.H. 2002. Fischer-Tropsch synthesis: Effect of water on the catalytic properties of a ruthenium promoted Co/TiO$_2$ catalyst. *Appl. Catal.* 233:255–62.

4. Li, J.L., Zhan, X.D., Zhang, Y.Q., Jacobs, G., Das, T.K., and Davis, B.H. 2002. Fischer-Tropsch synthesis: Effect of water on the deactivation of Pt promoted Co/Al$_2$O$_3$ catalysts. *Appl. Catal.* 228:203–12.

5. Dalai, A.K., Das, T.K., Chaudhari, K.V., Jacobs, G., and Davis, B. H. 2005. Fischer-Tropsch synthesis: Water effects on Co supported on narrow and wide-pore silica. *Appl. Catal.* 289:135–42.

6. Jacobs, G., Das, T.K., Patterson, P.M., Li, J.L., Sanchez, L., and Davis, B.H. 2003. Fischer-Tropsch synthesis XAFS: XAFS studies of the effect of water on a Pt-promoted Co/Al$_2$O$_3$ catalyst. *Appl. Catal.* 247:335–43.

7. Vada, S., Hoff, A., Ådnane, E., Schanke, D., and Holmen, A. 1995. Fischer-Tropsch synthesis on supported cobalt catalysts promoted by platinum and rhenium. *Topics Catal.* 2:155–62.

8. Claeys, M., and van Steen, E. 2002. On the effect of water during Fischer-Tropsch synthesis with a ruthenium catalyst. *Catal. Today* 71:419–27.

9. Schulz, H., Claeys, M., and Harms, S. 1997. Effect of water partial pressure on steady state Fischer-Tropsch activity and selectivity of a promoted cobalt catalyst. *Stud. Surf. Sci. Catal.* 107:193–200.

10. Das, T.K., Zhan, X.D., Li, J.L., Jacobs, G., Dry, M.E., and Davis, B.H. 2007. Fischer-Tropsch synthesis: Kinetics and effect of water for a Co/Al$_2$O$_3$ catalyst. *Stud. Surf. Sci. Catal.* 163:289–314.

11. Huff, G. A., Jr., and Satterfield, C. N. 1984. Intrinsic kinetics of the Fischer-Tropsch synthesis on a reduced fused-magnetite catalyst. *Ind. Eng. Chem. Process Des. Dev.* 23:696–705.

12. Das, T. K., Conner, W.A., Li, J.L., Jacobs, G., Dry, M.E., and Davis, B.H. 2005. Fischer-Tropsch synthesis: Kinetics and effect of water for a Co/SiO$_2$ catalyst. *Energy & Fuels* 19:1430–39.

13. Nettelhoff, H., Kokuun, R., Ledakpwicz, S., and Deckwer, W.D. 1985. Studies on the kinetics of Fischer-Tropsch synthesis in slurry phase. *Ger. Chem. Eng.* 8:177–85.

14. Krishnamoorthy, S., Tu, M., Ojeda, M. P., Pinna, D., and Iglesia, E. 2002. An investigation of the effects of water on rate and selectivity for the Fischer-Tropsch synthesis on cobalt-based catalysts. *J. Catal.* 211:422–33.

15. Minderhoud, J.K., Post, M.F.M., Sie, S.T., and Sudholter, E.J.R. 1986. U.S. Patent 4628133.

16. Bertole, C.J., Mims, C.A., and Kiss, G. 2002. The effect of water on the cobalt-catalyzed Fischer–Tropsch synthesis. *J. Catal.* 210:84–96.
17. Ma, W.P., Jacobs, G., Sparks, D.E., Spicer, R.L., Graham, U.M., and Davis, B. H. 2008. Comparison of the kinetics of the Fischer-Tropsch synthesis reaction between structured alumina supported cobalt catalysts with different pore size. *Prepr. Am. Chem. Soc. Div. Petro. Chem.* 53:99–102. (see Chapter 8 of this book.)
18. Jacobs, G., Ma, W.P., Ji, Y.Y., Khalid, S., and Davis, B.H. 2008. Characterization of CO/silica catalysts prepared by a novel NO calcination method. *Prepr. Am. Chem. Soc. Div. Petro. Chem.* 53:78–83.
19. Sietsma, J.R.A., van Dillen, A.J., de Jongh, P.E., and de Jong, K.P. 2008. PCT International Application WO 2008029177.
20. Iglesia, E. 1997. Design, synthesis, and use of cobalt-based Fischer-Tropsch synthesis catalysts. *Appl. Catal.* 161:59–78.
21. Zennaro, R., Tagliabue, M., and Bartholomew, C. 2000. Kinetics of Fischer-Tropsch synthesis on titania-supported cobalt. *Catal. Today* 58:309–19.
22. Chang, J., Teng, B.T., Bai, L., Chen, J.G., Zhang, R.L., Xu, Y.Y., Xiang, H.W., Li, Y.W., and Sun, Y.H. 2005. Detailed kinetic study of Fischer-Tropsch synthesis on $Co/ZrO_2/SiO_2$ catalyst. II. Construction and regression of kinetic models. *Cuihua Xuebao* 26:859–68.
23. Withers, H.P., Eliezer, K.F., and Mitchell, J.W. 1990. Slurry-phase Fischer-Tropsch synthesis and kinetic studies over supported cobalt carbonyl derived catalysts. *Ind. Eng. Chem. Res.* 29:1807–14.
24. Li, C., Cao, F.H., Ying, W.W., and Fang, D.Y. 2006. Intrinsic kinetics of ZrO_2 modified Co-Ru/γ-Al_2O_3 catalyst for F-T synthesis. *Huadong Ligong Daxue Xuebao Ziran Kexueban* 32:1253–57.
25. Iglesia, E., Reyes, S.C., and Soled, S.L. 1993. *Computer-aided design of catalysts*, ed. R.E. Becker and C.J Pereira. New York: Marcel Dekker.
26. McDonald, M.A., Storm, D.A., and Boudart, M. 1986. Hydrocarbon synthesis from carbon monoxide-hydrogen on supported iron: Effect of particle size and interstitials. *J. Catal.* 102:386–400.

4 The Formation and Influence of Carbon on Cobalt-Based Fischer-Tropsch Synthesis Catalysts
An Integrated Review

Denzil J. Moodley, Jan van de Loosdrecht,
Abdool M. Saib, and Hans J. W. Niemantsverdriet

CONTENTS

Cobalt-based Fischer-Tropsch synthesis (FTS) catalysts are the systems of choice for use in gas-to-liquid (GTL) processes. As with most catalysts, cobalt systems gradually lose their activity with increasing time on stream. There are various mechanisms that have been proposed for the deactivation of cobalt-based catalysts during realistic FTS conditions. These include poisoning, sintering, oxidation, metal support compound formation, restructuring of the active phase, and carbon deposition. Most of the recent research activities on cobalt catalyst deactivation during the FTS have focused on loss of catalyst activity due to oxidation of the metal and support compound formation. Relatively few recent studies have been conducted on the topic of carbon deposition on cobalt-based FTS catalysts. The purpose of this review is to integrate the existing open and patent literature with some of our own work on the topic of carbon deposition to provide a clearer understanding on the role of carbon as a deactivation mechanism.

4.1 INTRODUCTION AND SCOPE

The Fischer-Tropsch synthesis is a process that converts synthesis gas into mixtures of higher molecular weight hydrocarbons.[1] The FTS is at the heart of the GTL process, which converts natural gas to "clean" synfuels.[2] This approach is attractive due to rising oil prices and the need to comply with more stringent legislation on the quality of liquid fuels.[3]

The two catalytically active metals for FTS that are used in industry are iron (fused or precipitated) and cobalt (supported). Iron catalysts display higher water gas shift activities ($CO + H_2O \rightarrow CO_2 + H_2$) and are more suitable for use with coal- and biomass-derived synthesis gas feeds, which have lower hydrogen content.[4] Cobalt catalysts exhibit high per pass activities, have low water gas shift activity, which leads to improved carbon utilization, and are suitable for use on synthesis gas produced via reforming of natural gas.[5] Cobalt FTS catalysts yield mainly straight-chain hydrocarbons. Since cobalt is much more expensive than iron, dispersing the ideal concentration and size of metal nanoparticles onto a support can help reduce catalyst costs while maximizing activity and durability. However, as with almost all catalysts, cobalt FTS catalysts gradually deactivate with time on stream.

Various mechanisms have been proposed for the deactivation of cobalt-based catalysts during realistic FTS conditions. These include the following.

4.1.1 Oxidation of Active Phase and Support Compound Formation

The oxidation of cobalt metal to inactive cobalt oxide by product water has long been postulated to be a major cause of deactivation of supported cobalt FTS catalysts.[6-10] Recent work has shown that the oxidation of cobalt metal to the inactive cobalt oxide phase can be prevented by the correct tailoring of the ratio P_{H_2O}/P_{H_2} and the cobalt crystallite size.[11] Using a combination of model systems, industrial catalyst, and thermodynamic calculations, it was concluded that Co crystallites > 6 nm will not undergo any oxidation during realistic FTS, i.e., P_{H_2O}/P_{H_2} = 1–1.5.[11-14] Deactivation may also result from the formation of inactive cobalt support compounds (e.g., aluminate). Cobalt aluminate formation, which likely proceeds via the reaction of CoO with the support, is thermodynamically favorable but kinetically restricted under typical FTS conditions.[6]

4.1.2 Poisoning by Contaminants in the Synthesis Gas Feed

One of the causes of deactivation is the strong chemisorption of poisons on the metallic cobalt phase. According to Bartholomew,[15] poisons may (1) block active sites for the reaction, (2) electronically modify the metals nearest neighbor, affecting chemisorption and dissociation of CO, and (3) cause reconstruction of the catalyst surface, resulting in a more stable configuration. Sulfur,[16] halides, and NH_3/HCN[17-20] are generally the major poisons for cobalt catalysts during FTS. Poisoning is synthesis gas feed related and can therefore be minimized through synthesis gas purification steps; for example: (1) ZnO guard beds reduce sulfur levels significantly[21] and (2) a synthesis gas washing step with an aqueous solution of alkaline ferrous sulfate promotes the absorption of the above-mentioned impurities.[22]

4.1.3 Sintering of the Cobalt Active Phase

To prepare a good catalyst in terms of activity and cost, cobalt nanoparticles have to be well dispersed on a support that typically consists of alumina, silica, or titania. Small metal particles have a high surface free energy and tend to minimize this by either changing shape or agglomerating together (sintering). Sintering results in deactivation via the loss of catalytic surface area and has previously been reported during FTS on cobalt catalysts.[8,23,24] Sintering may occur via crystallite migration and coalescence or by atom migration/Oswald ripening.[15] The Hüttig temperature of cobalt, at which atoms at defects become mobile, is 253°C,[25] close to temperatures employed for realistic FTS conditions, supporting the above evidence for sintering. It should also be noted that sintering of the active phase may be facilitated by reaction water[15] and the formation of mobile subcarbonyl species.[26]

4.1.4 COBALT RECONSTRUCTION

It has been observed that cobalt may undergo large-scale reconstruction under a synthesis gas environment.[27] Reconstruction is a thermodynamically driven process that results in the stabilization of less reactive surfaces. Recent molecular modeling calculations have shown that atomic carbon can induce the clock reconstruction of an fcc cobalt (100) surface.[28] It has also been postulated and shown with *in situ* x-ray adsorption spectroscopy (XAS) on cobalt supported on carbon nanofibers that small particles (<6 nm) undergo a reconstruction during FTS that can result in decreased activity.[29]

4.1.5 FOULING BY PRODUCT WAX AND DEPOSITION OF CARBON

Although the FTS is considered a carbon in-sensitive reaction,[30] deactivation of the cobalt active phase by carbon deposition during FTS has been widely postulated.[31–38] This mechanism, however, is hard to prove during realistic synthesis conditions due to the presence of heavy hydrocarbon wax product and the potential spillover and buildup of inert carbon on the catalyst support. Also, studies on supported cobalt catalysts have been conducted that suggest deactivation by pore plugging of narrow catalyst pores by the heavy ($>C_{40}$) wax product.[39,40] Very often, regeneration treatments that remove these carbonaceous phases from the catalyst result in reactivation of the catalyst.[32] Many of the companies with experience in cobalt-based FTS research report that these catalysts are negatively influenced by carbon (Table 4.1).

The purpose of this review is to integrate the literature on this topic, along with some of the work we have performed, to provide a clearer understanding on the role of carbon as a deactivation mechanism. The minimization of carbon by promotion, regeneration of catalysts, and some selectivity implications will also be briefly discussed.

4.2 FORMATION OF CARBON DEPOSITS ON COBALT CATALYSTS DURING FTS AND IMPLICATIONS FOR ACTIVITY

Carbonaceous species on metal surfaces can be formed as a result of interaction of metals with carbon monoxide or hydrocarbons. In the FTS, where CO and H_2 are converted to various hydrocarbons, it is generally accepted that an elementary step in the reaction is the dissociation of CO to form surface carbidic carbon and oxygen.[1] The latter is removed from the surface through the formation of gaseous H_2O and CO_2 (mostly in the case of Fe catalysts). The surface carbon, if it remains in its carbidic form, is an intermediate in the FTS and can be hydrogenated to form hydrocarbons. However, the surface carbidic carbon may also be converted to other less reactive forms of carbon, which may build up over time and influence the activity of the catalyst.[15]

TABLE 4.1

Carbon Deactivation Postulated for Industrial Cobalt Catalysts

Company	Catalyst	Typical Conditions	Comments	Ref.
BP	Co/ZnO	218°C, 29 bar, $H_2/CO = 2$	Deactivation due to the formation of small amounts of inert carbon species on the cobalt active phase; regeneration required to maintain activity	35
Conoco-Phillips	Co/Al$_2$O$_3$	225°C, 24 bar, $H_2/CO = 2$	Regeneration process by steam needed due to coking of the catalyst caused by high support acidity or high temperatures in particles resulting from high initial conversions	38
ExxonMobil	Co/TiO$_2$	225°C, 20 bar, $H_2/CO = 2$	Regeneration process that is necessary due to the deposition of carbon or coke on catalyst	36
Shell	Co/Zr/SiO$_2$	220°C, 25 bar, $H_2/CO = 2$	Regeneration process needed to remove heavy products and carbonaceous deposits that diminish activity	37
Syntroleum	Co/Al$_2$O$_3$	220°C, 20 bar, $H_2/CO = 2$	Accumulation of unreactive polymeric carbon with time on stream resulting in deactivation	34

There are a number of ways that carbon may interact with a cobalt catalyst to affect its performance during FTS:

1. Carbon deposits or heavy hydrocarbons ($>C_{100}$) may block the catalyst pores causing diffusion problems.[39]

2. Carbon may adsorb on the metal surface irreversibly, therefore acting as a poison.[35] This irreversibly bonded carbon could also affect the adsorption and dissociation of neighboring species such as CO.

3. Carbon could also go subsurface and play a role in electronic inhibition of activity by affecting the adsorption and dissociation of CO.[41]

4. Carbon may bind to a metal surface and induce a surface reconstruction whereby a more active metal plane is transformed to one with a lower activity.[28]

5. At higher temperatures, out of the typical FT regime, carbon could encapsulate the active metal, thereby blocking access to reactants. In extreme cases carbon filaments can also be formed that can result in the breakup of catalyst particles.[42]

4.3 CLASSIFICATION OF CARBON TYPES ON COBALT FTS CATALYSTS

Figure 4.1 summarizes the different routes that can potentially lead to carbon deposition during FTS: (a) CO dissociation occurs on cobalt to form an adsorbed atomic carbon, which is also referred to as surface carbide, which can further react to produce the FT intermediates and products. The adsorbed atomic carbon may also form bulk carbide or a polymeric type of carbon. Carbon deposition may also result (b) from the Boudouard reaction and (c) due to further reaction and dehydrogenation of the FTS product (what is commonly called coke), a reaction that should be limited at typical FT reaction conditions. Carbon formed on the surface of cobalt can also spill over or migrate to the support. This is reported to readily occur on Co/Al_2O_3 catalysts.[43] The chemical nature of the carbonaceous deposits during FTS will depend on the conditions of temperature and pressure, the age of the catalyst, the chemical nature of the feed, and the products formed.

It would be fitting at this stage to define in detail the various carbon species for this review, as often different terms are used in the literature. A representation of the various carbon species is shown in Figure 4.2. Surface carbide or atomic carbon can be defined as isolated carbon atoms with only carbon-metal bonds, resulting from CO dissociation or disproportionation, the latter of which is not favored on cobalt at normal FTS conditions. Recent theoretical and experimental work has indicated that the CO dissociation is preferred at the step sites, so absorbed surface carbide is expected to be located near these sites.[44–46]

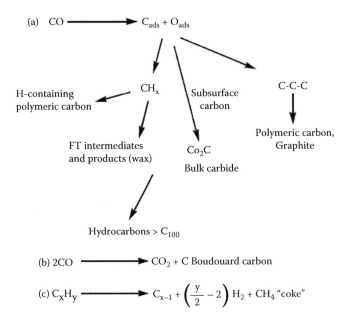

FIGURE 4.1 Possible modes of carbon formation during FTS on cobalt catalysts.

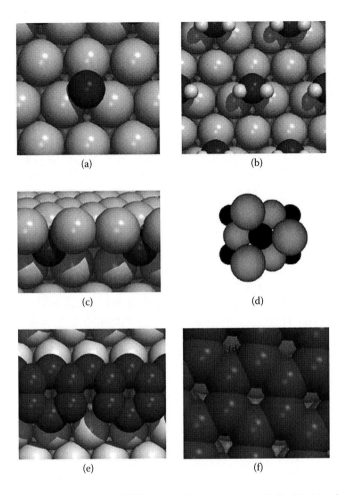

FIGURE 4.2 Representation of different carbon types on cobalt. (a) Atomic carbon/ surface carbide in a threefold hollow site. (b) CH_x species located in threefold hollow sites. (c) Subsurface carbon lying in octahedral positions below the first layer of cobalt. (d) Cobalt carbide (Co_2C) with an orthorhombic structure. (e) Polymeric carbon on a cobalt surface. (f) A sheet of graphene lying on a cobalt surface. The darker spheres represent carbon atoms in all the figures.

CH_x and hydrocarbon wax are, respectively, the active intermediates formed by the hydrogenation of surface carbide and products of FTS formed by chain growth and hydrogenation of CH_x intermediates. The hydrocarbon wax can contain molecules with the number of carbon atoms in excess of 100. Bulk carbide refers to a crystalline Co_xC structure formed by the diffusion of carbon into bulk metal. Subsurface carbon may be a precursor to these bulk species and is formed when surface carbon diffuses into an octahedral position under the first surface layer of cobalt atoms.

Polymeric carbon refers to chains of carbon monomers (surface carbide) that are connected by covalent bonds. It has been shown recently[47] that the barrier for C–C coupling on flat surfaces (1.22 eV) is half that for a step site (2.43 eV), and may indicate that the growth of these polymeric species is favored on terraces. Polymeric carbon may also refer to carbon chains that contain hydrogen. In the case of CO hydrogenation on ruthenium catalysts, polymeric carbon has been identified as a less reactive carbon that forms from polymerization of CH_x and has an alkyl group structure.[48]

Graphene is a single layer of carbon atoms densely packed into a benzene-ring structure and may be considered a precursor to graphite. In graphite each carbon atom is covalently bonded to three other surrounding carbon atoms. The flat sheets of carbon atoms are bonded into hexagonal structures, which are layered. These graphitic species (or free carbon, as they are often called) have strong carbon–carbon bonds and weaker bonds to the metal surface. The Boudouard reaction ($2CO \rightarrow C + CO_2$) at FTS temperatures (around 230°C) has been reported on cobalt catalysts and also results in the deposition of atomic carbon and its transformation to polymeric or graphitic forms of carbon on the surface.[49] Typically at high temperature Boudouard carbon can diffuse in cobalt to form metastable bulk carbide species.[50] The decomposition of the bulk carbide results in the formation of filaments and other forms of carbon on the surface. Filaments consist of stacked cone-segment (frustum)-shaped graphite basal plane sheets, grow with a catalyst particle at their tip and, as can be expected, lead to the breakup of the catalyst.[50] Another graphitic nanomaterial produced by carbon deposition is encapsulated metal nanoparticles.[50] These are roughly spherical formations, consisting of catalyst particles surrounded by graphitic carbon.

These different types of carbon tend to have different reactivities toward gases such as hydrogen, oxygen, or steam. Hence, a relatively simple technique such as temperature-programmed hydrogenation or oxidation can be used to classify them. Table 4.2 summarizes different reactivities of carbon species toward hydrogen.

4.4 FACTORS THAT GENERALLY INFLUENCE CARBON DEPOSITION ON CATALYSTS FOR CO HYDROGENATION

4.4.1 Temperature and Pressure

Temperature plays an important role in determining the amount and type of the carbon deposit. Generally during FTS at higher temperatures the amount of carbon deposited will tend to increase,[30,31] but the case is often not so straightforward. An example of temperature dependence on the rate of carbon deposition and deactivation is the case of nickel CO hydrogenation catalysts, as studied by Bartholomew.[56] At temperatures below 325°C the rate of surface carbidic carbon removal by hydrogenation exceeds that of its formation, so no carbon is deposited. However, above 325°C, surface carbidic carbon accumulates on the surface

TABLE 4.2

Examples of Various Carbon Species on Cobalt FTS Catalysts along with Their Hydrogenation Temperatures

| | | Reaction Conditions in Study | | | | |
Carbon Species	Catalyst	Temp. (°C)	$H_2/$ CO	Pressure (bar)	T_{hyd} (°C)	Ref.
CH$_x$ fragments	Co (0001)	220	1	1	<100	51,52
Surface carbide	Co/Al$_2$O$_3$	250	CO only[a]	1	180–200	31
Bulk carbide	Na-Co/Al$_2$O$_3$	240	2	50	<250	53
Hydrocarbons, paraffinic wax	Co/Al$_2$O$_3$	225	2	24	250–350	34
Polymeric carbon[b]	Co/Al$_2$O$_3$	225	2	24	>350	31,34
Graphite or graphene	Co/SiO$_2$	200	2	1	>620	54,55

[a] Surface carbide can be a product of both CO dissociation and disproportionation and can be formed from a mixture of H$_2$/CO as well.

[b] Polymeric carbon includes both carbon with only C–C bonds and hydrogen containing polymeric species.

since the rate of surface carbidic carbon formation is greater and exceeds that of its hydrogenation. As surface carbidic carbon accumulates (at 325–400°C), it is converted to a polymeric type of carbon that deactivates the nickel catalyst; however, above 425°C the rate of polymeric carbon hydrogenation exceeds that of formation and no deactivation occurs.

Higher temperatures will also aid in the transformation of surface carbon species into more stable species that will have decreased reactivity toward H$_2$. Nakamura et al.[49] showed that at 230°C, carburization of a Co/Al$_2$O$_3$ catalyst by CO results in formation of mainly carbidic carbon. Such carbidic carbon converts to graphitic carbon if the temperature is raised to around 430°C.[49] Increasing the exposure time to CO will also result in the formation of more stable carbon species.[57] If the catalyst is exposed to a too high temperature during FTS, undesired carbonaceous phases will be formed, which may damage the structural integrity of the catalyst (for example, carbon fibers or filaments).

Carbon deposition is a strong function of partial pressures of CO and H$_2$ in the gas phase. Rostrup-Nielsen showed that the amount of carbon deposited on the catalyst uniformly increases with the combined hydrogen and carbon monoxide pressure.[58] Moeller and Bartholomew[59] showed that the amount of carbon deposited on Ni catalysts was proportional to the partial pressure of CO. However, greater conversion at higher temperatures results in a corresponding decrease in P_{CO} and P_{H_2}, and may therefore lead to smaller amounts of carbon on the catalyst.[60] Higher conversions also lead to high water partial pressures, which can also influence carbon deposition, for example, by gasification of carbon or enhanced desorption of deleterious carbon/coke precursors. According to Dry, the formation rate of Boudouard carbon is a function of pressure for Fe catalysts.[61] He showed that at higher total pressure and lower $P_{H_2}/(P_{CO})^2$ ratio, the rate of carbon formation decreased.

4.4.2 Size and Crystallographic Nature of Metal Surface

Two studies[31,58] have suggested that carbon deposition rates are greater on smaller metal particles. This is most likely due to the presence of a higher concentration of defects on the small particles, which is known to enhance CO dissociation. Furthermore, in the case of Co/Al$_2$O$_3$ catalysts it was found that carbon formation rate and subsequent deactivation were higher for smaller cobalt particles.[31] The dissociation of adsorbed CO on a cobalt catalyst is also sensitive to the crystallographic structure of the surface, and it is known that dissociation of CO occurs readily on more open surfaces. The dissociation of CO$_{ads}$ and formation of surface carbidic carbon occurs preferentially on Co (1012) and (1120) than on Co (0001) and (1010) planes.[57,62] It is argued that CO adsorption at step sites (which are widely available on high index surfaces) weakens the C–O bond, which enables dissociation at lower temperatures. However, the carbon formed at these highly reactive cobalt sites may have enhanced stability (i.e., be strongly bound), and therefore may act as a poison. Hence, the optimum cobalt site is one that dissociates CO rapidly without leading to irreversible bonding of carbon.

4.4.3 Surface Coverage of Carbon

As the surface coverage of carbon increases, the deposited carbon becomes less reactive, as suggested by Koerts.[52] Using temperature-programmed hydrogenation, he showed that the formation of reactive surface carbidic carbon decreased from 70% to 10% as the surface coverage of carbon was increased toward 100%. Agrawal et al.[33] showed on Co/Al$_2$O$_3$ that greater CO concentrations resulting in an increased surface carbon concentration led to more rapid bulk carburization and rapid deactivation. Hence, the balance between dissociation and hydrogenation (i.e., the production and removal of carbon) must be maintained.

Molecular modeling work performed by Sasol researchers on fcc cobalt (100) shows that increased coverage of 50% atomic carbon will induce a clock type reconstruction (Figure 4.3) similar to that observed for the classic case of Ni (100).[28] The adsorption energy of the carbon is stabilized by 15 kJ/mol compared to the unreconstructed surface, resulting in a more stable surface.[28] The reconstruction results in a shorter distance between the carbon and cobalt but also an increase in coordination of the cobalt atoms and, thus, fewer broken bonds. The barrier for the carbon-induced clock reconstruction was found to be very small (1 kJ/mol), which suggested that the process is not kinetically hindered. The

FIGURE 4.3 Left: The unreconstructed surface of 50% C/fcc Co (100). Right: The clock reconstructed surface of 50% C/fcc Co (100). The darker spheres represent cobalt atoms and the lighter ones (in the fourfold hollow sites) represent carbon atoms. (Reprinted from Ciobica, I. M., van Santen, R. A., van Berge, P. J., and van de Loosdrecht, J., "Adsorbate Induced Reconstruction of Cobalt Surfaces," *Surface Science*, 602, 17–28. Copyright © 2008, with permission from Elsevier.)

atomic carbon may be hydrogenated in several bars of hydrogen that exist at FT conditions and the reconstruction can be lifted.

4.4.4 NATURE OF GAS FEED

The presence of a high concentration of H_2, i.e., high H_2/CO ratios, during FTS will make the formation of carbon deposits less favorable since the rate of hydrogenation of carbonaceous intermediates will be increased. This has been demonstrated by Sasol using model FTS tests (230°C, 1 bar, $H_2/CO = 0.5–2$) and subsequent temperature-programmed hydrogenation (TPH) experiments.[63] TPH analysis of the postreaction catalysts tested at higher H_2/CO ratios showed smaller amounts of less reactive carbonaceous phases, which may be detrimental to catalyst activity (Figure 4.4). A transformation of reactive carbon to more stable species was favored at low hydrogen concentration. Traces of poisons in the syngas feed also play a role in the deposition of carbon. The presence of low levels, i.e., ppm quantities, of sulfur in the feed stream resulted in a decrease in carbon deposition on Co/Al_2O_3 catalysts, probably via an ensemble site blocking effect.[64] It has been shown that increased water concentrations result in a decreased formation of carbon on nickel methanation catalysts.[65]

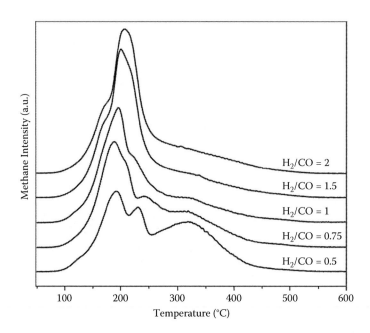

FIGURE 4.4 TPH profile of 20 wt% Co/Al$_2$O$_3$ catalysts tested in model FTS at different H$_2$/CO ratios (230°C, 1 bar). The P_{CO} was kept constant at 0.33 bar, while P_{H_2} was varied between 0.17 and 0.66 bar.[63]

4.5 STUDIES INVOLVING CARBON FORMATION ON COBALT CATALYSTS

4.5.1 Studies on Model Cobalt Systems at Model FT Conditions (CO + Syngas)

Carbon deposition from CO on a cobalt catalyst at low pressures is known to be a structure-sensitive process. CO is adsorbed molecularly on the low index surfaces (Co (0001)), but its dissociation occurs on the Co (1012), Co (1120), and polycrystalline surfaces.[57,62] Deposition of carbon on Co (1012) and the probable formation of Co$_3$C have been established by Auger emission spectroscopy (AES) and low-energy electron diffraction (LEED) techniques.[66]

Two forms of carbon (carbidic and graphitic) have been observed by x-ray photoelectron spectroscopy (XPS) on polycrystalline cobalt foil during the disproportionation of CO by Nakamura et al.[57] The dissociation of adsorbed CO occurred at temperatures higher than 60°C, and carbidic carbon and adsorbed oxygen were formed on the cobalt surface. After the surface is covered with adsorbed carbon and oxygen, no further dissociation of adsorbed CO occurs. Contrary to the dissociation of adsorbed CO, the deposition of carbon by the concerted Boudouard reaction continues on the carbidic carbon-deposited surface. The deposition of carbon increases with increasing exposure time, and there is an increase in its transformation to graphitic carbon.[57]

Johnson et al.[67] studied CO hydrogenation on bimetallic catalysts consisting of cobalt overlayers on W (100) and (110) single crystals at 200°C, 1 bar at a H_2/CO ratio of 2. AES spectra showed the postreaction Co/W surfaces to have high coverages of both carbon and oxygen, with carbon line shapes characteristic of bulk carbidic carbon.[67] The catalytic activity apparently could not be correlated with surface carbon level.[67]

Lahtinen et al.[68,69] studied CO hydrogenation on polycrystalline cobalt foil at various temperatures at 1 bar and H_2/CO ratio of 1.24. The cobalt surface was then characterized by AES immediately after the reaction without any further sample treatment. The C/Co ratio was almost constant as temperature was increased to 252°C. No significant deactivation for CO hydrogenation was observed on the foils at these conditions. At 297°C the C/Co ratio was significantly higher. From the peak shape of the carbon KLL Auger lines it was deduced that carbon formed at 297°C is in the graphitic form. Deactivation of the cobalt surface by carbon was observed at 276°C. On these metal foils the hydrogenation of CO occurs in the presence of an active carbidic overlayer. The transformation of this overlayer into graphite leads to a decrease in the catalytic activity of the metal surfaces.

The activity for CO hydrogenation was studied on Co (1120)- and (1012)-oriented single cobalt crystals by Geerlings et al.[51] The height of cobalt Auger peak decreased, while that of carbon increased due to the carbonaceous species on the surface. On the grooved Co (1120) surfaces long-chain hydrocarbon fragments grew; however, on the stepped Co (1012) surface long-chain fragments were not observed. The authors state that under FTS conditions, the step sites, which are very reactive for CO dissociation under UHV conditions, are poisoned by carbon. As a result of very strong binding of carbon atoms to these sites, efficient hydrogenation seems improbable. Hence, certain sites can aid carbon deposition and should be minimized. Beitel et al.[70] studied CO hydrogenation on Co (0001) at 250°C, 1 bar at a H_2/CO ratio of 2. They showed that the activity of a sputtered surface was greater than that of an annealed surface. However, the activity of both surfaces declined over time. They proposed that this could be due to the blocking of CO dissociation active sites by carbon deposition or by blocking of CO dissociation by hydrocarbons and water at defects. They conducted the experiments with clean syngas, and sintering could be eliminated for the most compact surface, and hence the observed deactivation could only be due to carbon.

It has been shown that it is favorable for surface carbon to go into the first subsurface layer of cobalt.[71] Diffusion to octahedral sites of the first subsurface layer is thermodynamically preferred by 50 to 120 kJ/mol and the corresponding activation energy is low. Theoretical calculations on the conversion of surface carbidic to subsurface carbon on Co (0001) found that the electron withdrawing power, and therefore the poisoning effect on potential CO adsorption, is maximal for subsurface carbon.[41] Metal d_{xz} orbitals are less likely to accept electrons from the CO 5σ orbital, and thus metal-CO bonding will weaken. The d_{xz} orbital will in turn be less able to back-donate into the CO 2π orbital, resulting in additional metal-CO bond weakening as well as reduced C–O bond weakening. The net result is that the presence of subsurface carbon is likely to reduce both CO adsorption and dissociation processes on nearby atoms. This electronic effect may be related to experimental work by Choi et al.,[72] who investigated the surface properties of 5 wt% Co/Al$_2$O$_3$ catalysts, exposed to CO at 250°C, by employing infrared (IR) and temperature-programmed desorption (TPD) techniques. They found that a carbon-deposited cobalt catalyst adsorbs CO more weakly, as evidenced by a new IR band at 2,073 cm^{-1}.

4.5.2 STUDIES ON SUPPORTED CATALYSTS AT MORE REALISTIC CONDITIONS

Lee et al. deposited carbon by CO disproportionation on Co/Al$_2$O$_3$ catalysts with different loadings (2–20 wt% Co) at different CO deposition temperatures (250–400°C).[31] Two forms of carbon where observed upon temperature-programmed surface reaction with hydrogen: an atomic or surface carbidic carbon (hydrogenated at ~190°C) and polymeric carbon (hydrogenated at 430°C). A fraction of the carbon was also resistant to hydrogenation at 600°C. They found that with increasing temperature of deposition, the amount of carbon deposited increased and surface carbidic carbon appeared to be transformed into polymeric and graphitic carbon (Figure 4.5a). These catalysts where carbon was artificially deposited were tested in the FTS at 250 to 300°C, H$_2$/CO = 2, and 1 bar, and exhibited lower activities when compared to the fresh catalyst (Figure 4.5b). The loss of activity was ascribed to the blockage of active sites by polymeric or graphitic carbon, which is irreversibly bound to the metal surface. Bulk carbide was not observed by AES, and as such, the authors argue that the deactivation was not due to an electronic effect. This experiment clearly establishes that stable carbon species generated from CO can be a poison in FTS.

Agrawal et al.[33] performed studies of Co/Al$_2$O$_3$ catalysts using sulfur-free feed synthesis gas and reported a slow continual deactivation of Co/Al$_2$O$_3$ methanation catalysts at 300°C due to carbon deposition. They postulate that the deactivation could occur by carburization of bulk cobalt and formation of graphite deposits on the Co surface, which they observed by Auger spectroscopy.

Thermogravimetric techniques such as thermogravimetric analysis–mass spectrometry (TGA-MS) have been used to show that polymeric or graphitic carbon deposits may form on catalysts during realistic FTS (215–232°C, 19–28 bar, H$_2$/CO = 1.98–2.28) in a two-stage slurry bubble column reactor.[34] The deposits are resistant to hydrogenation at temperatures well above those typical for FT (350°C). Gruver et al.[34] have shown that there is an increase in the amount of carbon resistant to hydrogen on Co/Al$_2$O$_3$ catalysts with an increase in time online (9 to 142 days), which could be related to catalyst activity, as indicated in Figure 4.6a. The carbon formed was even resistant to a regeneration procedure under O$_2$, indicating that it is quite stable. After 142 days the amount of hydrogen-resistant carbon formed on the catalyst was 1 wt%, which was sufficient to block the available surface cobalt atoms. Chemisorption measurements showed a linear decrease in H$_2$ chemisorption capacity with an increase in amount of residual carbon remaining

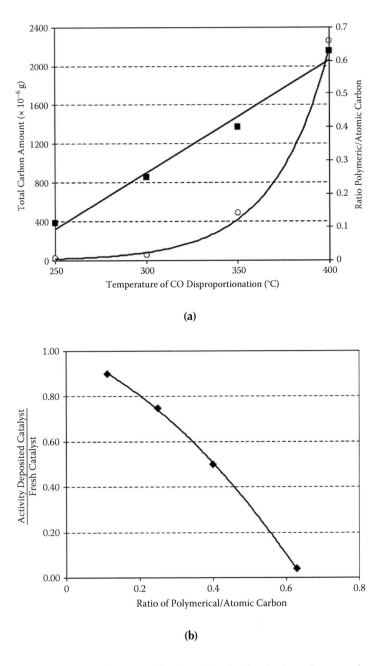

FIGURE 4.5 (a) The total amount of carbon (■) and ratio of polymeric to atomic carbon (o) deposited by the disproportionation of CO on Co/Al$_2$O$_3$ catalysts at various temperatures. (b) A loss of FTS activity (250°C, H$_2$/CO = 2, 1 bar) compared to a fresh catalyst is noted with increasing amounts of polymeric carbon on Co/Al$_2$O$_3$ catalysts. (Drawn from data provided in Lee et al.[31])

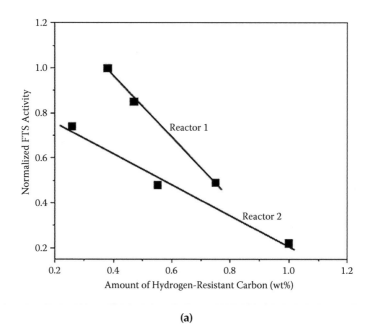

(a)

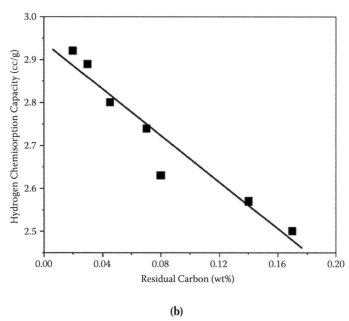

(b)

FIGURE 4.6   (a) Correlation between the amount of H$_2$-resistant carbon and loss of FTS activity (215–232°C, 19–28 bar, H$_2$/CO = 1.98–2.28) in a two-stage slurry bubble column using a Co/Al$_2$O$_3$ catalyst. (b) Correlation between amounts of residual carbon after O$_2$ treatment and H$_2$ chemisorption capacity. (Drawn from data provided in Gruver et al.[34])

after a regeneration step (Figure 4.6b). The deposited polymeric carbon was proposed as one of the causes of deactivation in the FTS. The authors do not make mention of the effect on sintering and poisons on the chemisorption capacity, nor did they determine whether the hydrogen-resistant carbon was located on the support or on cobalt. No bulk cobalt carbide was detected by x-ray diffraction (XRD). The slow accumulation of small amounts of deactivating stable carbon species on the cobalt active phase was also reported for Co/ZnO catalysts[35] tested in extended runs (218°C, 29 bar, $H_2/CO = 2$).

Carbon formation/deposition is a difficult deactivation mechanism to characterize on cobalt-based FTS catalysts. This is due to the low quantities of carbon that are responsible for the deactivation (<0.5 m%) coupled with the presence of wax that is produced during FTS. Furthermore, carbon is only detrimental to the FT performance if it is bound irreversibly to an active site or interacts electronically with it. Hence, not all carbon detected will be responsible for deactivation, especially if the carbon is located on the support.

Recently, we also undertook a study on carbon deposition of cobalt-based catalysts tested in a 100 bbl/day demonstration unit slurry bubble column operating at realistic FTS conditions.[73] Catalyst samples were removed from the reactor operated over a period of 6 months. The catalyst samples were wax extracted with tetrahydrofuran (THF) under argon at mild conditions (around 65°C) and transferred into a glove box in a protected environment, for passivation. In order to follow the accumulation and reactivity of the carbon on the passivated catalysts, a combination of TPH and TPH/TPO was used. TPH of the used catalysts showed the presence of three broad types of carbonaceous species (Figure 4.7). The least reactive species toward hydrogen (peak 3, ~430°C) was identified as polymeric carbon based on previous literature[31] and measurements of reference carburized compounds.[73] Coupled TPH/TPO experiments were performed to remove the less reactive carbon species and showed an increase in the polymeric type of carbon in amounts that were significant to block the available cobalt surface (Figure 4.8).

As mentioned earlier, the location of the inert polymeric carbon is believed to be a key issue in determining its effect on catalyst activity. Using a combination of hydrogen chemisorption, carbon and cobalt mapping with energy-filtered transmission electron microscopy (EFTEM), and low-energy ion scattering (LEIS), we were able to determine that the polymeric carbon was located on both cobalt and the alumina support.[73] The polymeric carbon located on the cobalt was postulated as one of the causes of activity decline in the extended FTS run. Indeed, regeneration of the catalyst to remove the polymeric carbon resulted in a dramatic recovery in the FTS activity.[73]

Barbier et al.[54] employed temperature-programmed hydrogenation on the carbon species on used Co/SiO$_2$ FTS catalysts (200°C, $H_2/CO = 2$, 1 bar) and showed that resulting methane evolution could be resolved into four peaks, representing different types of carbon, which vary in reactivity toward hydrogen. They showed that the formation of easily hydrogenated carbon decreased with increasing time on stream, while the carbon that was hydrogenated at higher temperatures increased with time on stream. This observation points to the fact that during the course of the

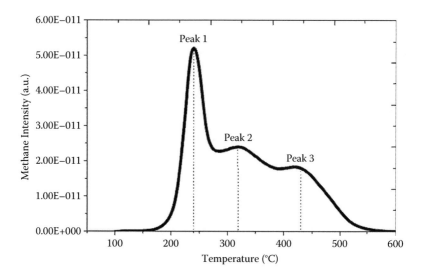

FIGURE 4.7 Methane TPH profile of a wax-extracted 20 wt% Co/Al_2O_3 FTS catalyst taken from a 100 bbl/day slurry bubble column operated at realistic FTS conditions.[73]

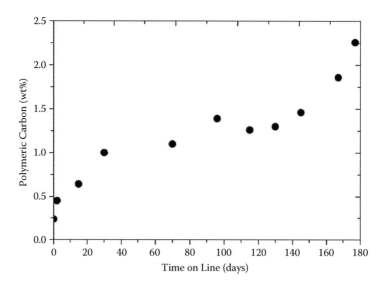

FIGURE 4.8 The correlation between polymeric carbon amount, as determined by TPO, with TOS for 20 wt% Co/Al_2O_3 catalysts taken from the slurry bubble column operated at realistic FTS conditions.[73]

reaction the slow formation of carbon phases that are resistant to H_2 occurs. They postulated that the nature of this carbon may be polymeric or even graphitic.

Pore blockage by carbon or heavy products may cause a loss in activity over time. Niemela and Krause[39] reported a loss of turnover frequency for Co/SiO_2 FTS catalysts due to preferential blocking of the narrowest catalyst pores by carbon. Puskas[74] found unusually high amounts of wax in the pores on a Co/Mg/diatomaceous earth catalyst tested in the FTS at 190°C, 1–2 bar, $H_2/CO = 2.55$ for 125 days. In a separate study it was concluded that pore plugging by the waxy products resulted in a fast deactivation of such catalysts.[75]

From the work reported in literature it can be thus concluded that there will be various forms of carbonaceous species, which vary in reactivity, that exist on the catalyst or support during FTS. Some forms of this carbon are active (atomic surface carbide and CH_x species) and even considered as intermediate species in FTS. However, it is also clear that especially during extended runs there may be a build up/transformation to less reactive forms of carbon (e.g., polymeric carbon). The amounts of these species may be small, but depending on their location, they may be responsible for a part of deactivation observed on cobalt-based FTS catalysts. The electronic interaction of carbon with the catalyst surface may also result in decreased activity.

4.6 BULK CARBIDE FORMATION IN THE FTS?

The formation and influence of bulk cobalt carbide during FTS has been a topic of interest for many research groups.[76–79] There is a general trend of decreasing bulk carbide stability as one goes from the left to the right of the periodic table through the transition metals. It has been shown that the activation energy for the diffusion of carbon into cobalt (145 kJ/mol) is much higher than that for iron (44–69 kJ/mol). This translated to a 10^5 times slower diffusion of carbon into cobalt than into iron.[80] Thus, it is reasonable to expect that cobalt will have a lesser tendency to form carbides than iron. Two forms of cobalt carbide are generally known for cobalt: Co_2C, which has an orthorhombic structure, and Co_3C, which has structure similar to that of cementite.

The formation of bulk cobalt carbide is quite a slow process since it requires the diffusion of carbon into the cobalt bulk. It was reported that the full conversion of unsupported and reduced Co to Co_2C only occurred after 500 h of exposure to pure CO at 230°C. Increasing the reaction temperature resulted in a faster rate of carburization.[81] Bulk cobalt carbides are considered to be thermodynamically metastable species, and therefore Co_2C will decompose to hcp cobalt and graphite, while Co_3C will decompose to fcc cobalt and methane. Thermal decomposition of bulk carbides under an inert atmosphere is believed to occur at 400°C.[81] Hydrogenation of the bulk carbides is believed to be a fast process and occurs around 200°C.[82,83]

Early work at the Bureau of Mines on $Co/ThO_2/$kieselguhr catalysts showed that bulk carbide was not an intermediate in the FTS, nor was it catalytically active.[82] Excessive amounts of carbides, produced by CO exposure prior to the reaction, were found to severely inhibit the FTS activity. Carbiding of $Co/ThO_2/$ kieselguhr catalysts in CO at 208°C had a dramatic effect on catalyst activity, decreasing conversion by 20% and increasing the formation of lighter hydrocarbons. The BET surface area of the catalysts remained constant; however, the CO chemisorption capacity decreased to 30% of the initial values before carbiding. In some cases a fourfold increase in activity was noticed after the hydrogenation of the carbide at 150–194°C. Also, it should be noted that XRD still showed the presence of bulk cobalt carbide postreaction in the case of the precarbided catalysts exposed to synthesis gas. This indicates that while the bulk carbide can be readily hydrogenated in pure hydrogen, it is stable for a considerable amount of time in synthesis gas mixtures at FTS conditions.

Recent work done by Xiong et al.[84] on Co/AC (activated carbon) catalysts showed that a Co_2C species formed during the catalyst reduction in hydrogen at 500°C. Evidence for the carbide in the Co/AC catalysts was obtained by x-ray diffraction and XPS measurements, and the formation of this Co_2C species reduced the FTS activity over the Co-based catalysts. The presence of bulk carbide also seems to enhance alcohol selectivity.[85]

Several workers have reported that bulk carbide does not form readily during normal FTS conditions.[76,82] Bureau of Mines work, using laboratory XRD measurements, showed that detectable amounts of bulk carbide were not formed under synthesis conditions.[82]

Work by Syntroleum on its Co/Al_2O_3 proprietary catalyst showed that bulk carbide is formed during FTS in a continuously stirred tank reactor (CSTR) reactor (216°C and 37 bar) in the presence of CO only for a period of 8 h (upset conditions).[76] The performance of the catalyst was severely affected when standard H_2/CO ratio (2) was reintroduced as the CO conversion dropped more than half and the methane selectivity doubled. An interesting observation was that the bulk carbide was hydrogenated to hexagonal cobalt at 225°C by treatment in a pure hydrogen stream. In general, small supported and reduced cobalt particles (<40 nm) are cubic in nature.[76,85,86] Similarly, recent work at Sasol[63] has shown that even exposure to pure CO for 2 h can be detrimental to activity and selectivity (Figure 4.9). Even several hours of operating with a standard H_2/CO ratio of 2, thereafter, could not recover the activity or methane selectivity. XRD analysis of the spent catalyst showed the presence of bulk cobalt carbide (Figure 4.10). The above experiment shows that if bulk cobalt carbide forms, it can be stable in a syngas environment for a considerable period of time.

Pankina et al.[87] performed *ex situ* postreaction TPH/magnetic studies on wax-extracted cobalt alumina catalysts tested in FTS and stated that methane evolution at 250°C corresponds to an increase in magnetization, which indicates the hydrogenation of cobalt carbide. The reduction of CoO was excluded as a cause of the increased magnetization. They argued that although cobalt carbide is said to be thermodynamically metastable during the FTS, it could be stable for small Co crystallites. This is due to the contribution of the surface free energy of small Co crystallites to the overall thermodynamic calculations.

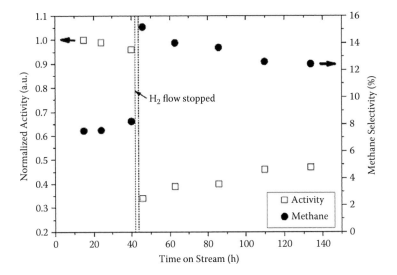

FIGURE 4.9 Activity profile and methane selectivity of an FTS run at 230°C, H_2/CO = 2, 20 bar showing drop in activity after stopping H_2 flow for 2 h. The H_2 flow was reintroduced thereafter.[63]

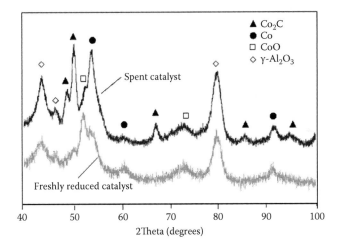

FIGURE 4.10 X-ray diffractograms of the spent 20 wt% Co/Al_2O_3 catalyst after the run (in Figure 4.9) compared to a freshly reduced catalyst in wax. The wax was melted *in situ* in nitrogen at 120°C to remove interfering diffraction patterns.[63]

Co_2C is rarely observed in the FTS by *ex situ* techniques (see Table 4.3). Ducreux et al.[88] observed the formation of Co_2C on Co/Al_2O_3 and $Co/Ru/TiO_2$ FTS catalysts by *in situ* XRD techniques and related it with a deactivation process (230°C, 3 bar, H_2/CO = 9; no wax). Machocki[89] also showed the formation of Co_2C on Co/SiO_2 catalysts after an initial 20 h induction period in the FTS

TABLE 4.3

An Overview of Reported Claims of Bulk Cobalt Carbide Being Observed after/when Performing Fischer-Tropsch Synthesis over Supported Cobalt-Based Catalysts

Catalyst	Reaction Conditions			Technique	Effect on Activity	Ref.
	H_2/CO Ratio	Temp. (°C)	Pressure (bar)			
Co/Pt/Al$_2$O$_3$	2	220	18	Synchrotron XRD	n.s.	90
Co/Ru/Al$_2$O$_3$	2	220	20	XRD	n.s.	91
Co/Al$_2$O$_3$ and TiO$_2$	9[a]	230	3	*In situ* XRD	↓	88
Co/ThO$_2$/ZSM-5/ Al$_2$O$_3$	1	280	21	XRD	↓	32
Co/SiO$_2$	1.1	275	1	XRD	↓	89
Co/Al$_2$O$_3$	2	220	1	TPH with magnetic measurements	n.s.	87
Fe/Co metal/oxide composite	1	230	10	XRD, TEM, and XPS	n.s.	79
Co/Al$_2$O$_3$[a]	9	400	1	AES line shape	↓	33
Na-Co/Al$_2$O$_3$	2	240	50	TPH with magnetic measurements, XPS, and XRD	n.s.	53

[a] Methanation conditions; n.s., not specified in study.

(275°C, 1 bar, H_2/CO = 1.1). This induction period is apparently needed to form a stable carbide nucleus. However, he also noted that bulk carburization occurs more readily on iron catalysts due to the stronger Fe–C bond, and that the hydrogenating ability of cobalt considerably decreases the amount of surface carbon that can migrate into bulk cobalt metal. Jacobs et al.,[90] employing synchrotron XRD, detected a small amount of Co$_2$C that may also have been formed during the synthesis (220°C, 18 bar, H_2/CO = 2). In a recent study Tavasoli et al.[91] showed the presence of cobalt carbide in used 30 wt% Co/Ru/Al$_2$O$_3$ catalysts that were tested at 220°C, 20 bar, and a H_2/CO ratio of 2 for over 40 days.

Pennline et al.[32] used bifunctional Co/ThO/ZSM-5 catalysts at 280°C, 21 bar, H_2/CO = 1 in the FTS. XRD of the used catalyst indicated that cobalt carbide is present. They found that the relative amount of the carbide species is larger on the used catalyst operated at 280°C than on the used catalyst operated at 320°C. They argued that this is because cobalt carbide begins to decompose around 300°C. Since this catalyst lacked high water gas shift activity, and a low feed gas ratio of

$H_2/CO = 1$ was used, the usage ratio of hydrogen to carbon monoxide was always greater than the feed ratio, and thus the catalyst was uniformly exposed to a low H_2/CO ratio, which increases the chance for carbide formation.

To summarize, from literature there does not seem to be much consensus on whether bulk cobalt carbide forms during realistic FTS conditions. Bulk carbide is generally considered a metastable species. However, it is clear that it may form under upset conditions. Furthermore, there is strong evidence to show that if bulk cobalt carbide is present, it is deleterious in terms of both catalyst activity and selectivity. With this in mind, it would be prudent to operate the catalyst in a regime (sufficiently high H_2/CO ratio) where bulk carbide formation is avoided.

4.7 MINIMIZATION OF CARBON DEPOSITS ON COBALT FTS CATALYSTS BY PROMOTION

Additives such as rare earth or noble metals are generally introduced into industrial cobalt FTS catalysts as structural or reduction promoters.[92] The addition of various promoters to cobalt catalysts has also been shown to decrease the amount of carbon produced during the FTS.[84,87,93,94] Also, the addition of promoter elements may decrease the temperature of regeneration, preventing the possible sintering of supported cobalt particles during such treatments.[92]

In the case of cobalt foils it has been found that 0.1 monolayer of potassium coverage reduces the formation of graphite at high FTS temperatures (307°C).[93] It is not exactly clear how the potassium reduced the formation of graphitic carbon deposits in this study. It is known that alkali ad-atoms on a transition metal surface exist in a partially ionic state, resulting in a work function decrease of the metal.[95] Potassium promotion results in a weakening of the C–O bond and an increase in the CO dissociation rate, resulting in increased coverage of active surface carbidic carbon.[96] Besides having an electronic effect, potassium could play a structural role in preventing graphite formation according to Wesner et al.[97] They argue that the epitaxial growth of graphite is favored on clean hexagonal cobalt and the promotion of the cobalt with potassium will disrupt the formation of epitaxial graphite islands by site blocking. However, already at the early stage of FT research it was shown that for supported cobalt catalyst, potassium was a poison.[98] The mobility of potassium during FTS conditions could also result in being distributed on the support.

Also, manganese added to cobalt on activated carbon catalysts resulted in a decrease in bulk carbide formation during reduction and a decrease in the subsequent deactivation rate.[84] Magnesium and yttrium added to the support in alumina-supported cobalt catalysts showed a lower extent of carburization. This was explained by a decrease in Lewis acidity of the alumina surface in the presence of these ions.[87]

It has also been postulated using molecular modeling and proven experimentally using temperature-programmed techniques that promotion with boron inhibits detrimental carbon formation.[71] *Ab initio* calculations indicate that boron behaves rather similarly to carbon and prefers to adsorb in the octahedral sites of the first subsurface layer of cobalt. The boron thus forces the carbon to remain on the surface in an active form. Additionally, boron present in the first subsurface layer reduces the surface carbon binding energy, lowering the carbon coverage, and may prevent the nucleation of graphene islands.[71]

It is also known that the common reduction promoters (e.g., Pt and Ru) aid in carbon gasification. Iglesia et al.,[94] using thermogravimetry and XPS, showed that in the case of ruthenium-promoted cobalt on titania catalysts, the promoter inhibits the deposition of carbon during FTS. Ruthenium may promote hydrogenolysis during the reaction, and the intimate association of ruthenium with cobalt might allow carbon deposits on the catalyst to be gasified via hydrogenolysis at lower temperatures, as opposed to carbon gasification via combustion with oxygen.

4.8   REGENERATION PROCESSES TO REMOVE CARBON DEPOSITS

Regeneration of cobalt-based Fischer-Tropsch synthesis catalysts is a cost-effective way to increase the life of the cobalt catalyst. In fact, workers at BP reported that the only way to manage activity decline and ensure a 4-year catalyst life was to regenerate their catalyst *in situ*.[35] This is important due to the high cost of cobalt, which can be a considerable proportion of the overall operating cost. In most cases the regeneration process relies on the effective removal of carbon.[36–38] The deleterious carbonaceous deposits can be removed by gasification with O_2, H_2O, and H_2,[15] which makes regeneration feasible. The order of decreasing reaction rate of carbon as $O_2 > H_2O > H_2$ can be generalized.

Already in the early stages of the industrial application of cobalt catalysts it was noted that to secure longer catalyst lifetime, regeneration with hydrogen was required.[99] Over the next few years new regeneration technologies were developed and improved for cobalt-based FTS catalysts. The regeneration can be performed in a few manners: (1) reductive regeneration, (2) oxidative regeneration, and (3) steam/water regeneration. All these regeneration procedures focus on the removal of deleterious carbon types, i.e., polymeric and graphitic.

Carbon-deactivated FTS catalysts can be rejuvenated or regenerated by treatment in hydrogen.[99–101] This can be done in both an *in situ*[99] and an *ex situ* manner.[101] Nondesorbing reaction products (heavy waxes) can also be removed from catalysts by treatment with hydrogen, or gases or vapors containing hydrogen.[101,102] Often it is necessary to remove wax and hydrocarbons from a spent catalyst before exposing it to regenerating gas such as oxygen, in order to limit exotherms that may result in damage to catalyst integrity. Various patents and publications claim that carbon-deactivated catalysts can best be regenerated by conventional wax removal, oxidation, and re-reduction techniques.[35–37,103,104] Steam regeneration can also be used to remove carbon from deactivated catalysts. Steam reacts with the carbon on the catalyst surface and forms CO and H_2, thus cleaning the surface.[38] The important message is that regeneration efforts focus largely on the removal of carbon.

4.9 THE EFFECT OF CARBON ON FTS SELECTIVITIES

Along with catalyst activity, product selectivity is a key issue in cobalt-based FTS.[1] For GTL processes the preferred product is long-chain waxy hydrocarbons. It is well known that FT reaction conditions have an important effect on product selectivities. High temperatures and H_2/CO ratios are associated with higher methane selectivity, lower probability of hydrocarbon chain growth, and lower olefinicity in the products.[105]

The deposition of the different types of inactive carbon species during FTS may have different influences on the product selectivities. It has been shown with CO adsorption studies on cobalt and molecular modeling that the presence of carbon will affect the CO adsorption strength and therefore the CO dissociation rate.[41,72] Consequently, the surface coverage of active carbon may decrease, leading to shorter-chained hydrocarbons. Indeed, Bertole et al.,[106] using isotopic transient experiments, showed that an increase in the amount of surface active carbon (surface carbide) will result in a higher chain growth probability and, thus, an increase in desired selectivity for Co/SiO_2 FTS catalysts. Furthermore, it has been shown that the presence of bulk cobalt carbide results in a dramatic increase in methane selectivity during the FTS.[76,82,84] It is also plausible that as the carbon becomes more stable, i.e., graphitic, the interaction with the metal would decrease and it would have a lesser effect on the product distribution.

Co/Al_2O_3 catalysts that contain higher amounts of less reactive polymeric carbon not only exhibited enhanced deactivation when tested in FTS when compared to the fresh catalyst, but also showed an increase in selectivity to olefinic products.[31] The authors postulated that this was probably due to the reduction in hydrogenation ability of the carbon deposited catalyst to convert primarily formed olefins into the corresponding paraffins.

Iglesia et al. showed that the pore diameter of the catalyst is an important parameter for tailoring selectivity.[107] Hence, carbon deposition leading to physical blocking of the pores could have an influence on the selectivity. Niemela and Krause[39] reported that for a Co/SiO_2 catalyst the relative turnover number for the C_2+ species may increase significantly during the initial phase of carbon deactivation due to preferential blocking of the narrowest catalyst pores. Puskas[74] also showed a decrease in the hydrocarbon growth rate of Co/Mg/diatomaceous earth catalyst with increasing time online, which they ascribed to pore blocking.

4.10 INTEGRATED UNDERSTANDING OF CARBON IN COBALT-BASED FTS UNDER REALISTIC SYNTHESIS CONDITIONS

Using the available literature[34,35,90,91] and our own work we have integrated our opinion on the presence of various carbon species on supported cobalt-based FTS catalysts operated under realistic FTS conditions (200–240°C, 10–40 bar, $H_2/CO = \sim 1-2$), and this is shown in Table 4.4. It is clear that only a few papers on supported cobalt catalysts tested at realistic FT conditions have been published on this topic.[34,35,90,91] It is believed that CH_x, surface carbide, paraffinic wax, and polymeric carbon are present during realistic conditions. The presence of small amounts of bulk carbide, although not observed in our work by XRD, cannot be excluded. Similarly, a graphite formation is expected to be unfavored at realistic FTS conditions. Boudouard carbon would be formed only in small amounts due to the relatively low temperature at realistic conditions and the presence of several bars of hydrogen. The amount of CO_2, which is a product of the Boudouard reaction, is also small over cobalt-based catalysts. Coke formation generally consists of dehydrogenation and cyclization reactions over acidic sites. The coke formation should be minimized taking into account the strong hydrogenation tendency of cobalt; the support is selected due to its low acidity and reaction temperatures under 250°C.[63] The primary products of the reaction are olefins but they are rapidly transformed to paraffins due to the hydrogenation ability of cobalt. It is well known that coking rates are lowest for paraffins.[30]

Additionally, the following factors are believed to have an increase in the amount of carbon deposited on cobalt catalysts: higher reaction temperature,[39,63] lower H_2/CO ratio,[63] higher CO partial pressures,[59] and cleaner (sulfur-free) synthesis gas.[33,64]

4.11 CONCLUSIONS

It is clear that the FTS over cobalt catalysts occurs in the presence of an active surface carbidic overlayer and in the presence of various hydrocarbon products. The conversion of this active surface carbidic carbon to other inactive forms over time results in deactivation of the catalyst. This integrated review has shown that the following carbon species can form on supported cobalt catalysts during realistic FTS: (1) CH_x species, (2) surface carbide, (3) wax, and (4) polymeric carbon. On the other hand, the following species are believed to be absent/minimal during FTS: (1) bulk cobalt carbide, (2) graphite or graphene, (3) carbon from the Boudouard reaction, and (4) carbon from coke formation. Additionally, it is evident that nondesorbing, heavy hydrocarbon wax can lead to pore plugging and deactivation. From the available literature and regeneration patents it does seem that deactivation by carbon deposits is an important deactivation pathway for cobalt-based FTS catalysts under realistic conditions that warrants further study.

TABLE 4.4

The Presence of Various Carbon Species in Supported Cobalt FTS Catalysts as Operated under Realistic FTS Conditions (200–240°C, 10–40 bar, H₂/CO = 1–2)

Carbon Species	Catalyst	Reaction Conditions in Study				Amount of Carbon (m%)	Observed in Study	Ref.
		Temp. (°C)	H₂/CO	Pressure (bar)	T_{hyd} (°C)			
CH$_x$ fragments	20 wt% Co/Pt/Al₂O₃	230	1.0–2.0	20	<100	—	No	Own work
Surface carbide	20 wt% Co/Pt/Al₂O₃	230	1.0–2.0	20	180–200	—	Peak 1	Own work
Bulk carbide	20 wt% Co/Pt/Al₂O₃	230	1.0–2.0	20	<250	—	No	Own work
	Co/Al₂O₃	215–230	2.0–2.3	19–28		Minor	No	34
	20 wt% Co/Pt/Al₂O₃	220	2.0	18		Minor	Yes	90
	30wt% Co/Pt/Al₂O₃	220	2.0	20			Yes	91
Hydrocarbons, paraffinic wax	20 wt% Co/Pt/Al₂O₃	230	1.0–2.0	20	250–350	—	Peak 2	Own work
	Co/Al₂O₃	215–230	2.0–2.3	19–28		—		34
Polymeric carbon	20 wt% Co/Pt/Al₂O₃	230	1.0–2.0	20	>350	2.0	Peak 3	Own work
	Co/Al₂O₃	215–230	2.0–2.3	19–28		1.0		34
	Co/ZnO	218	2.0	29		Small		35
Graphite or graphene	20 wt% Co/Pt/Al₂O₃	230	1.0–2.0	20	>620	—	No	Own work
	Co/Al₂O₃	215–230	2.0–2.3	19–28		1.0	Possibly	34

REFERENCES

1. Steynberg, A. P. 2004. Introduction to Fischer-Tropsch technology. *Stud. Surf. Sci. Catal.* 152:1–63.
2. Eilers, J., Posthuma, S. A., and Sie, S. T. 1990. The Shell middle distillate synthesis process (SMDS). *Catal. Lett.* 7:253–69.
3. Gas-to-liquids: Peering into the crystal ball. 2005. *Catalyst Review Newsletter*, p. 4.
4. Rao, V. U. S., Stiegel, G. J., Cinquegrane, G. J., and Srivastava, R. D. 1992. Iron-based catalysts for slurry-phase Fischer-Tropsch process: Technology review. *Fuel Process. Technol.* 30:83–107.
5. van Berge, P. J., Barradas, S., van de Loodsrecht, J., and Visagie, J. L. 2001. Advances in the cobalt catalyzed Fischer-Tropsch synthesis. *Erd. Erdgas Kohle* 117:138–42.
6. van Berge, P. J., van de Loosdrecht, J., Barradas, S., and van der Kraan, A. M. 2000. Oxidation of cobalt based Fischer-Tropsch catalysts as a deactivation mechanism. *Catal. Today* 58:321–34.
7. Jacobs, G., Das, T. K., Patterson, P. M., Li, J., Sanchez, L., and Davis, B. H. 2003. Fischer-Tropsch synthesis XAFS: XAFS studies of the effect of water on a Pt-promoted Co/Al$_2$O$_3$ catalyst. *Appl. Catal. A* 247:335–43.
8. Kiss, G., Kliewer, C. E., DeMartin, G. J., Culross, C. C., and Baumgartner, J. E. 2003. Hydrothermal deactivation of silica-supported cobalt catalysts in Fischer-Tropsch synthesis. *J. Catal.* 217:127–40.
9. Hilmen, A. M., Schanke, D., Hanssen, K. F., and Holmen, A. 1999. Study of the effect of water on alumina supported cobalt Fischer-Tropsch catalysts. *Appl. Catal. A* 186:169–88.
10. Li, J., Zhan, X., Zhang, Y., Jacobs, G., Das, T., and Davis, B. H. 2002. Fischer-Tropsch synthesis: Effect of water on the deactivation of Pt promoted Co/Al$_2$O$_3$ catalysts. *Appl. Catal. A* 228:203–12.
11. Saib, A. M., Borgna, A., van de Loosdrecht, J., van Berge, P. J., and Niemantsverdriet, J. W. 2006. XANES study of the susceptibility of nano-sized cobalt crystallites to oxidation during realistic Fischer-Tropsch synthesis. *Appl. Catal. A* 312:12–19.
12. van de Loosdrecht, J., Balzhinimaev, B., Dalmon, J.-A., Niemantsverdriet, J. W., Tsybulya, S. V., Saib, A. M., van Berge, P. J., and Visagie, J. L. 2007. Cobalt Fischer-Tropsch synthesis: Deactivation by oxidation? *Catal. Today* 123:293–302.
13. Saib, A. M., Borgna, A., van de Loosdrecht, J., van Berge, P. J., and Niemantsverdriet, J. W. 2006. In situ surface oxidation study of a planar Co/SiO$_2$/Si(100) model catalyst with nanosized cobalt crystallites under model Fischer-Tropsch synthesis conditions. *J. Phys. Chem. B* 110:8657–64.
14. Saib, A. M., Borgna, A., van de Loosdrecht, J., van Berge, P. J., Geus, J. W., and Niemantsverdriet, J. W. 2006. Preparation and characterization of spherical Co/SiO$_2$ model catalysts with well-defined nano-sized cobalt crystallites and a comparison of their stability against oxidation with water. *J. Catal.* 239:326–39.
15. Bartholomew, C. H. 2001. Mechanisms of catalyst deactivation. *Appl. Catal. A* 212:17–60.
16. van Berge, P. J., and Everson, R. C. 1997. Cobalt as an alternative Fischer-Tropsch catalyst to iron for the production of middle distillates. *Stud. Surf. Sci. Catal.* 107:207–12.
17. Baumann, P. R. F., Degeorge, C. W., and Leviness, S. C. 1997. WO 98/50485, to Exxon.
18. Behrmann, W. C., and Leviness, S. C. 1997. WO 98/50486, to Exxon.
19. Leviness, S. C., and Mitchell, W. N. 1997. WO 98/50488, to Exxon.
20. Chang, M., Stephen, J., and Mart, C. J. 1997. WO 98/50489, to Exxon.

21. Rostrup-Nielsen, J. R. 1984. In *Catalysis science and technology*, ed. J. R. Anderson and M. Boudart. Berlin: Springer-Verlag, 1–117.
22. Posthuma, S. A., and de Graaf, J. D. 1989. Great Britain Patent 2 231 581, to Shell.
23. Bian, G.-Z., Fujishita, N., Mochizuki, T., Ning, W.-S., and Yamada, M. 2003. Investigations on the structural changes of two Co/SiO_2 catalysts by performing Fischer-Tropsch synthesis. *Appl. Catal. A* 252:251–60.
24. Overett, M. J., Breedt, B., du Plessis, E., Erasmus, W., Maloka, J., and van de Loosdrecht, J. 2008. Sintering as a deactivation mechanism for an alumina supported cobalt Fischer Tropsch synthesis catalyst. *Prepr. Pap.-Am. Chem. Soc. Div. Pet. Chem.* 53:126–28.
25. Moulijn, J. A., van Diepen, A. E., and Kapteijn, F. 2001. Catalyst deactivation: Is it predictable? What to do? *Appl. Catal. A* 212:3–16.
26. Agnelli, M., Swaan, H. M., Marquez-Alvarez, C., Martin, G. A., and Mirodatos, C. 1998. CO hydrogenation on a nickel catalyst. II. A mechanistic study by transient kinetics and infrared spectroscopy. *J. Catal.* 175:117–28.
27. Wilson, J., and de Groot, C. 1995. Atomic-scale restructuring in high-pressure catalysis. *J. Phys. Chem.* 99:7860–66.
28. Ciobica, I. M., van Santen, R. A., van Berge, P. J., and van de Loosdrecht, J. 2008. Adsorbate induced reconstruction of cobalt surfaces. *Surf. Sci.* 602:17–27.
29. Bezemer, G. L., Bitter, J. H., Kuipers, H. P. C. E., Oosterbeek, H., Holewijn, J. E., Xu, X., Kapteijn, F., van Dillen, A. J., and de Jong, K. P. 2006. Cobalt particle size effects in the Fischer-Tropsch reaction studied with carbon nanofiber supported catalysts. *J. Am. Chem. Soc.* 128:3956–64.
30. Menon, P. G. 1990. Coke on catalysts—Harmful, harmless, invisible and beneficial types. *J. Mol. Catal.* 59:207–20.
31. Lee, D.-K., Lee, J.-H., and Ihm, S.-K. 1988. Effect of carbon deposits on carbon monoxide hydrogenation over alumina-supported cobalt catalysts. *Appl. Catal.* 36:199–207.
32. Pennline, H. W., Gormley, R. J., and Schehl, R. R. 1984. Process studies with a promoted transition metal-zeolite catalyst. *Ind. Eng. Chem. Prod. Res. Dev.* 23:388–93.
33. Agrawal, P. K., Katzer, J. R., and Manogue, W. H. 1981. Methanation over transition metal catalysts. II. Carbon deactivation of cobalt/alumina in sulfur-free studies. *J. Catal.* 69:312–26.
34. Gruver, V., Young, R., Engman, J., and Robota, H. J. 2005. The role of accumulated carbon in deactivating cobalt catalysts during FT synthesis in a slurry-bubble-column reactor. *Prepr. Pap.-Am. Chem. Soc. Div. Pet. Chem.* 50:164–66.
35. Font Freide, J. J. H. M., Gamlin, T. D., Hensman, J. R., Nay, B., and Sharp, C. 2004. Development of a CO_2 tolerant Fischer-Tropsch catalyst: From laboratory to commercial-scale demonstration in Alaska. *J. Nat. Gas Chem.* 13:1–9.
36. Soled, S. L., Iglesia, E., Fiato, R.M. and Ansell, G. B. 1995. U.S. Patent 5 397 806, to Exxon.
37. van der Burgt, M. J., and Ansorge, J. 1988. Great Britain Patent 2 222 531, to Shell.
38. Wright, H. A. 2002. U.S. Patent 6 486 220, to Conoco.
39. Niemela, M. K., and Krause, A. O. I. 1996. The long-term performance of Co/SiO_2 catalysts in CO hydrogenation. *Catal. Lett.* 42:161–66.
40. Pennline, H. W., and Pollack, S. S. 1986. Deactivation and regeneration of a promoted transition-metal-zeolite catalyst. *Ind. Eng. Chem. Prod. Res. Dev.* 25:11–14.
41. Zonnevylle, M. C., Geerlings, J. J. C., and van Santen, R. A. 1990. Conversion of surface carbidic to subsurface carbon on cobalt(0001): A theoretical study. *Surf. Sci.* 240:253–62.

42. Borko, L., Horvath, Z. E., Schay, Z., and Guczi, L. 2007. The role of carbon nanospecies in deactivation of cobalt based catalysts in CH$_4$ and CO transformation. *Stud. Surf. Sci. Catal.* 167:231–36

43. Boskovic, G., and Smith, K. J. 1997. Methane homologation and reactivity of carbon species on supported Co catalysts. *Catal. Today* 37:25–32.

44. Gong, X.-Q., Raval, R., and Hu, P. 2005. CH$_x$ hydrogenation on Co(0001). A density functional theory study. *J. Chem. Phys.* 122:24711/1–11/6.

45. Ge, Q., and Neurock, M. 2006. Adsorption and activation of CO over flat and stepped Co surfaces: A first principles analysis. *J. Phys. Chem. B* 110:15368–80.

46. Geerlings, J. J. C., Zonnevylle, M. C., and De Groot, C. P. M. 1991. Studies of the Fischer-Tropsch reaction on cobalt(0001). *Surf. Sci.* 241:302–14.

47. Cheng, J., Gong, X.-Q., Hu, P., Lok, C. M., Ellis, P., and French, S. 2008. A quantitative determination of reaction mechanisms from density functional theory calculations: Fischer-Tropsch synthesis on flat and stepped cobalt surfaces. *J. Catal.* 254:285–95.

48. Winslow, P., and Bell, A. T. 1985. Studies of the surface coverage of unsupported ruthenium by carbon- and hydrogen-containing adspecies during carbon monoxide hydrogenation. *J. Catal.* 91:142–54.

49. Nakamura, J., Tanaka, K., and Toyoshima, I. 1987. Reactivity of deposited carbon on cobalt-alumina catalyst. *J. Catal.* 108:55–62.

50. Nolan, P. E., Lynch, D. C., and Cutler, A. H. 1998. Carbon deposition and hydrocarbon formation on Group VIII metal catalysts. *J. Phys. Chem. B* 102:4165–75.

51. Geerlings, J. J. C., Zonnevylle, M. C., and De Groot, C. P. M. 1990. The Fischer-Tropsch reaction on a cobalt (0001) single crystal. *Catal. Lett.* 5:309–14.

52. Koerts, T. 1992. The reactivity of surface carbonaceous intermediates. PhD thesis, Eindhoven University of Technology, The Netherlands.

53. Mirodatos, C., Brum Pereira, E., Gomez Cobo, A., Dalmon, J. A., and Martin, G. A. 1995. CO hydrogenation over Ni- and Co-based catalysts: Influence of alkali addition on morphological and catalytic properties. *Top. Catal.* 2:183–92.

54. Barbier, A., Tuel, A., Arcon, I., Kodre, A., and Martin, G. A. 2001. Characterization and catalytic behavior of Co/SiO$_2$ catalysts: Influence of dispersion in the Fischer-Tropsch reaction. *J. Catal.* 200:106–16.

55. Potoczna-Petru, D. 1991. The interaction of model cobalt catalysts with carbon. *Carbon* 29:73–79.

56. Bartholomew, C. H. 1982. Carbon deposition in steam reforming and methanation. *Catal. Rev.-Sci.* 24:67–111.

57. Nakamura, J., Toyoshima, I., and Tanaka, K. 1988. Formation of carbidic and graphitic carbon from carbon monoxide on polycrystalline cobalt. *Surf. Sci.* 201:185–94.

58. Rostrup-Nielsen, J. R. 1974. Coking on nickel catalysts for steam reforming of hydrocarbons. *J. Catal.* 33:184–201.

59. Moeller, A. D., and Bartholomew, C. H. 1982. Deactivation by carbon of nickel, nickel-ruthenium, and nickel-molybdenum methanation catalysts. *Ind. Eng. Chem. Proc. Des. Dev.* 21:390–97.

60. Mukkavilli, S., Wittmann, C., and Tavlarides, L. L. 1986. Carbon deactivation of Fischer-Tropsch ruthenium catalyst. *Ind. Eng. Chem. Proc. Des. Dev.* 25:487–94.

61. Dry, M. E. 1982. Sasol's Fischer-Tropsch experience. *Hydrocarb. Process.* 61:121–24.

62. Hooker, M. P., and Grant, J. T. 1977. The use of Auger electron spectroscopy to characterize the adsorption of carbon monoxide transition metals. *Surf. Sci.* 62:21–30.

63. Moodley, D. J. 2008. Phd thesis, Eindhoven University of Technology, The Netherlands.

64. Kim, M. S., Rodriguez, N. M., and Baker, R. T. K. 1993. The interplay between sulfur adsorption and carbon deposition on cobalt catalysts. *J. Catal.* 143:449–63.
65. Gardner, D. C., and Bartholomew, C. H. 1981. Kinetics of carbon deposition during methanation of carbon monoxide. *Ind. Eng. Chem. Prod. Res. Dev.* 20:80–87.
66. Prior, K. A., Schwaha, K., and Lambert, R. M. 1978. Surface chemistry of the non-basal planes of cobalt: The structure, stability, and reactivity of C$_o$ (101̄2)-CO. *Surf. Sci.* 77:193–208.
67. Johnson, B. G., Bartholomew, C. H., and Goodman, D. W. 1991. Role of surface structure and dispersion in carbon monoxide hydrogenation on cobalt. *J. Catal.* 128:231–47.
68. Lahtinen, J., Anraku, T., and Somorjai, G. A. 1993. Carbon monoxide hydrogenation on cobalt foil and on thin cobalt film model catalysts. *J. Catal.* 142:206–25.
69. Lahtinen, J., Anraku, T., and Somorjai, G. A. 1994. C, CO and CO2 hydrogenation on cobalt foil model catalysts: Evidence for the need of CoO reduction. *Catal. Lett.* 25:241–55.
70. Beitel, G. A., de Groot, C. P. M., Oosterbeek, H., and Wilson, J. H. 1997. A combined in-situ PM-RAIRS and kinetic study of single-crystal cobalt catalysts under synthesis gas at pressures up to 300 mbar. *J. Phys. Chem. B* 101:4035–43.
71. Xu, J., Tan, K. F., Borgna, A., and Saeys, M. 2006. First principles based promoter design for heterogeneous catalysis. Poster 476ae, AIChE Annual meeting, San Francisco, November. http://aiche.confex.com/aiche/2006/techprogram/S2113.htm
72. Choi, J. G., Rhee, H. K., and Moon, S. H. 1985. IR and TPD study of fresh and carbon-deposited aluminum oxide-supported cobalt catalysts. *Appl. Catal.* 13:269–80.
73. Moodley, D. J., van de Loosdrecht, J., Saib, A. M., Overett, M. J., Datye, A. K., and Niemantsverdriet, J. W. 2009. Carbon deposition as a deactivation mechanism of cobalt-based Fischer-Tropsch synthesis catalysts under realistic conditions. *Appl. Catal. A*, 354:102–10.
74. Puskas, I. 1993. Unusual reactions on a cobalt-based Fischer-Tropsch catalyst. *Catal. Lett.* 22:283–88.
75. Puskas, I., Meyers, B. L., and Hall, J. B. 1993. A fast deactivating Fischer-Tropsch catalyst. *Prepr. Pap.-Am. Chem. Soc. Div. Pet. Chem.* 38:905–8.
76. Gruver, V., Zhan, X., Engman, J., Robota, H. J., Suib, S. L., and Polverejan, M. 2004. Deactivation of a Fischer-Tropsch catalyst through the formation of cobalt carbide under laboratory slurry reactor conditions. *Prepr. Pap.-Am. Chem. Soc. Div. Pet. Chem.* 49:192–94.
77. Hofer, L. J. E., Cohn, E. M., and Peebles, W. C. 1949. Isothermal decomposition of the carbide in a carburized cobalt Fischer-Tropsch catalyst. *J. Phys. Coll. Chem.* 53:661–69.
78. Weller, S., Hofer, L. J. E., and Anderson, R. B. 1948. The role of bulk cobalt carbide in Fischer-Tropsch synthesis. *J. Am. Chem. Soc.* 70: 799–801.
79. Tihay, F., Pourroy, G., Richard-Plouet, M., Roger, A. C., and Kiennemann, A. 2001. Effect of Fischer-Tropsch synthesis on the microstructure of Fe-Co-based metal/spinel composite materials. *Appl. Catal. A* 206:29–42.
80. Niemantsverdriet, J. W., and van der Kraan, A. M. 1981. On the time-dependent behavior of iron catalysts in Fischer-Tropsch synthesis. *J. Catal.* 72:385–88.
81. Hofer, L. J. E., and Peebles, W. C. 1947. X-ray diffraction studies of the action of carbon monoxide on cobalt-thoria-kieselguhr catalysts. I. *J. Am. Chem. Soc.* 69:2497–500.
82. U.S. Bureau of Mines. 1948–1959. *Synthetic liquid fuels from hydrogenation of carbon monoxide*, 19. Bulletin 578. Washington, DC: U.S. Government Print Office. http://www.fischer-tropsch.org.

83. Weller, S. E. 1947. Kinetics of carbiding and hydrocarbon synthesis with cobalt Fischer-Tropsch catalysts. *J. Am. Chem. Soc.* 69:2432–36.

84. Xiong, J., Ding, Y., Wang, T., Yan, L., Chen, W., Zhu, H., and Lu, Y. 2005. The formation of Co_2C species in activated carbon supported cobalt-based catalysts and its impact on Fischer-Tropsch reaction. *Catal. Lett.* 102:265–69.

85. Leclercq, L., Almazouari, A., Dufour, M., and Leclercq, G. 1996. Carbide-oxide interactions in bulk and supported tungsten carbide catalysts for alcohol synthesis. In *Chemistry of transition metal carbides and nitrides*, ed. S. T. Oyama, 345–61. Glasgow: Blackie.

86. Kitakami, O., Sato, H., Shimada, Y., Sato, F., and Tanaka, M. 1997. Size effect on the crystal phase of cobalt fine particles. *Phys. Rev. B* 56:13849–54.

87. Pankina, G. V., Chernavskii, P. A., Lermontov, A. S., and Lunin, V. V. 2002. Study of carbon deposits on the surface of supported cobalt catalysts in the Fischer-Tropsch process. *Petrol. Chem.* 42:217–20.

88. Ducreux, O., Lynch, J., Rebours, B., Roy, M., and Chaumette, P. 1998. In situ characterization of cobalt based Fischer-Tropsch catalysts: A new approach to the active phase. *Stud. Surf. Sci. Catal.* 119:125–30.

89. Machocki, A. 1991. Formation of carbonaceous deposit and its effect on carbon monoxide hydrogenation on iron-based catalysts. *Appl. Catal.* 70:237–52.

90. Jacobs, G., Patterson, P. M., Zhang, Y., Das, T., Li, J., and Davis, B. H. 2002. Fischer-Tropsch synthesis: Deactivation of noble metal-promoted Co/Al$_2$O$_3$ catalysts. *Appl. Catal. A* 233:215–26.

91. Tavasoli, A., Malek Abbaslou, R. M., and Dalai, A. K. 2008. Deactivation behavior of ruthenium promoted Co/α-Al$_2$O$_3$ catalysts in Fischer–Tropsch synthesis. *Appl. Catal. A* 346:58–64.

92. Morales, F., and Weckhuysen, B. M. 2006. Promotion effects in Co-based Fischer-Tropsch catalysis. *Catalysis* 19:1–40.

93. Lahtinen, J., and Somorjai, G. A. 1998. The effects of promoters in carbon monoxide hydrogenation on cobalt foil model catalysts. *J. Mol. Catal. A* 130:255–60.

94. Iglesia, E., Soled, S. L., Fiato, R. A., and Via, G. H. 1993. Bimetallic synergy in cobalt-ruthenium Fischer-Tropsch synthesis catalysts. *J. Catal.* 143:345–68.

95. Campbell, C. T., and Goodman, D. W. 1982. A surface science investigation of the role of potassium promoters in nickel catalysts for carbon monoxide hydrogenation. *Surf. Sci.* 123:413–26.

96. Snoeck, J.-W., Froment, G. F., and Fowles, M. 2002. Steam/CO$_2$ reforming of methane. Carbon formation and gasification on catalysts with various potassium contents. *Ind. Eng. Chem. Res.* 41:3548–56.

97. Wesner, D. A., Linden, G., and Bonzel, H. P. 1986. Alkali promotion on cobalt: Surface analysis of the effects of potassium on carbon monoxide adsorption and Fischer-Tropsch reaction. *Appl. Surf. Sci.* 26:335–56.

98. Pichler, H. 1952. Twenty-five years of synthesis of gasoline by catalytic conversion of carbon monoxide and hydrogen. *Adv. Catal.* 4: 271–341.

99. Arcuri, K. B., and Leviness, S. C. 2003. The regeneration of hydrocarbon synthesis catalyst, a partial review of the related art published during 1930 to 1952. Paper presented at the AIChE Spring National Meeting, New Orleans, April 2. www.fischer-tropsch.org.

100. Roelen, O., Heckel, H., and Hanisch, F. 1942. U.S. Patent 2 289 731, to Hydrocarbon Synthesis Corporation.

101. Iglesia, E., Soled, S. L., and Fiato, R. 1988. U.S. Patent 4 738 948, to Exxon.

102. Feisst, W., and Roelen, O. 1945. U.S. Patent 2 369 956, to Hydrocarbon Synthesis Corporation.

103. Huang, R., Agee, K. L., Arcuri, B. A., and Schubert, P. F. 2002. U.S. Patent 6 812 179, to Syntroleum Corporation.
104. Zhan, X., Arcuri, K., Huang, R., Agee, K., Engman, J., and Robota, H. J. 2004. Regeneration of cobalt-based slurry catalysts for Fischer-Tropsch synthesis. *Prep. Pap.-Am. Chem. Soc. Div. Pet. Chem.* 49:179–81.
105. Espinoza, R. L., Steynberg, A. P., Jager, B., and Vosloo, A. C. 1999. Low temperature Fischer-Tropsch synthesis from a Sasol perspective. *Appl. Catal. A* 186:13–26.
106. Bertole, C. J., Kiss, G., and Mims, C. A. 2004. The effect of surface-active carbon on hydrocarbon selectivity in the cobalt-catalyzed Fischer-Tropsch synthesis. *J. Catal.* 223:309–18.
107. Iglesia, E., Soled, S. L., Baumgartner, J. E., and Reyes, S. C. 1995. Synthesis and catalytic properties of eggshell cobalt catalysts for the Fischer-Tropsch synthesis. *J. Catal.* 153:108–22.

5 Catalytic Performance of Ru/Al$_2$O$_3$ and Ru/Mn/Al$_2$O$_3$ for Fischer-Tropsch Synthesis

Mohammad Nurunnabi, Kazuhisa Murata, Kiyomi Okabe, Megumu Inaba, and Isao Takahara

CONTENTS

Mn-modified Ru/Al$_2$O$_3$ catalyst was investigated for catalytic activity, selectivity, and stability for Fischer-Tropsch synthesis in a continuous stirred tank reactor (CSTR) under pressurized conditions. Without Mn, the Ru/Al$_2$O$_3$ catalyst showed low CO conversion and the deactivation rate was clearly observed at low (493 K) reaction temperature. In contrast, a small amount of Mn addition (Mn/Al = 1/19) on Ru/Al$_2$O$_3$ enhanced both CO conversion and C$_5{}^+$ selectivity for Fischer-Tropsch synthesis. Under pressurized conditions (20 to 60 bar), high catalytic activity and high resistance to catalyst deactivation with time on stream were observed over Ru/Mn/Al$_2$O$_3$. At 60 bar, equilibrium CO conversion was estimated to be about 96%. Characterization results—BET surface area x-ray diffraction (XRD), temperature-programmed reduction (TPR), transmission electron microscopy (TEM), and x-ray photoelectron spectroscopy (XPS)—indicate that the addition of Mn increases the concentration of metallic Ru active species on the catalyst surface by removing chlorine atoms from RuCl$_3$, thus increasing catalytic activity while inhibiting catalyst deactivation.

5.1 INTRODUCTION

A sustainable energy future requires a combination of factors, such as renewable resources and advanced energy technologies. Biomass has recently received increased attention as one potential source of renewable energy. Among all biomass conversion processes, the biomass-to-liquid (BTL) process is one of the most promising ways to utilize biomass in remote or local areas to form sulfur-free transportation fuels.[1] The development of catalysts for the Fischer-Tropsch (FT) synthesis is one of the key technologies required for the BTL process. It is well known that Ru, Co, and Fe can be utilized for FT synthesis in any way.[2,3] However, Ru metal, from $RuCl_3 \cdot nH_2O$, improves middle distillate yields and higher wax formation, which can then be cracked into middle distillates by using suitable catalyst supports. The problem with slurry phase FT reactors (which inhibit hot spot formation) is catalyst deactivation.[4] Therefore, new catalysts with high activity and stability are needed. Mn addition can promote FT activity. The effects of Mn addition strongly depend on the supports and on the preparation methods used to generate a catalyst with improved stability, activity, and selectivity toward carbon chain growth probability.[5,6] Previous reports suggested that a γ-Al_2O_3 support is more effective in the presence of Mn for FT synthesis.[6,7] In this paper, we have studied the catalytic performance (activity, selectivity, and stability) for Ru/Al_2O_3 and Ru/Mn/Al_2O_3 catalysts in the FT reaction under different Mn to Al ratios and pressures.

5.2 EXPERIMENTAL

5.2.1 CATALYST PREPARATION

Ru catalysts were prepared by the impregnation method using γ-Al_2O_3 (Soekawa Chemicals, Japan) and an aqueous solution of $RuCl_3 \cdot nH_2O$ (Soekawa Chemicals, Japan). After removal of the solvent by heating, the catalyst was dried in an oven overnight at 383 K. Subsequently, the catalyst was calcined in air at 673 K for 4 h. The prepared catalyst is denoted as Ru/Al_2O_3. The addition of Mn to Ru/Al_2O_3, denoted as Ru/Mn/Al_2O_3, was prepared in two steps. At first the support was impregnated with an aqueous solution of $Mn(NO_3)_2 \cdot 6H_2O$ (Wako Pure Chemical Industries Ltd., Japan) and then removed the solvent by evaporation at 348 K. After drying at 383 K overnight, the sample was calcined in air at 873 K for 5 h. In this study, Mn/Al molar ratios of 1/2 to 1/19 were used. In the second step, the Ru impregnation procedure was the same as the one used for the reference Ru/Al_2O_3. The loading amount of Ru used was 5 wt%. The BET surface area of these catalysts (from nitrogen porosimetry) is listed in Table 5.1.

5.2.2 CATALYTIC REACTION

FT synthesis was carried out in a 0.3 L autoclave-type slurry continuous stirred tank reactor under pressurized conditions. The catalyst was reduced with hydrogen

TABLE 5.1

Properties of BET, TPR, and XPS on Ru/Al$_2$O$_3$ and Ru/Mn/Al$_2$O$_3$[a]

Catalyst	BET (m^2/g)	H$_2$ Consumption[b] (10^{-8} mol/g)	Surface Atomic Concentrations in XPS (%)[c]				
			Ru	Mn	Cl	O	Al
Ru/Al$_2$O$_3$	81	82	1.3	—	0.9	60.0	37.8
Ru/Mn/Al$_2$O$_3$	80	111	2.3	0.3	0.7	60.0	36.7

[a] Mn to Al ratio was 1/19.

[b] Amount of hydrogen consumption on Ru reduction in TPR profiles at 380–600 K, where H$_2$-TPR reaction was Ru^{4+} + 2H$_2$ → Ru^0 + 4H$^+$.

[c] H$_2$ reduction at 473 K for 3 h with XPS experiment.

at 473 K, 20 bar for 5 h. After the pretreatment, the catalyst was suspended by using 80 g hexadecane as a solvent under atmospheric pressure and room temperature. The amount of catalyst used was 2.5 g and the partial pressure ratio of reactants was H$_2$/CO/Ar = 6/3/1. The reactions were carried out at T = 493 K, P = 10–60 bar, GHSV = 1,800 h^{-1}. The product gas was periodically analyzed with Shimadzu online gas chromatographs (GC-2014) in both flame ionization detector (FID) and thermal conductivity detector (TCD) detector, and the liquid hydrocarbons were analyzed by another Shimadzu gas chromatograph (GC-8A). Experimental details have been described in a previous report.[7]

5.2.3 Catalyst Characterization

The surface area of the catalysts was determined by the BET method using a Gemini (Micromeritics) instrument. X-ray diffraction (XRD) was measured by using a Mac Science M18XHF22-SRA diffractometer with Cu Kα radiation at 40 kV and 150 mA. The reducibility of the catalysts was characterized by a temperature-programmed reduction with H$_2$ (H$_2$-TPR) technique. Before the TPR measurement, the samples were heated at 773 K for 20 min (heating rate, 10 K/min) under Ar gas flow of 30 ml/min in order to remove any adsorbed species such as CO$_2$. After the sample was cooled down to room temperature under Ar flowing, the reactor was heated again from room temperature to 773 K at a heating rate of 10 K/min, and then the temperature was maintained for 0.5 h in a 5% H$_2$/Ar mixture with a gas flow of 35 ml/min. The temperature was measured by using a thermocouple located in the catalyst bed. The H$_2$ consumption was continuously monitored by TCD-GC. X-ray photoelectron spectroscopy (XPS) data were estimated by using a Shimadzu ESCA-850 with Mg-K$_α$ irradiation (8 kV, 30 mA) and without exposure to open air after reduction treatment in H$_2$ flow at 473 K for 3 h within the prechamber of the apparatus. The binding energies of XPS were referred to contamination C on the surface of the sample as the internal standard with C$_{1s}$ level at 284.6 eV. A transmission electron microscope (TEM) image was obtained on the H9000NAR (HITACHI) operated at 300 kV. The catalyst was reduced with H_2 at 473 K for 5 h under 20 bar. After this reduction, the TEM sample was deposited as dry powder on a microgrid mounted on a copper grid. A lot of metallic particles were observed over the catalyst. The average particle size (d) was estimated using the equation of

$$d = \sum_i n_i d_i^3 \Big/ \sum_i n_i d_i^2$$

(n_i, number of the particles; d_i, particle size).

5.3   RESULTS AND DISCUSSION

Figure 5.1 shows the effect of Ru loading on Ru/Mn/Al$_2$O$_3$ with CO conversion, C$_5^+$ selectivity, and CH$_4$ selectivity during FT synthesis. CO conversion increased with increasing Ru loading up to 5 wt%. For this loading, CH$_4$ selectivity and C$_5^+$ selectivity were almost constant, and at 5 wt% Ru loading the catalyst showed much higher CO conversion. When Ru increased from 5 to 10 wt%, CO

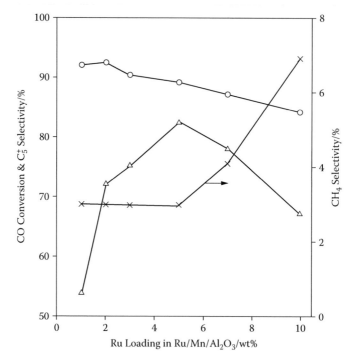

FIGURE 5.1   Effect of Ru loading on Ru/Mn/Al$_2$O$_3$ with CO conversion, C$_5^+$ selectivity, and CH$_4$ selectivity for FT synthesis. (Δ) CO conversion, (○) C$_5^+$ selectivity, (×) CH$_4$ selectivity. Reaction conditions: catalyst weight = 2.5 g, T = 493 K, P = 20 bar, H$_2$/CO = 2, GHSV = 1,800 h^{-1}, H$_2$ reduction = 473 K.

conversion and C$_5^+$ selectivity decreased while CH$_4$ selectivity increased almost linearly (Figure 5.1). These results indicate that, at these conditions, 5 wt% Ru on Ru/Mn/Al$_2$O$_3$ is the preferred loading for this type of FT synthesis.

Figure 5.2 shows the effect of Mn concentration (in Ru/Mn/Al$_2$O$_3$) on CO conversion, C$_5^+$ selectivity, and deactivation rate for FT synthesis at 493 K and 20 bar. Ru/Al$_2$O$_3$ exhibited moderate CO conversion, and a deactivation rate was clearly observed. On the other hand, the addition of small amounts of Mn to Ru/Al$_2$O$_3$ enhanced the catalyst activity significantly. When a Mn to Al ratio between 1/19 and 1/4 was used, the Ru/Mn/Al$_2$O$_3$ catalyst showed higher CO conversion than Ru/Al$_2$O$_3$, and its deactivation rate was negligible with time on stream. In these Mn/Al ranges, the values of CO conversion, C$_5^+$ selectivity, CH$_4$ selectivity, and carbon chain growth probability were about 84%, 95%, 3%, and 0.92, respectively. When Mn loading increased from 1/4 to 1/2 in Ru/Mn/Al$_2$O$_3$, CO conversion and C$_5^+$ selectivity dramatically decreased and deactivation rate increased almost linearly. The Ru/Mn/Al$_2$O$_3$ catalyst with Mn/Al = 1/2 showed a much lower catalytic activity than Ru/Al$_2$O$_3$. These results suggest that small amounts

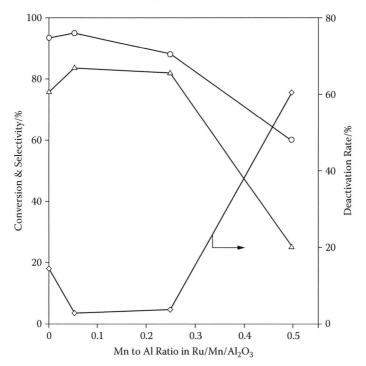

FIGURE 5.2 Effect of Mn concentration on Ru/Mn/Al$_2$O$_3$ with CO conversion, C$_5^+$ selectivity, and deactivation rate for FT synthesis. (Δ) CO conversion, ($\bigcirc$) C$_5^+$ selectivity, ($\Diamond$) deactivation rate. Reaction conditions: catalyst weight = 2.5 g, T = 493 K, P = 20 bar, H$_2$/CO = 2, GHSV = 1,800 h^{-1}, H$_2$ reduction = 473 K.

of Mn addition can play an important role on catalyst properties and effects are controlled by Mn/Al levels.

Figure 5.3 shows the dependence of CO conversion, C_5^+ selectivity, CH_4 selectivity, olefin–paraffin ratio, and space–time yield over Ru/Mn/Al$_2$O$_3$ catalyst on reaction time. In this experiment, the amount of solvent used was 117 g and the Mn to Al ratio was 1/19. High catalytic activity, selectivity, and stability were observed over Ru/Mn/Al$_2$O$_3$ catalyst with long reaction times for FT synthesis at 493 K and 20 bar. On this catalyst, olefin–paraffin ratio increased with increasing the reaction time up to 40 h, and then the ratio was almost constant. CO conversion, C_5^+ selectivity, CH_4 selectivity, and space-time yield were constant with time on stream. The equilibrium level of CO conversion was estimated to be about 80%; CH_4 selectivity was negligible. At these conditions, selectivity for C_{10}–C_{20} and carbon chain growth probability were about 44% and 0.92, respectively; over this Ru/Mn/Al$_2$O$_3$ catalyst the space–time yield was about 0.24 mol/gh.

Figure 5.4 represents the dependence of CO conversion and C_5^+ selectivity on reaction time over the Ru/Mn/Al$_2$O$_3$ catalyst for FT synthesis under pressurized

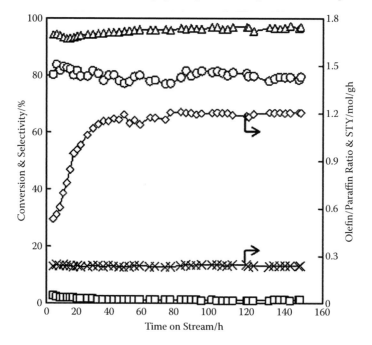

FIGURE 5.3 Results for FT synthesis with reaction time dependence of Ru/Mn/Al$_2$O$_3$ catalyst. (○) CO conversion, (△) C_5^+ selectivity, (◇) CH_4 selectivity, (◊) olefin-paraffin ratio in C_2–C_4 products, (×) space-time yield on CO conversion and C_5^+ selectivity. Ru content was 5 wt% and Mn to Al ratio was 1/19. Reaction conditions: catalyst weight = 2.5 g, T = 493 K, P = 20 bar, H_2/CO = 2, GHSV = 1,800 h^{-1}, H_2 reduction = 473 K.

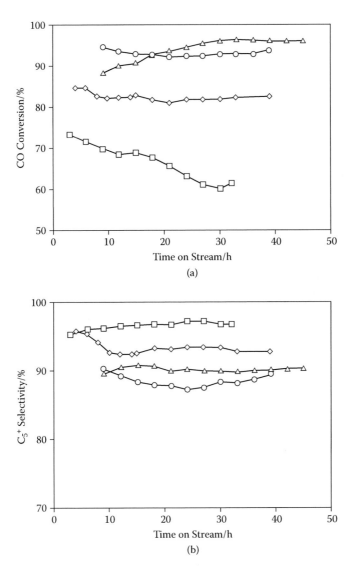

FIGURE 5.4 Reaction time dependence of (a) CO conversion and (b) C$_5^+$ selectivity over Ru/Mn/Al$_2$O$_3$ catalyst for FT synthesis under pressurized conditions. (□) 10 bar, (◇) 20 bar, (○) 40 bar, (△) 60 bar. Mn to Al ratio was 1/19. Reaction conditions: catalyst weight = 2.5 g, T = 493 K, H$_2$/CO = 2, GHSV = 1,800 h^{-1}, H$_2$ reduction = 473 K.

conditions. The Mn to Al ratio was 1/19 and the reaction temperature was 493 K. CO conversion increased and C$_5^+$ selectivity decreased with increasing pressure. Under high pressures of 20, 40, and 60 bar, CO conversion was very stable with time on stream. In particular, under 60 bar, high catalytic activity and high resistance to catalyst deactivation were observed in the CSTR used for FT synthesis.

Under this pressure, equilibrium CO conversion and C_5^+ selectivity were about 96 and 90%, respectively (Figure 5.4).

Figure 5.5 shows the TPR profiles of Mn/Al_2O_3, Ru/Al_2O_3, and $Ru/Mn/Al_2O_3$. It is clear that in Mn/Al_2O_3 there is a barely visible very weak and broad peak centered near 680 K representing the possible reduction of some Mn_2O_3 to Mn_3O_4 (Figure 5.5a). There is a weak other hydrogen reduction peak near 740 K that can be safely attributed to the reduction of intermediate Mn_3O_4 to MnO. This result suggests that higher reduction temperatures are needed for Mn due to a stronger Mn-support interaction. This observation is in agreement with previous reports.[8,9] On the other hand, in Figure 5.5b and c, Ru/Al_2O_3 and $Ru/Mn/Al_2O_3$ show only one hydrogen reduction peak at 480 and 510 K, respectively, assigned to the reduction of RuO_2 to Ru^0. Characterization results are listed in Table 5.1. The amount of hydrogen consumption in the TPR profiles increased with Mn addition to Ru/Al_2O_3, and this can be due to an increased ease of Ru reduction. The important point is that $Ru/Mn/Al_2O_3$ showed a delay reduction peak with respect to Ru/Al_2O_3, and reduction peaks attributed to the Mn species were not observed in Figure 5.5c. This suggests a possible Ru–Mn species interaction forcing the RuO_2 phase to be reduced at a higher temperature (see Figure 5.5b and c).

Figure 5.6 shows the XRD patterns of Ru/Al_2O_3 and $Ru/Mn/Al_2O_3$ after catalyst calcination, H_2 reduction, and FT reaction. After catalyst calcination, the diffraction peaks assigned to Mn_2O_3[10] were clearly observed in $Ru/Mn/Al_2O_3$,

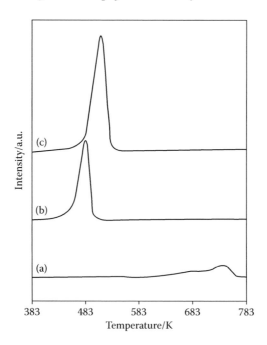

FIGURE 5.5 H_2-TPR profiles of (a) Mn/Al_2O_3, (b) Ru/Al_2O_3, and (c) $Ru/Mn/Al_2O_3$ at 300 to 800 K.

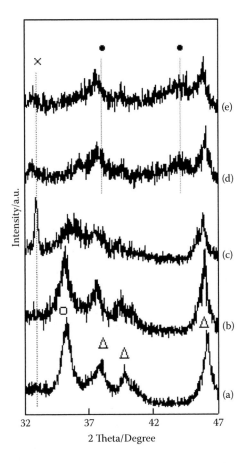

FIGURE 5.6 XRD patterns of (a) Ru/Al$_2$O$_3$ and (b–e) Ru/Mn/Al$_2$O$_3$ catalysts: (b–c) after calcination, (d) after H$_2$ reduction, and (e) after FT reaction. Mn to Al ratio was (a) 0/100, (b) 1/19, and (c–e) 1/2. (○) RuO$_2$, (△) Al$_2$O$_3$, (×) Mn$_2$O$_3$, (●) Ru^0.

Mn/Al = ½ (see Figure 5.6c). However, Mn species were not observed over Ru/Mn/Al$_2$O$_3$, Mn/Al = 1/19 (Figure 5.6b), possibly because at this low Mn concentration, Mn species are below the limit of detection by XRD. A comparison of Figure 5.6a and b indicates that Ru/Al$_2$O$_3$ and Ru/Mn/Al$_2$O$_3$ (Mn/Al = 1/19) yield similar diffractograms showing peaks attributable to Al$_2$O$_3$ and RuO$_2$ formation.[11] RuO$_2$ broad diffraction peaks with low intensity are also observed over Ru/Mn/Al$_2$O$_3$ (Mn/Al = 1/2; see Figure 5.6c). In Figure 5.6c, the diffraction peak at low diffraction angles is attributed to Mn$_2$O$_3$. This peak broadens and decreases significantly after reduction in H$_2$ and reaction (Figure 5.6d and e). The BET surface area (SA) data listed in Table 5.1 indicate that Mn addition has a negligible effect on SA in Ru/Al$_2$O$_3$ and Ru/Mn/Al$_2$O$_3$ (Mn/Al = 1/19) catalysts; additional details have been shown in previous reports.[6,12] Ru/Mn/Al$_2$O$_3$ (Mn/Al = 1/2) catalyst after H$_2$ reduction and FT reaction yields similar XRD diffractograms (Figure 5.6d and e). In this case, all RuO_2 was reduced and Ru^0 formed in both samples.[13] Using the Scherrer equation, the particle size of Ru after reduction and reaction was calculated to be about 8 nm. This value is also well in agreement with the TEM measurement, shown in Figure 5.7.

Figure 5.7 shows the TEM image of $Ru/Mn/Al_2O_3$ catalyst after H_2 reduction at 473 K. A lot of round-shaped particles that correspond to Ru metals can be observed on the catalyst surface. The average particle size (d) can be obtained using the equation

$$d = \sum_i n_i d_i^3 \Big/ \sum_i n_i d_i^2$$

(n_i, number of the particles; d_i, particle size), and it was estimated to be about 8 nm over the $Ru/Mn/Al_2O_3$ surface. We also measured the energy-dispersive x-ray (EDX) analysis of one metallic particle; the spot region of EDX analysis is also indicated in Figure 5.7. The elements of Mn, Al, and Cl as well as Ru were clearly observed in the spot region of EDX analysis.

Surface atomic concentrations by XPS results are listed in Table 5.1. $Ru/Mn/Al_2O_3$ shows higher Ru concentration than Ru/Al_2O_3, suggesting that the catalyst reducibility was improved significantly by the presence of Mn. This result is consistent with the TPR result in Table 5.1. Furthermore, Cl was found present on Ru/Al_2O_3 and $Ru/Mn/Al_2O_3$ surfaces. A previous report suggested that the residual chlorine ions are partitioned between the support and the metal, and that the chlorine on the surface could inhibit both CO and hydrogen chemisorption.[14]

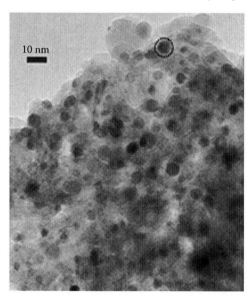

FIGURE 5.7   TEM image of $Ru/Mn/Al_2O_3$ catalyst after H_2 reduction at 473 K.

Therefore, it is necessary to evaluate the effect of Mn addition on the catalyst resistance to deactivation. From Table 5.1, the important data are the Cl/Mn ratio for FT synthesis, and the Cl/Mn ratio for our catalyst is about 2.3. This ratio can be related to the formation of MnCl$_2$ over Ru/Mn/Al$_2$O$_3$.[6,7] One possible explanation is that the addition of Mn can accelerate a removal of Cl atom from RuCl$_3$ forming MnCl$_2$, thus eliminating the deleterious effects of Cl on CO and H$_2$ chemisorption. As a result, the concentration of metallic Ru, associated with catalytic performance in the FT reaction, increases. Thus, Ru/Mn/Al$_2$O$_3$ catalyst shows high activity with high resistance to catalyst deactivation for FT synthesis under higher reaction pressure.

5.4   CONCLUSIONS

In contrast to Ru/Al$_2$O$_3$, Ru/Mn/Al$_2$O$_3$ catalysts showed high catalytic activity and high resistance to catalyst deactivation during FT synthesis. Under high pressure of 60 bar, the equilibrium CO conversion was estimated to be about 96% over Ru/Mn/Al$_2$O$_3$. TEM, XPS, and TPR results indicate that the addition of Mn to Ru/Al$_2$O$_3$ can accelerate removal of Cl atoms from RuCl$_3$ to form MnCl$_2$, thus increasing the density of metallic Ru species on the catalyst surface and with it catalytic activity.

REFERENCES

1. Luke, H.W. 2005. Biomass-to-liquids (BTL) fuels—A promising option for the future. *Erdol Erdgas Kohle* 121:3–5.
2. Iglesia, E., Soled, S.L., Fiato, R.A., and Via, G.H. 1993. Bimetallic synergy in cobalt ruthenium Fischer–Tropsch synthesis catalysts. *J. Catal.* 143:345–68.
3. Schulz, H. 1999. Short history and present trends of Fischer–Tropsch synthesis. *Appl. Catal. A Gen.* 186:3–12.
4. Dry, M.E. 2004. FT catalysts. *Stud. Surf. Sci. Catal.* 152:533–600.
5. Morales, F., de Groot, F.M.F., Gijzeman, O.L.J., Mens, A., Stephan, O., and Weckhuysen, B.M. 2005. Mn promotion effects in Co/TiO$_2$ Fischer–Tropsch catalysts as investigated by XPS and STEM-EELS. *J. Catal.* 230:301–8.
6. Nurunnabi, M., Murata, K., Okabe, K., Inaba, M., and Takahara, I. 2007. Effect of Mn addition on activity and resistance to catalyst deactivation for Fischer–Tropsch synthesis over Ru/Al$_2$O$_3$ and Ru/SiO$_2$ catalysts. *Catal. Commun.* 8:1531–37.
7. Nurunnabi, M., Murata, K., Okabe, K., Inaba, M., and Takahara, I. 2008. Performance and characterization of Ru/Al$_2$O$_3$ and Ru/SiO$_2$ catalysts modified with Mn for Fischer–Tropsch synthesis. *Appl. Catal. A Gen.* 340:203–11.
8. Richter, M., Trunschke, A., Bentrup, U., Brzezinka, K.W., Schreier, E., Schneider, M., Pohl, M.M., and Fricke, R. 2002. Selective catalytic reduction of nitric oxide by ammonia over egg-shell MnO$_x$/NaY composite catalysts. *J. Catal.* 206:98–113.
9. Kapteijn, F., van Langeveld, A.D., Moulijn, J.A., Andreini, A., Vuurman, M.A., Turek, A.M., Jehng, J.M., and Wachs, I.E. 1994. Alumina-supported manganese oxide catalysts. *J. Catal.* 150:94–104.
10. International Center for Diffraction Data (ICDDD), 78-0390.
11. International Center for Diffraction (ICDD), 88-0323.

6 Preparation of Highly Active Co/SiO$_2$ Catalyst with Chelating Agents for Fischer-Tropsch Synthesis
Role of Chelating Agents

Naoto Koizumi, Daichi Hongo, Yukiya Ibi, Yasuhiko Hayasaka, and Muneyoshi Yamada

CONTENTS

In order to investigate effects of complex formation between Co and chelating agents on Co-SiO$_2$ interaction at high Co loading, we have tried to prepare Co/SiO$_2$ catalysts having high Co loading using nitrilotriacetic acid (NTA) or trans-1,2-Cyclohexane-diamine-N,N,N′,N′-tetraacetic acid (CyDTA). The stepwise impregnation method, where Co nitrate solution and chelating agent solution were stepwise impregnated followed by drying and calcination, successfully increased Co loading to 20 mass%. The catalysts thus prepared showed higher Fischer-Tropsch synthesis (FTS) activities at high Co loadings (≤10 mass%) than the catalyst prepared without using chelating agents. CyDTA was more effective than NTA. The stepwise impregnation method with CyDTA led to higher dispersion and reducibility of Co at 20 mass%-Co, leading to higher metallic Co surface area than those of previously reported catalysts with various types of support and Co loadings. The effects of complex formation on the structure of Co species formed during drying and calcination were investigated by Fourier transform infrared (FT-IR), x-ray diffraction (XRD), extended x-ray absorption fine structure (EXAFS), and x-ray photoelectron spectroscopy (XPS).

6.1 INTRODUCTION

Because of increasing demands for high-quality diesel fuels, many studies on fundamental and technological aspects of Co-based FTS catalysts have been made in order to improve our understanding of their activity and selectivity property.[1–10]

From these studies, it is generally assumed that metallic Co particle is an FTS active species for Co-based catalysts. Iglesia[2] investigated the FTS activities of Co/Al$_2$O$_3$, Co/TiO$_2$, and Co/SiO$_2$ catalysts with various Co loadings. It was found that the turnover frequency (TOF) of these catalysts is independent of the particle size of metallic species (10–250 nm, determined by H$_2$ adsorption), even though the type of supports and Co loading were varied over wide ranges. This means that the rate of CO conversion (Co weight basis) increases with decreasing the cluster size in this range. On the other hand, Bezemer et al.[8] recently investigated the FTS activity of Co supported on carbon nanofiber (CNF) prepared by different procedures, and found that the TOF of Co/CNF catalysts linearly increases with the particle size (2–8 nm, determined by XPS). When the particle size is in 8 to 27 nm range, the TOFs are identical to those reported by Iglesia.[2] According to their results, the formation of too small metallic particles is not suitable for higher rate of conversions. However, the TOFs of Co/Al$_2$O$_3$, Co/TiO$_2$, and Co/SiO$_2$ catalysts remain unclear when the size of metallic Co particles is below 10 nm. Different type of supports may cause different types of Co-support interactions, leading to different types of particle size effects.

On the other hand, the dispersion and reducibility of Co has been investigated extensively for various catalysts with different types of Co precursor, support, and Co loading. For example, supported Co catalysts are usually prepared by the impregnation of Co nitrate solution followed by drying, calcinations, and H_2 reduction. As is well known, this preparation method leads to lower dispersion, but higher reducibility. The particle size of metallic species from Co nitrate precursor is reported to be 20 to 40 nm.[9,10] Besides, several authors reported that the use of Co acetate (Co/SiO$_2$),[9] Co acethylacetonate, Co oxalate (Co/TiO$_2$),[11] or Co nitrate + Co acetate mixed salt (Co/SiO$_2$),[9] instead of the conventional Co nitrate precursor, results in the formation of metallic particles with smaller sizes. The reducibility of Co on these catalysts is, on the other hand, somewhat lower compared with the catalyst prepared from nitrate precursor. These results suggest that Co precursor has a strong influence on the dispersion and reducibility of Co, that is, Co-support interaction.

In previous studies, the authors have tried to control Co-support interaction using novel types of Co precursor; a Co complex was chosen as precursor. Specifically, catalysts (5 mass%-Co) were prepared by co-impregnation of Co nitrate and chelating agent, or organic acid solutions (1:1 molar ratio), followed by drying and calcination. Chelating agents or organic acids with various Co^{2+} complex formation constants were used in catalyst preparation. It was found that Co/SiO$_2$ catalyst prepared with NTA shows three times higher FTS activity than the catalyst prepared with Co nitrate alone. Furthermore, volcano-type dependencies were found when CO conversion and H_2 uptake over these catalysts were plotted against their logarithmic complex formation constants.[12] Characterization results showed that chelating agents that form stable complexes in the impregnating solution are effective for the formation of fine metallic particles (less than 5 nm), whereas the formation of too stable complex results in lower reducibility of Co. Because 1:1 NTA-Co complex has an intermediate stability, both higher dispersion and reducibility were obtained with the catalyst prepared with NTA.[12–14] It was also found that the TOFs of these catalysts are constant in the 1 to 20 nm particle size range,[13] which is evidently different from that reported for Co/CNF catalyst. Because Co precursor that can attain both higher dispersion and reducibility has not been known so far, our results suggest that the complex formation creates a novel type of Co-support interactions, leading to higher FTS activity.

If such a kind of novel Co-support interactions can be realized at higher Co loadings, higher FTS activities are then to be expected. Therefore, it is important to investigate what kind of Co-support interaction is formed by the complex formation at higher Co loadings, and how it affects the FTS activity of Co/SiO$_2$ catalyst. However, unfortunately, Co loading by the above-mentioned co-impregnating method was limited to a maximum of 5 mass% because of lower solubility of chelating agents in the presence of higher concentration of Co. However, this disadvantage is expected to be overcome by the stepwise impregnation method, where separately prepared Co nitrate and chelating agent solutions were stepwise impregnated onto SiO$_2$ support.

In the present work, in order to investigate complex formation effects on Co-SiO$_2$ interaction at higher Co loading, we have prepared Co/SiO$_2$ catalysts (having higher Co loading) using chelating agents and the stepwise impregnation method. NTA (log K_{Co} = 10) and CyDTA (log K_{Co} = 19) were chosen as chelating agents, because they were both effective in the formation of fine metallic particles in the co-impregnation method. Their FTS activity and selectivity were investigated under high-pressure reaction conditions. The dispersion and reducibility of Co, and the structure of Co species at drying and calcination steps, were also investigated by several spectroscopic techniques to elucidate the effects of complex formation.

6.2   EXPERIMENTAL

6.2.1   CATALYST PREPARATION BY THE STEPWISE IMPREGNATION METHOD

Q-15 SiO$_2$ powder purchased from Fuji Silisia (BET surface area = 192 m^2 g^{-1}, pore volume = 1.03 ml g^{-1}, 150–250 µm) was used as support. Aqueous solution containing NTA (Dojindo Chemicals, purity > 99%) or CyDTA (Dojindo chemicals, purity > 99%) was first impregnated onto SiO$_2$ support followed by drying at 393 K. Then, aqueous Co(NO$_3$)$_2$·6H$_2$O (Wako Pure Chemicals, purity > 99.5%) solution was impregnated onto dried L/SiO$_2$ powder (L = NTA or CyDTA) followed by drying (393 K) and calcination (723 K). Co loading was varied from 5 to 20 mass% as metallic Co (SiO$_2$ weight basis), whereas NTA and CyDTA loading was fixed constant. Thus, the Co to L molar ratio is varied from 1 (5 mass%-Co) to 4 (20 mass%-Co) depending on Co loading. The catalysts thus prepared are denoted as Co(X)/L/SiO$_2$, where X stands for Co loading in mass% as metallic Co (SiO$_2$ weight basis).

6.2.2   ACTIVITY TEST

FTS activity and selectivity of these catalysts were investigated with a high-pressure fixed bed reactor system. The reactor consisted of a stainless steel tube with an internal diameter of 7 mm in an electronically heated oven. The gases, H$_2$ (purity > 99.995%) and 33% CO/62% H$_2$/5% Ar (purity > 99.99995%), were used in the reaction without purification. The calcined catalyst was reduced in a stream of H$_2$ at 773 K for 6 h. After H$_2$ reduction, the H$_2$ stream was changed into a high-pressure CO/H$_2$/Ar stream (1.1 MPa) at room temperature. Hereafter, the catalyst was heated to 453 K at the rate of 4 K min^{-1}, and then to 503 K at the rate of 0.2 K min^{-1} for the activity evaluation. Gaseous products were analyzed with online GC/TCD (Shimadzu, GC8A) and GC/FID (Shimadzu, GC14B), whereas liquid products were collected with an ice trap during on stream. After the reaction, liquid products were removed from the ice trap, and then subjected to offline GC/FID (Agilent Technologies, 6850 series GC) analysis. The chain growth probability (α) was calculated from molar fractions of C$_{10}$–C$_{20}$ hydrocarbon.

6.2.3 Catalysts Characterization

6.2.3.1 Dried Catalysts

Complex formation between Co and chelating agents was investigated by FT-IR spectroscopy after the second impregnation followed by drying. The dried sample was diluted with KBr powder and then formed into a self-supporting wafer. FT-IR measurements were carried out in a transmittance mode on an FTS6000 FT-IR spectrometer (Varian) with spectral resolution and accumulation time of 4 cm^{-1} and 1,024, respectively.

6.2.3.2 Calcined Catalysts

Physical properties of calcined catalysts were investigated by N_2 adsorption at 77 K with an AUTOSORB-1-C analyzer (Quantachrome Instruments). Before the measurements, the samples were degassed at 523 K for 5 h. Specific surface areas (S_{BET}) of the samples were calculated by multiplot BET method. Total pore volume (V_{tot}) was calculated by the Barrett-Joyner-Halenda (BJH) method from the desorption isotherm. The average pore diameter (D_{ave}) was then calculated by assuming cylindrical pore structure. Nonlocal density functional theory (NL-DFT) analysis was also carried out to evaluate the distribution of micro- and mesopores.

The effects of chelating agents on the coordination structure and the electronic state of Co species on calcined catalysts were investigated by XRD, Co K-edge EXAFS, and XPS. XRD patterns of calcined catalysts were measured on a MiniFlex diffractometer (Rigaku). Cu Kα radiation was used as an x-ray source, with the x-ray tube operating at 30 kV and 15 mA. Diffraction intensities were recorded from 20 to 90 degrees at the rate of 2.00 degrees min^{-1} with a sampling width of 0.02 degrees. The observed diffraction peaks were assigned by reference to JCPDS data. Co K-edge EXAFS measurements were carried out at BL-14B2 in SPring-8 synchrotron radiation facility (Harima, Japan). EXAFS spectra were measured by Quick EXAFS mode in a transmission setup, and analyzed in a conventional manner using REX2000 software (Rigaku). Polycrystalline Co_3O_4 (Wako Pure Chemicals, purity = 64–77% as Co) and α-Co_2SiO_4 were also subjected to EXAFS measurement as references. Polycrystalline α-Co_2SiO_4 was prepared by calcinations of stoichiometric mixture of a polycrystalline Co_3O_4 and amorphous SiO_2 powder at 1,523 K for 24 h. Backscattering amplitude and phase shift for Co-O and Co-Co coordination shells were extracted from the spectrum of polycrystalline Co_3O_4. XP spectra of the calcined catalysts were measured with an ESCA200 spectrometer (SIENTA). Monochromatized Mg K_α line was used as an excited source. Charge shift was corrected using BN powder as an internal standard.

6.2.3.3 Reduced Catalyst

TPR and hydrogen uptake measurements were carried out to investigate the effect of chelating agents on the formation of metallic particles. Apparatus and procedures for TPR measurement were described in our previous paper.[13] The reduction degree of Co after H_2 reduction (*%Reduction*) was calculated from TPR profiles of calcined and reduced (773 K for 6 h) catalysts using the following equation:

$$\%Reduction = \left[1 - \frac{Amount\ of\ H_2\ consumed\ during\ TPR\ of\ the\ reduced\ catalyst}{Amount\ of\ H_2\ consumed\ during\ TPR\ of\ the\ calcined\ catalyst} \right] \times 100 \quad (6.1)$$

Hydrogen uptake of reduced catalysts (X) was measured by volumetric method with an AUTOSORB-1-C analyzer (Quantachrome Instruments). Hydrogen adsorption was carried out at 373 K after *in situ* H_2 reduction at 773 K for 6 h in the adsorption cell. The dispersion and particle size of metallic Co were calculated by the following equations, assuming that the stoichiometry for hydrogen adsorption on the metallic site is unity:

$$Co^0\ dispersion = \frac{2X}{N_0 \dfrac{\%Reduction}{100}} \times 100 \quad (6.2)$$

$$Co^0\ particle\ size = \frac{82W \dfrac{\%Reduction}{100}}{X} \quad (6.3)$$

N_0 and W in these equations are the total number of Co atoms and the weight percentage of metallic Co, respectively.

6.3　RESULTS AND DISCUSSION

6.3.1　FTS ACTIVITY OF CATALYSTS PREPARED BY THE STEPWISE IMPREGNATION METHOD

As reported in previous papers,[12,13] Co loading of co-impregnated catalysts is limited to a maximum of 5 mass% because of lower solubility of the chelating agent. In contrast, it was found that Co loadings can be increased to 20 mass% by the stepwise impregnation method. The FTS activity of these catalysts prepared by stepwise impregnation method is investigated and reported below.

Figure 6.1 shows CO conversion (at 20 h on stream) over Co(X)/SiO$_2$ and Co(X)/L/SiO$_2$ catalysts as a function of their Co loading. This figure also includes the conversion over the catalysts prepared by the co-impregnation method (L-Co(5)/SiO$_2$) as references. CO conversion over Co(X)/SiO$_2$ catalyst linearly increases from 20% to 60% with increasing Co loading. A similar trend

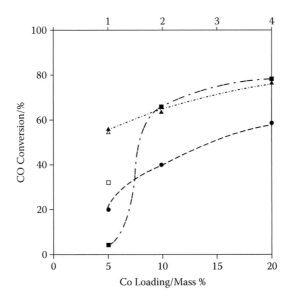

FIGURE 6.1   Effect of Co loading on CO conversions over Co(X)/SiO$_2$ (●), Co(X)/NTA/ SiO$_2$ (▲), and Co(X)/CyDTA/SiO$_2$ (■) catalysts. The conversions over NTA-Co(5)/SiO$_2$ (Δ) and CyDTA-Co(5)/SiO$_2$ (□) catalysts are also plotted in this figure. NTA and CyDTA loadings were fixed at 16 and 31 mass%, respectively, for the catalysts prepared by the stepwise impregnation method. Reaction conditions: 503 K, 1.1 MPa, 5.0 g-cat h mol^{-1}.

is observed over Co(X)/NTA/SiO$_2$; this catalyst always shows higher conversions than Co(X)/SiO$_2$. The conversion over Co(20)/NTA/SiO$_2$ catalyst is ca. 80%, which is more than two times greater than that over NTA-Co(5)/SiO$_2$ catalyst. In contrast, Co(5)/CyDTA/SiO$_2$ shows much lower conversion than the Co(5)/SiO$_2$ catalyst. An abrupt increase in conversion is observed over Co(X)/CyDTA/SiO$_2$ catalyst around 10 mass%-Co. Conversion further increases with Co loading, and reaches a level similar to that of the Co(20)/NTA/SiO$_2$ catalyst. Therefore, the stepwise impregnation method with these chelating agents successfully leads to higher activity at higher Co loadings. It is noted here, again, that the Co to L molar ratio is 4 when Co loading is 20 mass%. This means that only 5 mass% of Co formed a complex with the chelating agent. The remaining 15 mass% did not. In this relation, it is noteworthy that the conversion over Co(20)/CyDTA/SiO$_2$ catalyst is higher than the simple sum of the conversions over Co(5)/CyDTA/ SiO$_2$ (Co/L = 1 mol mol^{-1}) and Co(20)/SiO$_2$ catalysts. This suggests that there is some interaction between the Co-CyDTA complex and free Co species during the catalyst preparation steps.

In order to investigate effects of these chelating agents in more detail, FTS reactions were carried out at different conditions. Figure 6.2 shows CO conversion over Co(20)/SiO$_2$, Co(20)/NTA/SiO$_2$, and Co(20)/CyDTA/SiO$_2$ catalysts as a function of W/F. At 1.25 g h mol^{-1}, the conversion decreases in the following

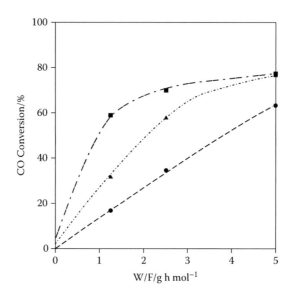

FIGURE 6.2 CO conversions over Co(20)/SiO$_2$ (●), Co(20)/NTA/SiO$_2$ (▲), and Co(20)/CyDTA/SiO$_2$ (■) catalysts as a function of W/F. Reaction conditions: 503 K, 1.1 MPa.

order: Co(20)/CyDTA/SiO$_2$ >> Co(20)/NTA/SiO$_2$ > Co(20)/SiO$_2$. Therefore, CyDTA is more effective than NTA.

6.3.2 SELECTIVITY AND SPACE-TIME YIELD OF LIQUID HYDROCARBON

Table 6.1 summarizes selectivity and space-time yield (STY) of liquid hydrocarbon obtained with the catalysts having 20 mass%-Co loading (W/F = 1.25 g h mol^{-1}), in comparison with those catalysts having 5 mass%-Co loading (W/F = 5 g h mol^{-1}). C$_{5+}$ selectivity for Co(20)/SiO$_2$ catalyst is ca. 80 C-mol%. The chain growth probability of C$_{10-20}$ hydrocarbon (α) for this catalyst is 0.85. Both C$_{5+}$ selectivity and α value for Co(20)/SiO$_2$ catalyst are identical to those for Co(5)/SiO$_2$ catalyst. Co(20)/SiO$_2$ catalyst yields 360 g kg-cat^{-1} h^{-1} of C$_{5+}$ hydrocarbons, which is ca. 2.5 times greater than that obtained with Co(5)/SiO$_2$ catalyst. Concerning effects of chelating agents, C$_{5+}$ selectivity for Co(20)/NTA/SiO$_2$ and Co(20)/CyDTA/SiO$_2$ catalysts is comparable with that for Co(20)/SiO$_2$ catalyst, whereas α values for the former catalysts are slightly lower. However, the use of these chelating agents greatly improves the STY of liquid hydrocarbon. With Co(20)/NTA/SiO$_2$ and Co(20)/CyDTA/SiO$_2$ catalysts, 608 and 1,500 g kg-cat^{-1} h^{-1} of C$_{5+}$ hydrocarbon are obtained, respectively. Furthermore, the STY of C$_{10-20}$ hydrocarbon (equivalent to the diesel fraction) reaches 815 g kg-cat^{-1} h^{-1} over Co(20)/CyDTA/SiO$_2$ catalyst. Ohtsuka and his coworkers[15] reported that SBA-15-supported Co catalyst (Co loading: 20 mass% as metallic Co) yields 710 and 350 g kg-cat^{-1} h^{-1} of C$_{5+}$ and C$_{10-20}$ hydrocarbons, respectively, at 503 K and 2.0 MPa with W/F of 2.4 g h mol^{-1}. As far as we know, this is champion data for

TABLE 6.1

Effects of Chelating Agents on the Selectivity and Space-Time Yield of Hydrocarbon[a]

| | W/F g-cat h mol-CO^{-1} | CO Conversion/% | Product Selectivity C-mol% | | | α^c | STY of C$_{10}$-C$_{20}$ g kg-cat^{-1} h^{-1} | STY of C$_{5+}$ g kg-cat^{-1} h^{-1} |
			CO$_2$	CH$_4$	C$_{5+}$			
Co(5)/SiO$_2$	5	19	0.7	8.4	79.2	0.86	145	46
Co(20)/SiO$_2$	1.25	19	—^{d}	9.8	80.2	0.85	360	209
NTA-Co(5)/SiO$_2$	5	53	0.7	12.7	74.2	0.82	348	110
Co(20)/NTA/ SiO$_2$ $^{b)}$	1.25	30	—^{d}	10.5	78.8	0.83	608	270
Co(20)/ CyDTA/SiO$_2$ $^{b)}$	1.25	60	0.5	14.1	76.7	0.81	1.500	815

^{a} Reaction conditions: 503K, 1.1 MPa.
^{b} Co^{2+}/NTA or CyDTA = 4 mol mol^{-1}.
^{c} Chain growth probability of C$_{10}$-C$_{20}$ hydrocarbon.
^{d} Not detected.

hydrocarbon productivity with Co catalysts. Co(20)/CyDTA/SiO$_2$ catalyst yields C$_{5+}$ and C$_{10-20}$ hydrocarbons more than two times greater than Co/SBA-15 catalyst, even under lower reaction pressure.

6.3.3 PHYSICAL PROPERTIES OF CALCINED CATALYSTS

To elucidate the origin of the promoting effect of CyDTA, we investigated at first the physical properties of calcined catalysts prepared with and without using CyDTA. N$_2$ adsorption isotherms of these catalysts were of type IV, indicating the presence of mesopore. Specific surface area (S_{BET}), total pore volume (V_{tot}), and average pore diameter (D_{ave}) of some calcined catalysts are summarized in Table 6.2. Specific surface area and total pore volume of calcined Co(20)/SiO$_2$ catalyst are ca. 80% of SiO$_2$ support. Because SiO$_2$ content of this catalysts is ca. 80%, surface area and pore volume normalized to catalyst weight should decrease to the same degree. In contrast, total pore volume of calcined Co(20)/CyDTA/SiO$_2$ catalyst is ca. 70% of that of SiO$_2$, although SiO$_2$ content of this catalyst is ca. 80%. Average pore diameter of this catalyst is evidently smaller than with other samples as well.

NL-DFT analysis was then carried out to obtain more accurate information of the distribution of mesopores. It was shown that smaller mesopores with 2 to 12 nm diameter are formed only on calcined Co(20)/CyDTA/SiO$_2$ catalyst, whereas mesopore diameter is distributed above 12 nm on other catalysts. In Table 6.2, cumulative pore volume (V_c) and specific surface area (S_c) of mesopores are tabulated. Because of the presence of smaller mesopores, cumulative specific surface area of calcined Co(20)/CyDTA/SiO$_2$ catalyst is larger than that of calcined Co(20)/SiO$_2$ catalyst, whereas cumulative pore volume of the former is smaller. The formation

TABLE 6.2
Physical Properties of Some Calcined Catalysts

	S_{BET}[a] m^2 g-cat^{-1}	V_{tot}[b] ml g-cat^{-1}	D_{ave}[c] nm	S_c[d] m^2 g-cat^{-1}	V_c[e] ml g-cat^{-1}
Co(5)/CyDTA/SiO$_2$	242.6	1.24	20.5	250.7	1.24
Co(20)/CyDTA/ SiO$_2$	206.7	0.87	16.8	199.5	0.85
Co(20)/SiO$_2$	189.4	0.97	20.5	195.2	0.95
SiO$_2$	232.3	1.23	21.2	242.8	1.19

[a] Specific surface area calculated by multiplot BET method.
[b] Total pore volume calculated by BJH method.
[c] Average pore diameter.
[d] Cumulative surface area of mesopore from NL-DFT analysis.
[e] Cumulative pore volume of mesopore from NL-DFT analysis.

of these smaller mesopores thus is reason for smaller average pore diameter of this catalyst, although their origin is unknown.

6.3.4 Co Reducibility

The effect of CyDTA on the reducibility of Co was then investigated by TPR. Figure 6.3 shows TPR profiles of calcined and reduced catalysts. During TPR of calcined Co(5)/SiO$_2$ catalyst, two strong peaks are observed below 700 K. These strong peaks are assigned to the reduction of Co$_3$O$_4$ to CoO (ca. 600 K), and CoO to the metallic Co (ca. 670 K), as already reported in previous studies.[16–18] A similar profile is obtained with a Co(20)/SiO$_2$ catalyst. After H$_2$ reduction of these catalysts at 773 K, no peak is observed in their TPR profiles, showing that most of Co on calcined catalysts is reduced by hydrogen under these conditions. In marked contrast with these profiles, TPR of calcined Co(5)/CyDTA/SiO$_2$

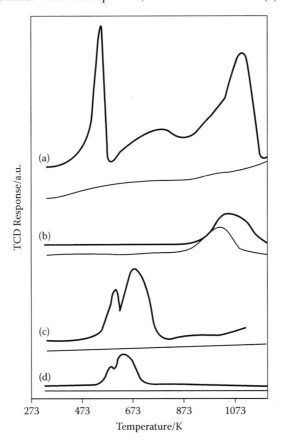

FIGURE 6.3 TPR profiles of calcined (thick line) and reduced (thin line) catalysts: (a) Co(20)/CyDTA/SiO$_2$, (b) Co(5)/CyDTA/SiO$_2$, (c) Co(20)/SiO$_2$, and (d) Co(5)/SiO$_2$. Heating rate: 5 K min^{-1}.

catalyst evolves only a broad peak above 1,000 K. A high temperature peak is observed even after H_2 reduction, whereas the peak position is slightly shifted toward lower temperature. Therefore, most of Co cannot be reduced under the conditions employed here. When Co loading is increased to 20 mass% (calcined Co(20)/CyDTA/SiO$_2$ catalyst), however, several peaks are evolved at lower temperatures. An intense peak is observed at ca. 473 K, which accompanies a broad peak ranging from 700 to 900 K. A relatively weak peak is also observed above 1,000 K in this TPR profile. These peaks completely disappear after H_2 reduction. Most of Co on this catalyst is reduced by hydrogen at 773 K. To obtain a better quantitative insight, the reduction degree of Co after H_2 reduction was calculated by Equation 6.1 and is summarized in Table 6.3. The reduction degree for Co(X)/CyDTA/SiO$_2$ increases from 17% to 93% for Co loading of 5 and 10 mass%. This steep increase is comparable with that observed for the CO conversion vs. Co loading relationship (Figure 6.1).

6.3.5 DISPERSION OF METALLIC CO

H_2 uptake of reduced catalysts is summarized in Table 6.3. The dispersion and particle size of metallic Co species were then calculated from H_2 uptake and reduction degree of Co using Equations 6.2 and 6.3 (Table 6.3). The dispersion of metallic species on the Co(X)/CyDTA/SiO$_2$ catalyst is two to three times higher than that on the Co(X)/SiO$_2$ catalyst. The highest dispersion is obtained with the Co(20)/CyDTA/SiO$_2$ catalyst even though this catalyst has the highest Co loading. The particle size of metallic species on this catalyst was calculated to be ca. 8 nm. Figure 6.4 shows the relationship between the reduction degree of Co and the dispersion of metallic species. At low Co loading, modification with CyDTA improves the dispersion of metallic species by a factor of 2, but it strongly depresses the reduction degree of Co. In contrast, at high Co loading, modification with CyDTA improves the dispersion of metallic species by a factor of 3 without loss of reduction degree of Co. In other words, both higher dispersion and reducibility of Co can be obtained by the stepwise impregnation method with CyDTA at high Co loading. Higher dispersion and reducibility of Co leads to

TABLE 6.3
Effects of CyDTA on the Reducibility of Co and Dispersion of Metallic Co Species

	H_2 Uptake 10^{-6} mol g-cat^{-1}	Dispersion of Co^0 %	Reduction Degree %	Co^0 Particle Size nm
Co(5)/CyDTA/SiO$_2$	6	8.9	17	12
Co(10)/CyDTA/SiO$_2$	64	9.0	93	11
Co(20)/CyDTA/SiO$_2$	173	12.7	96	8
Co(20)/SiO$_2$	57	4.2	96	23
Co(5)/SiO$_2$	20	4.7	85	17

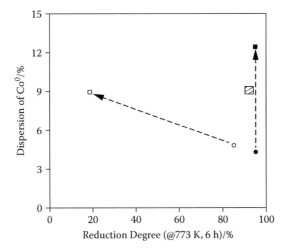

FIGURE 6.4 Relationship between the dispersion of metallic Co particle and reducibility of Co: Co(20)/CyDTA/SiO$_2$ (■), Co(10)/CyDTA/SiO$_2$ (▨), Co(20)/SiO$_2$ (●), and Co(5)/SiO$_2$ (○).

higher metallic Co surface area, which is why the Co(20)/CyDTA/SiO$_2$ catalyst shows higher FTS activity.

6.3.6 COMPARISON WITH PREVIOUS STUDIES (METALLIC CO SURFACE AREA)

To show the impact of the stepwise impregnation method with CyDTA on the formation of a metallic Co surface area, the metallic surface area (SA) on our catalyst was compared with literature results. Figure 6.5 compares the metallic SA of our catalysts prepared by co-impregnation and stepwise impregnation methods with literature data[8,9,19–31] that include SiO$_2$-, Al$_2$O$_3$-, TiO$_2$-, and CNF-supported catalysts prepared by various methods. Catalysts prepared with promoters are not included in this figure. As we can see in Figure 6.5, the reduced Co(20)/CyDTA/SiO$_2$ catalyst has a greater metallic surface area than any other catalysts shown here. Therefore, the stepwise impregnation method with CyDTA is effective for higher metallic SA generation and higher FTS activity.

6.3.7 ROLE OF CYDTA DURING DRYING AND CALCINATIONS STEPS

As mentioned above, the stepwise impregnation method with CyDTA at high Co loading leads to higher metallic Co surface area after H$_2$ reduction. CyDTA could form a complex with Co during the second impregnation step. This Co-CyDTA complex would be preserved during the following drying step because the Co-CyDTA complex has a higher stability (log K_{Co} = 19), but it is burnt off during the final calcination step. However, only part of Co could be involved in complex formation at high Co loading. This situation is evidently different from that

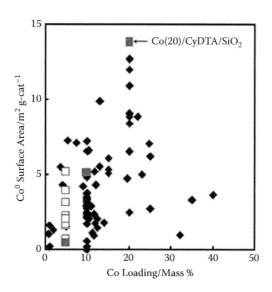

FIGURE 6.5 Impact of the stepwise impregnation method with CyDTA on the formation of higher Co^0 surface area: $Co(X)/CyDTA/SiO_2$ (■); $L-Co(5)/SiO_2$ (L = glycine, citric acid, aspartic acid, NTA, EDTA, CyDTA, TTHA) (□)[12,13]; and Co/SiO_2, Co/Al_2O_3, Co/TiO_2, and Co/CNF (◆).[18–30]

for the catalysts prepared by the co-impregnation method. So, the question is how the Co-CyDTA complex works during the drying and calcination steps in the stepwise impregnation method to create higher metallic surface area after H_2 reduction. The activity test used suggests that there is some interaction between the Co-CyDTA complex and the free Co species during the catalyst preparation steps. To investigate this from a different point of view, the structure of Co species on dried and calcined catalysts was investigated by spectroscopic techniques. Complex formation between Co and CyDTA was first investigated by FT-IR spectroscopy on the catalyst after the second impregnation followed by drying. Then, the effect of CyDTA on the coordination structure (dispersion) and electronic state (support interaction) of Co on calcined catalysts was investigated by XRD, Co K-edge EXAFS, and XPS.

6.3.8 COMPLEX FORMATION

The FT-IR spectrum of the $Co(5)/CyDTA/SiO_2$ catalyst after the second impregnation followed by drying is shown in Figure 6.6 and compared with those of dried $Co(5)/SiO_2$ and $CyDTA/SiO_2$ catalysts. In an aqueous solution, the complex formation between CyDTA and Co ion yields $[CoCyDTA]^{2-}$ and $[CoHCyDTA]^-$ complexes depending on pH. At higher pH, the Co ion is fully coordinated with CyDTA ($[CoCyDTA]^{2-}$ complex), whereas this complex is protonated to yield $[CoHCyDTA]^-$ complex at lower pH.[32,33] Their structures are shown in Figure 6.7.

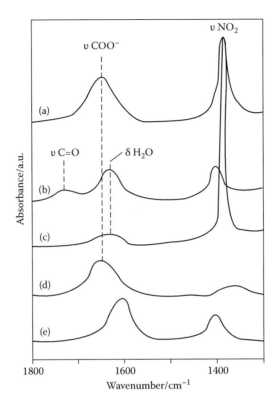

FIGURE 6.6   FT-IR spectra of the catalyst with low Co loading after the second impregnation followed by drying: (a) Co(5)/CyDTA/SiO$_2$, (b) CyDTA/SiO$_2$, (c) Co(5)/SiO$_2$, (d) [CoHCyDTA]$^-$, and (e) [CoCyDTA]$^{2-}$ complexes.

These complexes were precipitated from aqueous solutions using ethanol, and subjected to FT-IR measurement as references. Their structure can be easily distinguished by FT-IR spectroscopy because IR band frequencies of their v_{as}COO$^-$ are evidently different (see Figure 6.6). In the spectrum of dried Co(5)/SiO$_2$ catalyst, a strong band assigned to NO$_3^-$ is observed at 1,386 cm^{-1}. A weak and broad band is also observed at 1,631 cm^{-1} in this spectrum (δH$_2$O), showing the presence of adsorbed water even after the drying step. In the spectrum of the dried CyDTA/SiO$_2$ catalyst, three bands are observed at 1,729 cm^{-1}, 1,629 cm^{-1}, and 1,400 cm^{-1}. These bands are assigned to vC=O, δH$_2$O, and the scissor vibration of CH$_2$, respectively. When Co nitrate solution was impregnated onto dried CyDTA/SiO$_2$ followed by drying (dried Co(5)/CyDTA/SiO$_2$ catalyst), an intense band was evolved at 1,646 cm^{-1} as well as the band assigned to NO$_3^-$. IR band frequency of the former band (1,646 cm^{-1}) is evidently higher than those of the bands observed for dried Co/SiO (1,631 cm^{-1}) and dried CyDTA/SiO$_2$ (1,629 cm^{-1}). This IR band frequency is well consistent with that of the band for the [CoHCyDTA]$^{-1}$ complex. Therefore, protonated Co-CyDTA complex is a main species after the

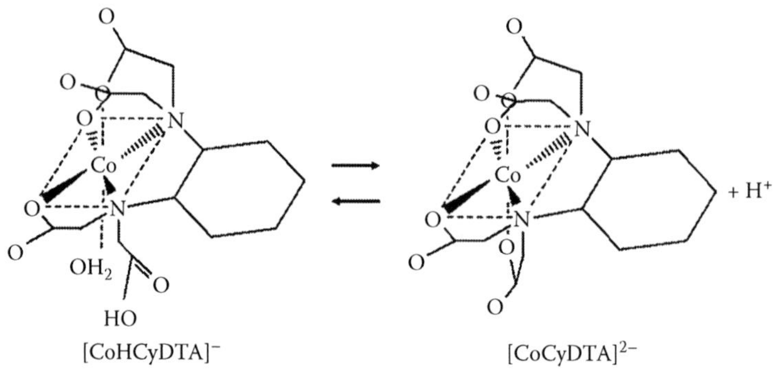

FIGURE 6.7 Structure of Co-CyDTA complexes.

second impregnation followed by drying at low Co loading, i.e., Co/CyDTA = 1 mol mol^{-1}.

FT-IR spectra of dried catalysts with higher Co loadings are shown in Figure 6.8. When the catalyst is prepared without CyDTA, the IR band of adsorbed nitrate species is more intense at high Co loading (20 mass%-Co). IR band intensity of this species increases as well with Co loading for dried Co(X)/CyDTA/SiO$_2$ catalyst. Furthermore, the higher-frequency band assigned to the [CoHCyDTA]$^-$ complex shifts toward lower frequency with Co loading. At 20 mass%-Co, this band is observed near the band of adsorbed water. These results indicate that both Co nitrate species and a small amount of [CoHCyDTA]$^-$ complex are present at higher Co loadings.

6.3.9 COORDINATION STRUCTURE OF CO

The coordination structure of Co on calcined catalysts was then investigated by XRD and Co K-edge EXAFS spectroscopy. Figure 6.9 shows XRD patterns of calcined Co(X)/CyDTA/SiO$_2$ catalysts in comparison with those of calcined Co(X)/SiO$_2$ catalysts. XRD patterns of polycrystalline Co$_3$O$_4$ and α-Co$_2$SiO$_4$ are also included in this figure as references. The XRD pattern of calcined Co(5)/ SiO$_2$ catalyst shows several diffraction peaks at ca. 37, 58, and 65 degrees. More intense peaks are observed at the same diffraction angles in the XRD pattern of calcined Co(20)/SiO$_2$ catalyst. By comparison with the patterns of reference compounds, these diffraction peaks are assigned to Co$_3$O$_4$ species. To the contrary, no diffraction peak is observed in the pattern of calcined Co(5)/CyDTA/ SiO$_2$ catalyst. Only a weak and broad peak is observed around 37 degrees in the pattern of calcined Co(20)/CyDTA/SiO$_2$ catalyst. Thus, the stepwise impregnation method with CyDTA leads to the formation of x-ray amorphous Co species after the calcination.

Figure 6.10 shows Fourier transforms of Co K-edge EXAFS spectra of these calcined catalysts and polycrystalline Co oxides. Three peaks are clearly observed in EXAFS spectra of calcined Co(X)/SiO$_2$ catalysts at almost the same

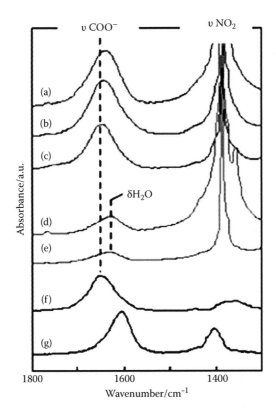

FIGURE 6.8 FT-IR spectra of the catalyst with different Co loadings after the second impregnation followed by drying: (a) Co(20)/CyDTA/SiO$_2$, (b) Co(10)/CyDTA/SiO$_2$, (c) Co(5)/CyDTA/SiO$_2$, (d) Co(20)/SiO$_2$, (e) Co(5)/SiO$_2$, (f) [CoHCyDTA]$^-$, and (g) [CoCyDTA]$^{2-}$ complexes.

positions as observed for polycrystalline Co$_3$O$_4$. These peaks are also observed in the spectra of calcined Co(X)/CyDTA/SiO$_2$; note that their intensities are evidently weaker than those in the spectra of the catalysts without CyDTA. Structural parameters of coordination shells calculated by curve-fitting analysis are summarized in Table 6.4. Co-O and Co-Co interatomic distances for calcined catalysts are consistent with those for polycrystalline Co$_3$O$_4$ within experimental error. Therefore, a Co$_3$O$_4$-like structure is formed on these calcined catalysts irrespective of CyDTA usage. Coordination numbers of Co-Co shells are between those observed for calcined Co(5)/SiO$_2$ and Co(20)/SiO$_2$ catalysts. Coordination numbers of Co-Co shells for the calcined Co(X)/CyDTA/SiO$_2$ are smaller than those for Co-loaded silica, indicating that the stepwise impregnation method with CyDTA leads to the formation of a dispersed Co$_3$O$_4$ phase. If we assume that the [CoHCyDTA]$^-$ complex and Co nitrate species are present without interacting during drying or calcination, then both XRD pattern and EXAFS spectra of calcined Co(20)/CyDTA/SiO$_2$ would be comparable to those of calcined Co(20)/

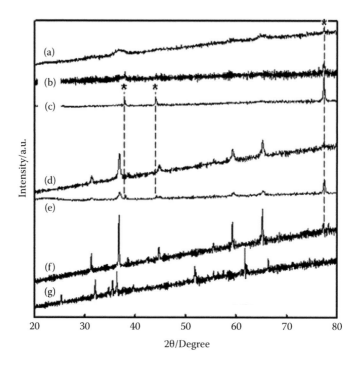

FIGURE 6.9 XRD patterns of calcined catalysts: (a) Co(20)/CyDTA/SiO$_2$, (b) Co(10)/CyDTA/SiO$_2$, (c) Co(5)/CyDTA/SiO$_2$, (d) Co(20)/SiO$_2$, (e) Co(5)/SiO$_2$, (f) polycrystalline Co$_3$O$_4$, and (g) polycrystalline α-Co$_2$SiO$_4$ (asterisks stand for the diffraction peak from Al sample folder).

SiO$_2$ catalyst because XRD and EXAFS peaks observed for calcined Co(5)/CyDTA/SiO$_2$ are negligible or much weaker than those for calcined Co(20)/SiO$_2$. However, this was not observed in the present study, and Co$_3$O$_4$ species were in a more dispersed state on the calcined Co(20)/CyDTA/SiO$_2$ catalyst.

6.3.10 ELECTRONIC STATE OF CO

The electronic state of Co on calcined catalysts was further investigated by XPS to make clear the effect of CyDTA on Co–SiO$_2$ interaction after calcination. Figure 6.11 compares XP spectra of calcined Co(X)/CyDTA/SiO$_2$ and Co(X)/SiO$_2$. The spectra of some reference compounds (polycrystalline Co$_3$O$_4$ and α-Co$_2$SiO$_4$) are also included in this figure. In the spectra of calcined Co(X)/SiO$_2$, Co $2p_{3/2}$ peaks are observed at 780.1 and 780.0 eV. The energy splitting between Co $2p_{1/2}$ and Co $2p_{3/2}$ is 15.1 (X = 5 mass%-Co) and 15.3 eV (X = 20 mass%-Co), respectively, which is comparable with polycrystalline Co$_3$O$_4$ (15.2 eV) rather than polycrystalline CoO (15.8 eV; not shown here). Thus, Co$_3$O$_4$ species with a weak support interaction is formed on these calcined catalysts. In contrast, the Co $2p_{3/2}$ peak appears at a higher-energy side (781.4 eV) in the spectrum of

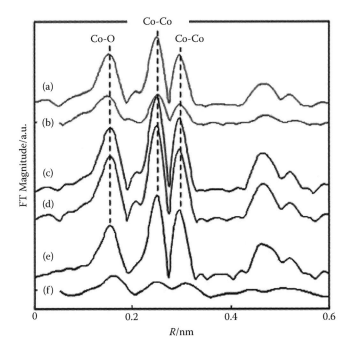

FIGURE 6.10 Fourier transform of Co K-edge EXAFS of calcined catalysts: (a) Co(20)/CyDTA/SiO$_2$, (b) Co(5)/CyDTA/SiO$_2$, (c) Co(20)/SiO$_2$, (d) Co(5)/SiO$_2$, (e) polycrystalline Co$_3$O$_4$, and (f) polycrystalline α-Co$_2$SiO$_4$.

calcined Co(5)/CyDTA/SiO$_2$ catalyst. This binding energy is comparable to the one for polycrystalline α-Co$_2$SiO$_4$. Furthermore, the satellite peak in this spectrum is clearly observed at 786.4 eV; for polycrystalline α-Co$_2$SiO$_4$ it is at 785.1 eV. All these results indicate that dispersed Co$_3$O$_4$ species, revealed by Co K-edge EXAFS spectroscopy, strongly interact with the SiO$_2$ support, making the electronic state of Co in this species comparable with that of Co in silicate species. With increasing Co loading, the Co $2p_{3/2}$ peak shifts toward a lower energy value. The binding energy of Co $2p_{3/2}$ peak for calcined Co(20)/CyDTA/SiO$_2$ is comparable to that of calcined Co(X)/SiO$_2$, indicating a weak interaction with the silica support.

6.3.11 THE ROLE OF CyDTA

Considering the characterization results presented above, the role of CyDTA can be interpreted as follows. At low Co loading, [CoHCyDTA]$^-$ complex was a main species after the second impregnation followed by drying. After the calcination step, this protonated complex is converted into dispersed Co$_3$O$_4$ species that strongly interact with the SiO$_2$ support. It is suggested that the [CoHCyDTA]$^-$ complex interacts with the SiO$_2$ surface by a hydrogen bonding between the

TABLE 6.4

Coordination Number (N) and Interatomic Distance (R) of Co-O and Co-Co Shells Calculated from Co K-edge EXAFS Spectra of Calcined Catalysts, and Polycrystalline Co_3O_4 and $\alpha\text{-}Co_2SiO_4$ (Figure 6.10)

	Co-O		Co-Co^{1st}		Co-Co^{2nd}	
	N	R/nm	N	R/nm	N	R/nm
Co(5)/CyDTA/SiO$_2$	2.7	0.189	1.2	0.286	1.4	0.338
Co(20)/CyDTA/SiO$_2$	5.5	0.191	3.1	0.287	5.6	0.337
Co(5)/SiO$_2$	6.7	0.193	4.1	0.288	7.9	0.337
Co(20)/SiO$_2$	6.8	0.193	4.2	0.287	8.0	0.337
Polycrystalline Co$_3$O$_4$	5.3	0.192	4.0	0.288	8.0	0.335

	Co-O		Co-Si		Co-Co^{1st}		Co-Co^{2nd}	
	N	R/nm	N	R/nm	N	R/nm	N	R/nm
Polycrystalline α-Co$_2$SiO$_4$	6	0.211	1.5	0.277	1	0.305	2	0.333

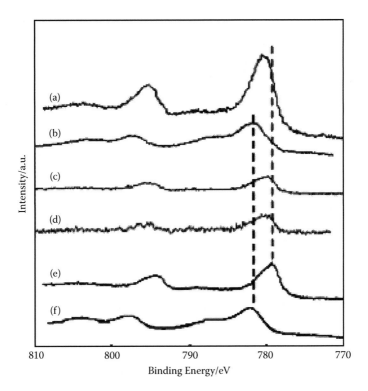

FIGURE 6.11 XP spectra in Co $2p$ region of calcined catalysts: (a) Co(20)/CyDTA/SiO$_2$, (b) Co(5)/CyDTA/SiO$_2$, (c) Co(20)/SiO$_2$, (d) Co(5)/SiO$_2$, (e) polycrystalline Co$_3$O$_4$, and (f) polycrystalline α-Co$_2$SiO$_4$.

–COOH group of the complex and SiOH groups of the SiO$_2$ surface. This interaction leads to the strong Co$_3$O$_4$–SiO$_2$ interaction during calcinations by condensation reaction between the –COOH and SiOH groups, which is responsible for the lower reducibility of this catalyst.

At high Co loadings, on the other hand, both Co nitrate species and a small amount of [CoHCyDTA]$^-$ complex are present after the second impregnation followed by drying. Considering the results for the catalyst having low Co loading, this complex would be converted after calcination to dispersed Co$_3$O$_4$ species that strongly interact with the silica support. However, a dispersed Co$_3$O$_4$ weakly bonded to the support was observed after calcinations; that is, the overall Co-support interaction is weaker than in the catalyst with low Co loading. This is probably due to the fact that most of the Co species is free from complex formation at the drying step. Therefore, these catalysts have higher reducibilities. Furthermore, the dispersion of this Co$_3$O$_4$ species is higher than that expected from the simple assumption that the [CoHCyDTA]$^-$ complex and Co nitrate species are present without interacting during drying or calcination. In other words, the dispersion of Co species free from the complex formation is improved after the calcination in the presence of a protonated complex. As a result, both higher dispersion and reducibility of Co can be obtained by the stepwise impregnation method with CyDTA at higher Co loadings. It is suggested that CyDTA affects Co-support interactions in direct (through complex formation) and indirect manners; the mechanism of an indirect manner is still unknown.

6.4   CONCLUSIONS

In order to investigate effects of complex formation on Co-SiO_2 interaction at high Co loading, the stepwise impregnation method was used to prepare Co/SiO_2 catalysts having high Co levels in the presence of NTA or CyDTA. Their FTS activity and selectivity, and the structure of Co species at each preparation step were also investigated. Important results can be summarized as follows:

1. The stepwise impregnation method successfully increased Co loading of Co/SiO_2 catalyst, prepared with NTA or CyDTA, to 20 mass%.
2. The stepwise prepared catalysts showed higher FTS activities at higher Co loadings ($\geq$10 mass%-Co) than the catalyst without chelating agents. CyDTA was more effective than NTA.
3. The catalyst prepared with CyDTA yielded 815 g kg-cat^{-1} h^{-1} of C_{10-20} hydrocarbon at 503 K and 1.1 MPa, which is much greater than yields reported previously.
4. TPR and hydrogen uptake measurements have shown that the stepwise impregnation method with CyDTA leads to higher dispersion and higher reducibility of Co at high Co loading, leading to higher metallic Co surface area than those of previously reported catalysts having various types of support (SiO_2, Al_2O_3, TiO_2, and CNF) and various Co loadings.
5. FT-IR measurement showed that both Co nitrate and a small amount of [CoHCyDTA]$^-$ complex are formed after the second impregnation, followed by drying at higher Co loadings ($\geq$10 mass%-Co). It was also revealed by Co K-edge EXAFS and XPS that this complex after calcination is converted into dispersed Co_3O_4 species strongly interacted with the SiO_2 support. The dispersion of Co species improved after calcinations, irrespective of the presence of the complex. It is suggested that CyDTA affects Co-support interactions in direct (through complex formation) and indirect manners, different from co-impregnated catalysts.

ACKNOWLEDGMENTS

This research has been supported by the Japan Society for the Promotion of Science (JSPS), Grant-in-Aid for Scientific Research (S), 17106011, 2005. The synchrotron radiation experiments were performed at the BL14B2 in the SPring-8 with the approval of the Japan Synchrotron Radiation Research Institute (JASRI) (Proposal 2008A1825).

REFERENCES

1. Dry, M. E. 1981. The Fischer-Tropsch synthesis. In *Catalysis science and technology*, ed. J. R. Anderson and M. Boudart, 195–300. Vol. 1. Berlin: Springer-Verlag.
2. Iglesia, E. 1997. Design, synthesis, and use of cobalt-based Fisher-Tropsch synthesis catalyst. *Appl. Catal.* 161:59–78.
3. Davis, B. H. 2001. Fischer–Tropsch synthesis: Current mechanism and futuristic needs. *Fuel Process. Technol.* 71:157–66.
4. Dry, M. E. 2002. The Fischer-Tropsch process: 1950–2000. *Catal. Today* 71:227–41.
5. Ming, J., Koizumi, N., Ozaki, T., and Yamada, M. 2001. Adsorption properties of cobalt and cobalt-manganese catalysts studied by in-situ diffuse reflectance FTIR using CO and CO+H$_2$ as probes. *Appl. Catal. A Gen.* 209:59–70.
6. Bian, G., Fujishita, N., Mochizuki, T., Ning, W., and Yamada, M. 2003. Investigations on the structural changes of two Co/SiO$_2$ catalysts by performing Fischer–Tropsch synthesis. *Appl. Catal. A Gen.* 252:251–60.
7. Bian, G., Mochizuki, T., Fujishita, N., Nomoto, H., and Yamada, M. 2003. Activation and catalytic behavior of several Co/SiO$_2$ catalysts for Fischer-Tropsch synthesis. *Energy & Fuels* 17:799–803.
8. Bezemer, G. L., Bitter, J. H., Kuipers, H. P. C. E., Oosterbeek, H., Holewijn, J. E., Xu, X., Kapteijn, F., von Dillen A. J., and de Jong, K. P. 2006. Cobalt particle size effects in the Fischer-Tropsch reaction studied with carbon nanofiber supported catalysts. *J. Am. Chem. Soc.* 128:3956–64.
9. Sun, S., Tsubaki, N., and Fujimoto, K. 2000. The reaction performance and characterization of Fischer-Tropsch synthesis Co/SiO$_2$ catalyst prepared from mixed Co salt. *Appl. Catal. A Gen.* 202:121–31.
10. Feller, A., Claeys, M., and van Steen, E. 1999. Cobalt cluster effects in zirconium promoted Co/SiO$_2$ Fischer-Tropsch catalysts. *J. Catal.* 185:120–30.
11. Kraum, M., and Baerns, M. 1999. Fischer-Tropsch synthesis: The influence of various cobalt compounds applied in the preparation of supported cobalt catalysts on their performance. *Appl. Catal. A Gen.* 186:189–200.
12. Mochizuki, T., Hara, T., Koizumi, N., and Yamada, M. 2007. Novel preparation method of highly active Co/SiO$_2$ catalyst for Fischer-Tropsch synthesis with chelating agents. *Catal. Lett.* 113:165–69.
13. Mochizuki, T., Hara, T., Koizumi, N., and Yamada, M. 2007. Surface structure and Fischer-Tropsch synthesis activity of highly active Co/SiO$_2$ catalysts prepared from the impregnating solution modified with some chelating agents. *Appl. Catal. A Gen.* 317:97–104.
14. Mochizuki, T., Koizumi, N., Hamabe, Y., Hara, T., and Yamada, M. 2007. Preparation of highly active Co/SiO$_2$ Fischer-Tropsch synthesis catalyst with chelating agents: Effect of chelating agents on the structure of Co species formed during the preparation steps. *J. Jpn. Petrol. Inst.* 50:262–71.
15. Ohtsuka, Y., Arai, T., Takasaki, S., and Tsubouchi, N. 2003. Fischer-Tropsch synthesis with cobalt catalysts supported on mesoporous silica for efficient production of diesel fuel fraction. *Energy & Fuels* 17:804–9.
16. Brown, R., Cooper, M., and Whan, D. 1982. Temperature programmed reduction of alumina-supported iron, cobalt and nickel bimetallic catalysts. *Appl. Catal.* 3:177–86.

17. Viswanathan, B., and Gopalkrishnan, R. 1986. Effect of support and promoter in Fischer-Tropsch cobalt catalysts. *J. Catal.* 99:342–48.

18. Sexton, B., Hughes, A., and Turney, T. 1986. An XPS and TPR study of the reduction of promoted cobalt-kieselguhr Fischer-Tropsch catalysts. *J. Catal.* 97:390–406.

19. Iglesia, E., Reyes, S. C., Madon, R. J., and Soled, S. L. 1993. Selectivity control and catalyst design in the Fischer-Tropsch synthesis: Sites, pellets, and reactors. *Adv. Catal.* 39:221–302.

20. Jacobs, G., Das, T. K., Zhang, Y., Li, J., Racoillet, G., and Davis, B. H. 2002. Fischer–Tropsch synthesis: Support, loading, and promoter effects on the reducibility of cobalt catalysts. *Appl. Catal. A Gen.* 233:263–81.

21. Jacobs, G., Chaney, J. A., Patterson, P. M., Das, T. K., and Davis, B. H. 2004. Fischer–Tropsch synthesis: Study of the promotion of Re on the reduction property of Co/Al$_2$O$_3$ catalysts by in situ EXAFS/XANES of Co K and Re L$_{III}$ edges and XPS. *Appl. Catal. A Gen.* 264:203–12.

22. Dalai, A. K., Das, T. K., Chaudhari, K. V., Jacobs, G., and Davis, B. H. 2005. Fischer–Tropsch synthesis: Water effects on Co supported on narrow and wide-pore silica. *Appl. Catal. A Gen.* 289:135–42.

23. Sun, S., Tsubaki, N., and Fujimoto, K. 2000. Characteristic feature of Co/SiO$_2$ catalysts for slurry phase Fischer-Tropsch synthesis. *J. Chem. Eng. Jpn.* 33:232–38.

24. Tsubaki, N., Sun, S., and Fujimoto, K. 2001. Different functions of the novel metals added to cobalt catalysts for Fischer-Tropsch synthesis. *J. Catal.* 199:236–46.

25. Tsubaki, N., Yoshii, K., and Fujimoto, K. 2002. Anti-ASF distribution of Fischer-Tropsch hydrocarbons in supercritical-phase reactions. *J. Catal.* 207:371–75.

26. Zhang, Y., Shinoda, M., and Tsubaki, N. 2004. Development of bimodal cobalt catalysts for Fischer-Tropsch synthesis. *Catal. Today* 93:55–63.

27. Zhang, Y., Koike, M., Yang, R., Hinchiranan, S., Vitidsant, T., and Tsubaki, N. 2005. *Appl. Catal. A Gen.* 292:252.

28. Zhang, Y., Hanayama, K., and Tsubaki, N. 2006. The surface modification effects of silica supported by organic solvents for Fischer-Tropsch synthesis catalysts. *Catal. Comm.* 7:251–54.

29. Liu, Y., Zhang, Y., and Tsubaki, N. 2007. The effect of acetic acid pretreatment for cobalt catalysts prepared from cobalt nitrate. *Catal. Comm.* 8:773–76.

30. Zhang, Y., Li, Y., Yang, G., Sun, S., and Tsubaki, N. 2007. Effects of impregnation solvent on Co/SiO$_2$ catalyst for Fisher-Tropsch synthesis: A highly active and stable catalyst with bimodal sized cobalt particle. *Appl. Catal. A Gen.* 321:79–85.

31. Sabi, A. M., Claeys, M., and van Steen, E. 2002. Silica supported Fischer-Tropsch catalysts: Effect of pore diameter of support. *Catal. Today* 71:395–402.

32. Martell, A. E. 1964. *Section II: Organic ligands in stability constant of metal-ion complexes.* Special Publication 17. London: The Chemical Society.

33. Martell, A. E., and Smith, R. M. 1982. *Critical stability constants.* Vol. 5, 1st suppl. New York: Plenum Press.

7 Fischer-Tropsch Synthesis Temperature-Programmed EXAFS/XANES Characterization of the Impact of Cu and Alkali Promoters to Iron-Based Catalysts on the Carbide Formation Rate

Gary Jacobs, Amitava Sarkar, Burtron H. Davis, Donald Cronauer, A. Jeremy Kropf, and Christopher L. Marshall

CONTENTS

Temperature-programmed reduction combined with x-ray absorption fine-structure (XAFS) spectroscopy provided clear evidence that the doping of Fischer-Tropsch synthesis catalysts with Cu and alkali (e.g., K) promotes the carburization rate relative to the undoped catalyst. Since XAFS provides information about the local atomic environment, it can be a powerful tool to aid in catalyst characterization. While XAFS should probably not be used exclusively to characterize the types of iron carbide present in catalysts, it may be, as this example shows, a useful complement to verify results from Mössbauer spectroscopy and other temperature-programmed methods. The EXAFS results suggest that either the Hägg or ε-carbides were formed during the reduction process over the cementite form. There appears to be a correlation between the α-value of the product distribution and the carburization rate.

7.1   INTRODUCTION

Iron-based Fischer-Tropsch synthesis catalysts are not only used commercially for high-temperature Fischer-Tropsch synthesis [1], but are increasingly becoming the focus for converting low H_2/CO ratio synthesis gas at lower temperature [2]. Such low-temperature processing yields hydrocarbon distributions with higher α-values and, as a consequence, much less light gas production (and especially less methane). Another benefit to the use of iron-based catalysts is that the product slate is richer in α-olefinic products, which are more valuable than the paraffinic products produced using cobalt-based catalysts. Iron-based catalysts are often used to convert low H_2/CO ratio syngas, because the catalysts can intrinsically adjust the syngas ratio upward by converting a fraction of CO with H_2O to produce H_2 and CO_2 via the water-gas shift reaction. If coal-to-liquids is going to be commercialized worldwide, environmental and political factors will likely dictate the necessity to sequester the CO_2 produced. This in turns means that the relative rates of Fischer-Tropsch synthesis and water-gas shift should be optimized to limit the amount of CO_2 formed. In this way, the burden of CO_2 sequestration from the Fischer-Tropsch synthesis step may be limited. There is increasing interest to better understand the catalyst structure-function relationships to achieve this goal, so that better formulations may be developed. Unlike cobalt catalysts, where the active sites are likely the metallic cobalt crystallites [3], making them relatively easier to characterize, working iron catalysts are often a complex mixture of iron carbide compounds [4], often containing an oxidic fraction.

The aim of this work was to apply combined temperature-programmed reduction (TPR)/x-ray absorption fine-structure (XAFS) spectroscopy to provide clear evidence regarding the manner in which common promoters (e.g., Cu and alkali, like K) operate during the activation of iron-based Fischer-Tropsch synthesis catalysts. In addition, it was of interest to compare results obtained by EXAFS with earlier ones obtained by Mössbauer spectroscopy to shed light on the possible types of iron carbides formed. To that end, model spectra were generated based on the existing crystallography literature for four carbide compounds of interest. The experimental data of the catalyst were qualitatively compared with the models.

7.2 EXPERIMENTAL

7.2.1 CATALYST PREPARATION, ACTIVATION, AND REACTION TESTING

A commercial ultrafine iron oxide catalyst (NANOCAT® Superfine Iron Oxide, Mach I, Inc., 3–5 nm particle size, surface area = 250 m²/g, and bulk density = 0.05 g/cm³) was used as the reference α-Fe_2O_3 catalyst. The promoted iron catalysts were prepared from a base catalyst with atomic composition 100Fe/4.6Si or 100Fe/5.1Si prepared by a precipitation method as previously reported [5]. The precipitated base iron catalyst was first prepared using a ferric nitrate solution obtained by dissolving $Fe(NO_3)_3$ in deionized water, and the amount of tetraethyl orthosilicate needed to make Si:Fe atomic ratio of 4.6:100 or 5.1:100 was added. The mixture was stirred vigorously until the tetraethyl orthosilicate was hydrolyzed. This mixture of tetraethyl orthosilicate and iron nitrate was then added to a continuously stirred tank reactor (CSTR) precipitator vessel together with a separate stream of 30% ammonium hydroxide that was added at a rate to maintain a pH of 9.0. By maintaining the slurry pH at 9 and an average residence time of 6 min, a base catalyst material with an iron to silicon molar ratio of 100:4.6 or 100:5.1 was obtained. The slurry was then filtered with a vacuum drum filter and washed twice with deionized water. The final filtration cake was dried in an oven with flowing air at 110°C for 24 h. The catalyst was crushed and calcined at 350°C under an airflow for 4 h. The Fe/Si catalyst base powder was impregnated with the proper amount of aqueous K_2CO_3 solution to produce a potassium-promoted catalyst with the desired composition (i.e., atomic ratios) of 100Fe/4.6Si/1.44K, 100Fe/4.6Si/1.5K, 100Fe/4.6Si/5.0K, or 100Fe/5.1Si/5.0K. Copper was added to some of these catalysts by wet impregnation with aqueous Cu(NO3)2,3H2O solution to produce the desired catalyst composition (atomic ratio) of 100Fe/4.6Si/1.44K/2.0Cu or 100Fe/4.6Si/5.0K/2.0Cu.

The FTS experiments were conducted in a 1 L CSTR equipped with a magnetically driven stirrer with turbine impeller, a gas-inlet line, and a vapor outlet line with a stainless steel (SS) fritted filter (7.0 μm) placed external to the reactor. A tube fitted with an SS fritted filter (0.5 or 2.0 μm opening) extends below the liquid level of the reactor for withdrawing reactor wax (rewax, solid at room temperature) maintaining a desired liquid level in the reactor. Separate mass flow controllers were used to control the flow of H_2 and CO at the desired rates. The gases were premixed in a mixing vessel before entering the reactor. CO was passed through a vessel containing lead oxide–alumina to remove any traces of iron carbonyl. The mixed gases entered the CSTR below the stirrer, which was operated at 750 rpm. The reactor temperature was maintained constant ($\pm$ 1°C) by a temperature controller.

Melted (150°C) Polywax 3000 (PW3000, polyethylene fraction with average molecular weight of 3,000, Baker Petrolite, Inc.) or Ethylflo 164 hydrocarbon oil

(C_{30} oil, homopolymer of 1-decene, Ethyl Corp., Inc.) served as the start-up solvent for the experiments. The catalyst (ca. 5–8 g) was added to start-up solvent (ca. 300 g) in the CSTR. The reactor temperature was then raised to 270°C at a rate of 1°C/min. The catalyst was activated using CO at a space velocity of 3.0 sl/h/g Fe at 270°C and 175 psig for 24 h. FTS was then started by adding synthesis gas mixture (H_2:CO ratio of 0.7) to the reactor at a space velocity of either 3.1 or 5.0 sl/h/g Fe. The conversions of CO and H_2 were obtained by gas chromatography (GC) analysis (HP Quad Series Micro-GC equipped with thermal conductivity detectors) of the product gas mixture. The reaction products were collected in three traps maintained at different temperatures—a hot trap (200°C), a warm trap (100°C), and a cold trap (0°C). The products were separated into different fractions (rewax, wax, oil, and aqueous) for quantification by GC analysis. However, the oil and the wax (liquid at room temperature) fractions were mixed prior to GC analysis.

7.2.2 CRYSTAL STRUCTURE AND DEVELOPMENT OF THE THEORETICAL EXAFS MODELS

EXAFS theoretical structural models were prepared using the Atoms [6] program to input crystal structure parameters obtained from the literature [7–9] into the FEFF software [10] to determine, based in part on the muffin tin potentials and scattering geometries, the path contributions that make up the theoretical EXAFS functions (i.e., χ(k)) for each carbide compound. The crystal parameters, provided in Tables 7.1 and 7.2, were used to construct three-dimensional views of the carbide compounds using the software CrystalOgraph [11]. The resulting single scattering path contributions for each carbidic compound are given in Tables 7.3a through 7.3d, provided in the Section 7.5. These paths' contributions served as theoretical models for the FEFFIT [12] input file. Theoretical χ(k) functions for each carbide, along with the resulting k^1-weighted theoretical Fourier transform magnitude spectra (i.e., radial distribution functions), were generated. The parameters used included the nominal coordination numbers set by the crystal structure, no lattice contraction factor (i.e., α), an amplitude reduction factor (i.e., S_0^2) of 0.90, a shift in E_0 of zero, and a comparison of spectra generated with no disorder (i.e., Debye-Waller factor of zero) or relatively high disorder (i.e., Debye-Waller factor of 0.010), as commonly found with iron FT catalysts.

7.2.3 TEMPERATURE-PROGRAMMED EXTENDED X-RAY ABSORPTION FINE-STRUCTURE (EXAFS) AND X-RAY ABSORPTION NEAR-EDGE (XANES) SPECTROSCOPIES

In situ CO-TPR XAFS studies were performed at the Materials Research Collaborative Access Team (MR-CAT) beam line at the Advanced Photon Source,

TABLE 7.1

Lattice Parameters for Carbide Compounds

Carbide Compound and Reference	Formula	Bravais Lattice	Space Group	a (Å)	b (Å)	c (Å)	α (deg)	β (deg)	γ (deg)
Cementite [7]	Fe_3C	Orthorhombic	P n m a (62)	4.5133	5.0679	6.7137	90	90	90
Hägg carbide [7]	Fe_5C_2	Monoclinic	C 2/c (15)	11.504	4.524	5.012	90	97.60	90
ε-carbide [8]	Fe_3C^*	Hexagonal	P 63 2 2 (182)	4.767	4.767	4.354	90	90	120
η-carbide [7]	Fe_2C^*	Orthorhombic	P n n m (58)	4.687	4.261	2.830	90	90	90

The structures of ε- and η-carbides are closely related and are more correctly written Fe_2C_{1-x}. At 29% carbon content, the ε-carbide is favored, while at 33% carbon content, the structure starts to form the η-carbide [6].

TABLE 7.2
Atomic Positions Using Cartesian Coordinates and Corresponding Wyckoff Letters

Carbide Compound	Atom	Color	Cartesian Coordinates
Cementite	Fe 4c	Red to White	$(x_1, y_1, z_1) = (0.036, 0.250, 0.852)$
	Fe' 8d	Orange to Gray	$(x_2, y_2, z_2) = (0.186, 0.063, 0.328)$
	C 4c	Blue to Black	$(x_3, y_3, z_3) = (0.890, 0.250, 0.450)$
Hägg carbide	Fe 8f	Red to White	$(x_1, y_1, z_1) = (0.092, 0.091, 0.421)$
	Fe' 8f	Orange to Light Gray	$(x_2, y_2, z_2) = (0.207, 0.577, 0.302)$
	Fe' 4e	Peach to Dark Gray	$(x_3, y_3, z_3) = (0.000, 0.566, 0.250)$
	C 8f	Blue fo Black	$(x_4, y_4, z_4) = (0.109, 0.300, 0.082)$
ε-carbide	Fe 6g	Red to White	$(x_1, y_1, z_1) = (0.333, 0.000, 0.000)$
	C 2c	Blue to Black	$(x_2, y_2, z_2) = (0.333, 0.667, 0.250)$
	C' 2d	Green to Gray	$(x_3, y_3, z_3) = (0.333, 0.667, 0.750)$
η-carbide	Fe 4g	Red to White	$(x_1, y_1, z_1) = (0.671, 0.250, 0.000)$
	C 2a	Blue to Black	$(x_2, y_2, z_2) = (0.000, 0.000, 0.000)$

Argonne National Laboratory. A cryogenically cooled Si(111) monochromator selected the incident energy and a rhodium-coated mirror rejected higher-order harmonics of the fundamental beam energy. The experiment setup was similar to that outlined by Jacoby [13]. A stainless steel multisample holder (4.0 mm i.d. sample wells) was used to monitor the *in situ* reduction of five samples and one reference during a single TPR run. The samples were diluted with silica (ultrapure silica gel) of the Silicycle Chemical Division (S10040T) at a level of 10:1 silica:catalyst. The silica and Fe catalysts were pulverized together using a mortar and pestle, and the Fe concentration only varied ±10% across the pellet diameter, indicating an acceptable degree of mixing.

The pellet charge was about 0.0065 g (0.0060 to 0.0068) into a cell with a 4.0 mm diameter. The holder was placed in the center of a quartz tube, equipped with gas and thermocouple ports and Kapton windows. The amount of sample used was optimized for the Fe K edge, considering the absorption by Si of the catalyst. The quartz tube was placed in a clamshell furnace mounted on the positioning table. Each sample cell was positioned relative to the beam by finely adjusting the position of the table to an accuracy of 20 μm (for repeated scans). Once the positions were fine-tuned, the samples were heated to about 120°C in 5% CO/He at 10°C/min. Then the samples were heated to about 270°C over a 3-hour period. They were held at this temperature for 4 h and then cooled.

The Fe K-edge spectra were recorded in the transmission mode and a metallic iron foil spectrum was measured simultaneously with each sample spectrum for energy calibration. X-ray absorption spectra for each sample were collected from 7,520 to 8,470 eV, with a step size of 0.40 eV and acquisition times of ca. 68 s per sample. Measuring each sample, in turn, and repeating

TABLE 7.3A

Path Parameters Generated by FEFF (Single Scattering) for Cementite

		Cementite	
Atom	Interaction	No. Degeneracies	Distance (C)
Fe2	Fe-C	1.0	1.8313
Fe2	Fe-Fe	1.0	1.8954
Fe2	Fe-Fe	1.0	2.0289
Fe1	Fe-Fe	2.0	2.0289
Fe1	Fe-C	1.0	2.0784
Fe2	Fe-C	1.0	2.2035
Fe1	Fe-Fe	2.0	2.2317
Fe2	Fe-Fe	1.0	2.2317
Fe2	Fe-C1	0.0	2.2868
Fe2	Fe-Fe	2.0	2.4878
Fe1	Fe-Fe	2.0	2.6397
Fe1	Fe-C	1.0	2.7782
Fe1	Fe-Fe	2.0	2.8300
Fe2	Fe-Fe	1.0	2.8300
Fe2	Fe-Fe	1.0	2.9258
Fe2	Fe-Fe	2.0	3.1276
Fe2	Fe-Fe	1.0	3.1725
Fe1	Fe-C	1.0	3.2043
Fe1	Fe-Fe	2.0	3.2366
Fe1	Fe-C	2.0	3.2481
Fe1	Fe-C	2.0	3.2624
Fe1	Fe-Fe	2.0	3.4013
Fe2	Fe-Fe	1.0	3.4013
Fe2	Fe-C	1.0	3.4154
Fe2	Fe-Fe	2.0	3.4655

allowed eighty-four scans to be collected for each sample over a 7-hour period. The sample's temperature change from the absorption edge through the end of the scan was then about 1.3°C, while each sample was measured approximately every 10°C.

7.2.4 XAS DATA REDUCTION AND FITTING

7.2.4.1 XANES Analysis

XANES spectra were processed using the WinXAS program [14]. For XANES processing, a simultaneous pre- and postedge background removal was carried

TABLE 7.3B
Path Parameters Generated by FEFF (Single Scattering) for the Hägg Carbide

| | The Hägg Carbide | | |
Atom	Interaction	No. Degeneracies	Distance (Å)
Fe2	Fe-C	1.0	1.9309
Fe1	Fe-C	1.0	1.9433
Fe1	Fe-C	1.0	1.9766
Fe2	Fe-C	1.0	1.9921
Fe3	Fe-C	2.0	2.0022
Fe3	Fe-C	2.0	2.0388
Fe2	Fe-C	1.0	2.3475
Fe3	Fe-Fe	2.0	2.3616
Fe2	Fe-Fe	1.0	2.3616
Fe1	Fe-C	1.0	2.4964
Fe1	Fe-Fe	1.0	2.4958
Fe1	Fe-Fe	1.0	2.4965
Fe3	Fe-Fe	2.0	2.4965
Fe1	Fe-Fe	1.0	2.5415
Fe2	Fe-Fe	2.0	2.5502
Fe3	Fe-Fe	2.0	2.5762
Fe1	Fe-Fe	1.0	2.5901
Fe3	Fe-Fe	2.0	2.5901
Fe2	Fe-Fe	2.0	2.6010
Fe2	Fe-Fe	1.0	2.6170
Fe1	Fe-Fe	2.0	2.6378
Fe1	Fe-Fe	1.0	2.6428
Fe2	Fe-Fe	1.0	2.6428
Fe1	Fe-Fe	1.0	2.6463
Fe2	Fe-Fe	1.0	2.6463
Fe1	Fe-Fe	1.0	2.6737
Fe2	Fe-Fe	1.0	2.6737
Fe1	Fe-Fe	1.0	2.6937
Fe3	Fe-Fe	2.0	2.6937
Fe1	Fe-Fe	1.0	2.6988
Fe2	Fe-Fe	1.0	2.6988
Fe1	Fe-Fe	1.0	2.7788
Fe2	Fe-Fe	1.0	2.7788
Fe1	Fe-C	1.0	2.8702
Fe1	Fe-C	1.0	3.4258
Fe2	Fe-Fe	1.0	3.4642
Fe3	Fe-Fe	2.0	3.4642
Fe2	Fe-C	1.0	3.4887

TABLE 7.3C
Path Parameters Generated by FEFF (Single Scattering) for ε-carbide

		ε-carbide	
Atom	Interaction	No. Degeneracies	Distance (Å)
Fe1	Fe-C	4.0	1.9259
Fe1	Fe-Fe	6.0	2.6951
Fe1	Fe-Fe	6.0	2.7520
Fe1	Fe-C	4.0	3.3594

TABLE 7.3D
Path Parameters Generated by FEFF (Single Scattering) for η-carbide

		η-carbide	
Atom	Interaction	No. Degeneracies	Distance (Å)
Fe1	Fe-C	1.0	1.8742
Fe1	Fe-C	2.0	1.9441
Fe1	Fe-Fe	4.0	2.6626
Fe1	Fe-Fe	2.0	2.6662
Fe1	Fe-Fe	4.0	2.7376
Fe1	Fe-Fe	2.0	2.8300
Fe1	Fe-C	1.0	3.3205
Fe1	Fe-C	2.0	3.3943

out using two degree 1 polynomials over the ranges of 6.90 to 7.01 keV and 7.22 to 8.09 keV, respectively, and the resulting spectra were normalized by dividing by the height of the absorption edge. The spectra were then calibrated after correcting the edge position for the Fe^0 reference, which typically resulted in an offset of ~1 eV, and sectioned over the range of interest up to ~50 eV above the edge jump. Once the spectra were processed, they were compared to reference compounds. All of the beginning spectra bore similarities to bulk Fe_2O_3. As the temperature trajectory was followed, it was clearly observed that a transition from Fe_2O_3 to Fe_3O_4 took place for all of the samples. Beyond this transition, the white line continued to increase slightly in intensity, while the edge shifted to lower energy, in agreement with the continuing removal of O from the Fe_3O_4 compound, resulting in what we describe as an oxygen-deficient form. The final spectra of the catalysts, those recorded after carburization in CO for 4 h at 270°C, were compared with spectra for Fe_3O_4 and the iron foil, in order to qualitatively assess the extent of reduction.

7.2.4.2 EXAFS Analysis

Data reduction of EXAFS spectra was performed using WinXAS [14]. The normalized spectra were analyzed over the k-range of 2.5 to 10 Å^{-1}. A square-weighted degree 7 spline was used to remove the background of the $\chi(k)$ function. Finally, the data in k-space were converted to R-space using a Bessel window to obtain the radial distribution function.

Li et al. in 2001 [15] found that used iron FT catalyst samples analyzed by EXAFS contained primarily an oxidic component resembling Fe_3O_4 and an iron carbidic fraction. Due in part to the nano-crystalline size of iron particles used in our work, and in part to the presence of promoters, the EXAFS and XANES spectra for the catalyst during the Fe_3O_4 transition of the TPR, though maintaining characteristics resembling the Fe_3O_4 reference, also contain enough differences to be of concern for applying the Fe_3O_4 reference as a basis for removing the oxidic fraction from the final catalyst spectra. Therefore, we alternately utilized the catalyst spectrum at the point in the TPR in which the spectrum most resembles Fe_3O_4 as a basis for a subtraction method [16]. $\chi(k)$ for the oxide spectrum was multiplied by a variable factor and subtracted from $\chi(k)$ of the catalyst spectrum until the peaks for the oxide fraction in the Fourier transform magnitude achieved a minimum. The Fourier transform of this difference $\chi(k)$ spectrum, as well as the $\chi(k)$ spectrum itself, represents estimates of the carbidic fraction.

7.3 RESULTS AND DISCUSSION

In the work of Nagakura [8], Fe films were studied by electron diffraction. The films were carburized in a CO stream at temperatures ranging from 140 to 500°C for times from 1 to 6 h. The ε-carbide was produced by carburization below 250°C, the carbide χ (i.e., the Hägg carbide [9]) by carburization at temperatures between 250 and 350°C, and the carbide θ (cementite Fe_3C) by carburization above 350°C. The cubic carbide was not formed in these experiments. The detailed crystal structure of the three carbides was obtained. Irreversible phase transitions were observed from ε to χ at 380 to 400°C and from χ to θ at 550°C. In these transitions, the composition did not change significantly, and no rings due to C were seen; the maximum change was from $Fe_{2.8}C$ to Fe_3C. The particle size of the ε-carbide, as estimated from the ring width, was the same as that of the original α-Fe, suggesting that ε-carbide forms by diffusion of C atoms into the α-Fe lattice along imperfections without recrystallization. The χ- and θ-carbides were found to have very complex crystal structures and much larger particle sizes. A closely related carbide to ε-carbide is the more recently discovered η-Fe_2C [7], which is metastable and forms prior to the formation of the Hägg carbide. The transition from ε-carbide to η-carbide occurs above 33% C, given the formula Fe_2C_{1-x}. There is only a slight difference in crystalline structure between the ε- and η-carbides.

Following from this discussion, and considering that carburization was carried out at 270°C in 5% CO/He, one would anticipate that the catalyst should likely

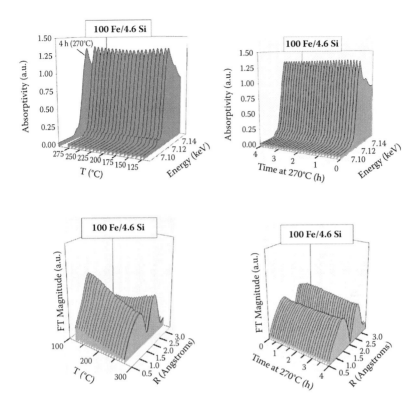

FIGURE 7.1   (Top) XANES and (bottom) EXAFS spectra (left) as a function of temperature and (right) as a function of time at 270°C in 5% CO/balance inert for the catalyst with an atomic ratio of 100Fe/4.6 Si.

evolve into one or more of the lower-temperature carbide forms (e.g., ε-, η-, or χ-carbides, and mixtures thereof).

Figures 7.1 through 7.5 display three-dimensional graphs of the XANES spectra (top) and the EXAFS spectra (bottom) as a function of (left) temperature and time at 270°C (right). Figure 7.6 selects the main XANES spectra of importance from those series of spectra for comparison against the reference of interest. Figure 7.6 (top left) displays the initial spectra at room temperature and compares these against the Fe_2O_3 reference. The line shape and pre-edge feature for each catalyst are quite similar to those of the Fe_2O_3 reference. The intensity and sharpness of the pre-edge feature suggest that some of the Fe is located in a tetrahedral environment. For α-Fe_2O_3, the crystal structure is a corundum structure in which the Fe ions occupy octahedral sites, while for γ-Fe_2O_3, the Fe^{III} ions are essentially randomly distributed between octahedral and tetrahedral sites. Therefore, the latter may be a better description for the material. All the catalysts eventually convert (top right) to an oxide form that is quite similar in line shape to the Fe_3O_4 reference. The peak maximum of

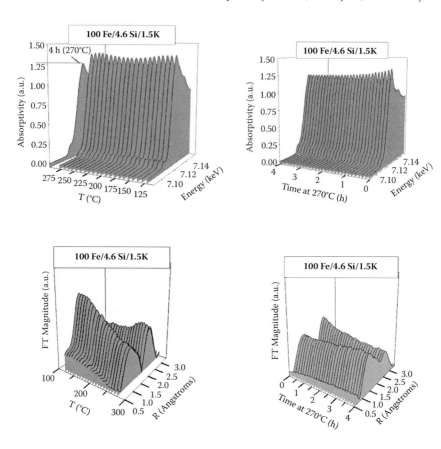

FIGURE 7.2 (Top) XANES and (bottom) EXAFS spectra (left) as a function of temperature and (right) as a function of time at 270°C in 5% CO/balance inert for the catalyst with an atomic ratio of 100Fe/4.6Si/1.5K.

the white line shifts to lower energy. In addition; the sharpness of the pre-edge diminishes slightly, indicating that a fraction of the Fe, though less, is still located in a tetrahedral environment. In Fe_3O_4, the Fe^{II} ions are located in octahedral sites, while the Fe^{III} ions are distributed between octahedral and tetrahedral sites. The range is quite narrow for all the catalysts (between 180 and 195°C). With continued heating, the white line intensity increases slightly, the peak maximum continues to shift to lower energy, and there is even less sharpness in the pre-edge feature. This is in agreement with continuing reduction of the Fe_3O_4, where the oxide is becoming more deficient in oxygen. One could describe the intermediate oxide as becoming more like FeO, where the Fe^{II} are situated in octahedral sites. The lack of sharpness in the pre-edge is in agreement with less tetrahedral symmetry. Isolating the maximum white line intensity for this oxide from the reduction profiles, one can see that the maximum intensity occurs at lower temperature with increasing level of promoters

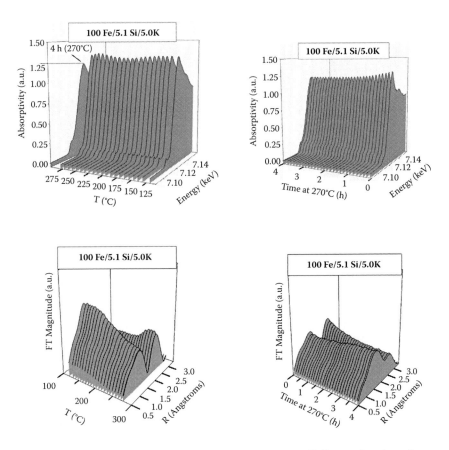

FIGURE 7.3 (Top) XANES and (bottom) EXAFS spectra (left) as a function of temperature and (right) as a function of time at 270°C in 5% CO/balance inert for the catalyst with an atomic ratio of 100Fe/4.6Si/5.0K.

present in the order: (1) 100Fe/4.6Si—270°C, (2) 100Fe/4.6Si/1.5K—256°C, (3) 100Fe/4.6Si/5.0K—249°C, (4) 100Fe/4.6Si/1.44K/2.0Cu—235°C, and (5) 100Fe/4.6Si/5.0K/2.0Cu—227°C. Finally, after heating the catalysts for 4 h at 270°C, it is clear that the catalysts with "more metallic character" are achieved with higher K and Cu promoter levels. Extent of reduction (i.e., lower white line intensity) followed the order: 100Fe/4.6Si ≪ 100Fe/4.6Si/1.5K < 100Fe/4.6Si/5.0K < 100Fe/4.6Si/1.44K/2.0Cu ≪ 100Fe/4.6Si/5.0K/2.0Cu (Figure 7.7).

The EXAFS spectra of interest were also extracted from the TPR experiment. Although we detected Fe_2O_3 and Fe_3O_4 oxides along the trajectory of the TPR, the corresponding EXAFS spectra (Figure 7.8, top left and top right) presented some ambiguities in comparison with the reference spectra as to the oxide type present. When we achieve 4 h of carburization in 5% CO at 270°C (Figure 7.8, bottom), the contribution from the residual oxide is still present. Nevertheless, it is also obvious

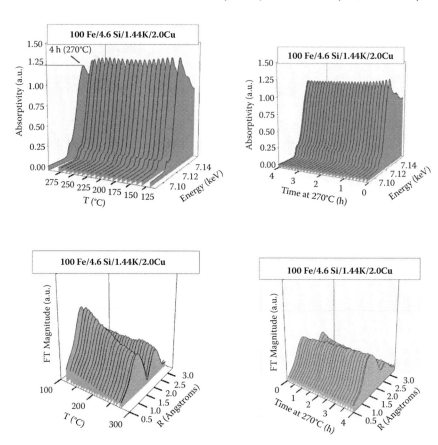

FIGURE 7.4 (Top) XANES and (bottom) EXAFS spectra (left) as a function of temperature and (right) as a function of time at 270°C in 5% CO/balance inert for the catalyst with an atomic ratio of 100Fe/4.6Si/1.44K/2.0Cu.

that an asymmetric peak corresponding to iron carbide is developing in the region around ~2 Å (Figure 7.9). To extract the carbidic fraction, the oxide fraction was subtracted from the catalyst $\chi(k)$ spectra. However, instead of utilizing the Fe_3O_4 reference compound as a basis for oxide subtraction, we chose the spectrum in the TPR of the catalyst itself that most resembles Fe_3O_4 (see Figure 7.8 top right), as demonstrated by the examples provided in Figure 7.10. Using this spectrum ameliorates the effect of elevated temperature as well as morphological and particle size differences of the sample compared to a bulk oxide. By extracting the iron carbide from the four catalysts containing an obvious signal indicating the presence of carbide, the plots in Figure 7.10 were thus obtained. In agreement with the XANES trend of increasing iron carbide content with increasing promoter content, the iron carbide Fe-Fe first shell coordination peak intensities increase in the order 100Fe/Si (not plotted, but assumed from Figure 7.6) << 100Fe/4.6Si/1.5K < 100Fe/4.6Si/5.0K < 100Fe/4.6Si/1.44K/2.0Cu << 100Fe/4.6Si/5.0K/2.0Cu.

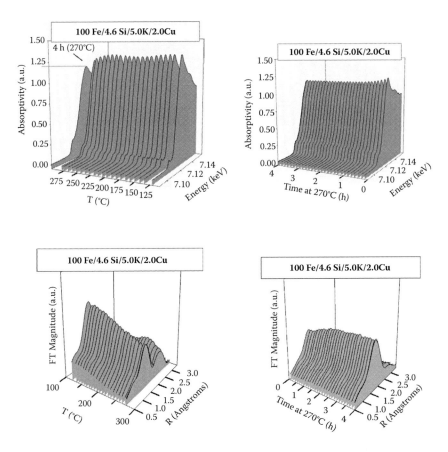

FIGURE 7.5   (Top) XANES and (bottom) EXAFS spectra (left) as a function of temperature and (right) as a function of time at 270°C in 5% CO/balance inert for the catalyst with an atomic ratio of 100Fe/4.6Si/5.0K/2.0Cu.

The models of the crystalline structures constructed using the CrystalOgraph software package are provided in Figure 7.11. Results of theoretical EXAFS modeling are plotted in terms of the Fourier transform magnitude spectra. One can observe from Figure 7.12 that there is, in addition to an obvious attenuation of the coordination peaks, a shift to lower distance with increasing degree of disorder, as modeled by increasing the values of the Debye-Waller factor from 0 to 0.010. In comparison with the experimental results, it is clear that the ε-carbide and Hägg carbide forms (i.e., real and imaginary components, and magnitude of the Fourier transform spectra) are in better agreement qualitatively than the cementite form.

Catalytic performance testing of both unpromoted and promoted catalysts for the FTS reaction was conducted. Table 7.4 summarizes the catalyst compositions and reaction conditions employed. Variations in catalytic activity, as well as differences in selectivities among the catalysts listed in Table 7.4, are reported in Figure 7.13. Comparison of CO conversion levels among different catalysts, as

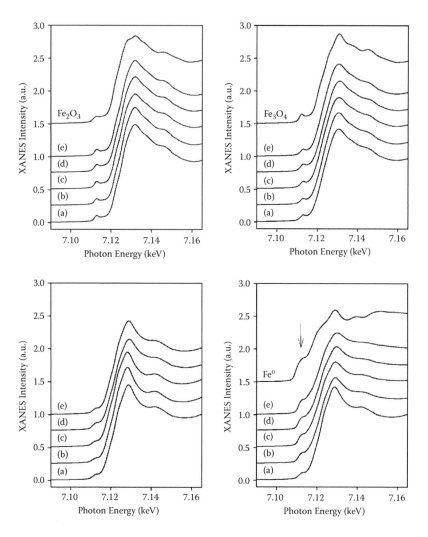

FIGURE 7.6   XANES spectra in flowing 5% CO, including: (top left) initial room temperature spectra that resemble Fe_2O_3; (top right) after transition to oxide resembling Fe_3O_4 reference; (bottom left) after transition to an oxide-deficient form of Fe_3O_4; and (bottom right) final spectrum of each catalyst after 4 h at 270°C and comparison to Fe^0 reference. Arrow indicates peak representative of increasing metal content. Catalysts include: (a) 100Fe/4.6Si, (b) 100Fe/4.6Si/1.5K, (c) 100Fe/4.6Si/5.0K, (d) 100Fe/4.6Si/1.44K/2.0Cu, and (e) 100Fe/4.6Si/5.0K/2.0Cu.

shown in Figure 7.13a, indicates that unpromoted α-Fe_2O_3 results in the highest initial conversion rate (about 85%) among the six catalysts studied; however, the rate of deactivation was very fast for the unpromoted catalyst (~3.12% CO conversion per day). In the case of the unpromoted catalyst, CO conversion changed from 85% to 10% over a period of 300 h. The α-value of the unpromoted catalyst

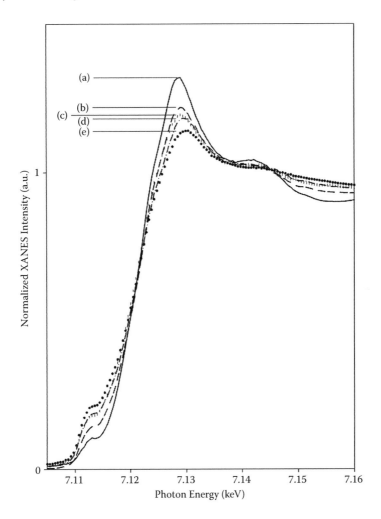

FIGURE 7.7 Overplots of XANES spectra for catalysts after 4 h of carburization at 270°C, including (a) 100Fe/4.6Si, (b) 100Fe/4.6Si/1.5K, (c) 100Fe/4.6Si/5.0K, (d) 100Fe/4.6Si/1.44K/2.0Cu, and (e) 100Fe/4.6Si/5.0K/2.0Cu.

(calculated from carbon number 5 and onward) was found to be 0.85. Addition of Si (100Fe/4.6Si) resulted in a slow increase in the CO conversion rate from about 52.5% to 65% over 300 h of synthesis time. The α-value for the 100Fe/4.6Si catalyst was found to be 0.89. Addition of K resulted in an induction period and significantly affected the CO conversion rate, although the magnitude of the resulting effect depended on the Fe/K atomic ratio. The induction period was defined to be the region where the conversion increases initially from a low level to a maximum value, before declining to attain a lower stable activity at the same reaction conditions. The catalyst with a composition of 100Fe/4.6Si/1.5K displayed an induction period of 150 h and a stable CO conversion rate of about 85%, the

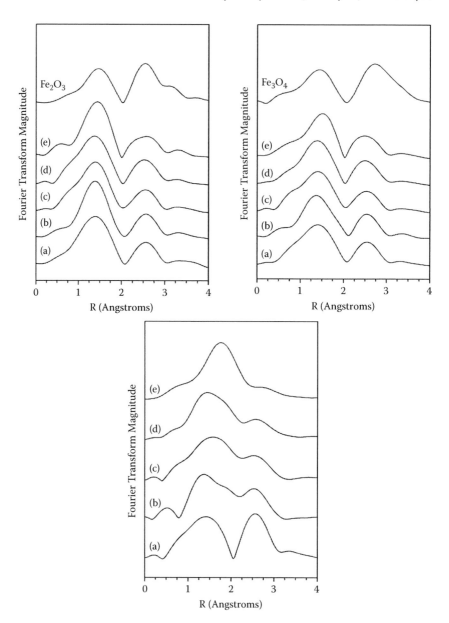

FIGURE 7.8 EXAFS spectra in flowing 5% CO, including (top left) initial room temperature spectra that resemble (f) Fe_2O_3; (top right) after transition to oxide resembling (f) Fe_3O_4 reference; and (bottom) final spectrum of each catalyst after 4 h at 270°C. Catalysts include (a) 100Fe/4.6Si, (b) 100Fe/4.6Si/1.5K, (c) 100Fe/4.6Si/5.0K, (d) 100Fe/4.6Si/1.44K/2.0Cu, and (e) 100Fe/4.6Si/5.0K/2.0Cu.

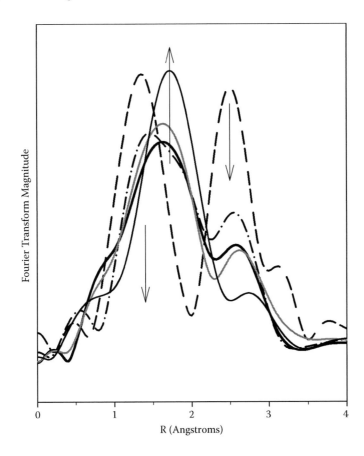

FIGURE 7.9   Final spectra after carburization at 270°C and cooling to ambient conditions. Spectra indicate that increasing promoter content improves the carburization rate. Arrows indicate loss of oxidic fraction and increase in carbidic fraction. Samples include: (solid, dashed) α-Fe$_2$O$_3$, (dash-dotted) 100Fe/4.6Si/1.5K, (thick-solid) 100Fe/4.6Si/5.0K, (gray) 100Fe/4.6Si/1.44K/2.0Cu, and (thin solid) 100Fe/4.6Si/5.0K/2.0Cu.

highest among all six catalysts studied. However, the catalyst with a higher Si and K content (100Fe/5.1Si/5.0K) resulted in a more stable, but lower CO conversion rate (about 45%), in agreement with earlier results [17]. High K content in the promoted catalyst was found to decrease the CO conversion rate in the present study, and was also observed by Luo et al. [17]. The α-values of the 100Fe/4.6Si/1.5K and 100Fe/5.1Si/5.0K catalysts were found to be 0.91 and 0.93, respectively. In the case of the Cu-promoted catalyst, a comparison of the CO conversion rates between the catalysts 100Fe/4.6Si/1.44K/2.0Cu and 100Fe/4.6Si/5.0K/2.0Cu indicates that the catalyst with composition 100Fe/4.6Si/5.0K/2.0Cu resulted in an induction period and a higher CO conversion rate (about 45%) relative to the 100Fe/4.6Si/1.44K/2.0Cu catalyst, which showed an initial decrease in conversion followed by a stable CO conversion rate of about 20%. The α-values for the

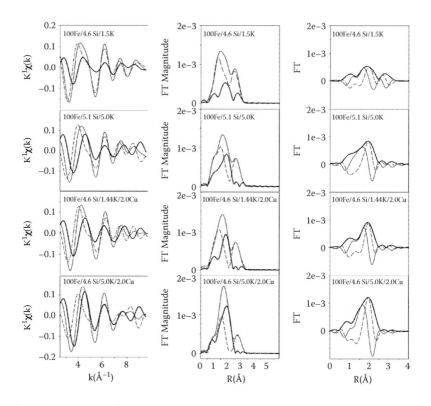

FIGURE 7.10   (Left) $k^1 \cdot \chi(k)$ versus k spectra and (center) corresponding Fourier transform magnitude spectra for (solid light) the catalyst, (dashed) the fraction of the catalyst that is oxide, and (solid heavy) the difference, which represents the extracted signal of the carbidic fraction. (Right) Extracted carbidic fraction, including (thick solid light) the Fourier transform magnitude, and the (solid light) real and (dashed) imaginary parts of the Fourier transform.

100Fe/4.6Si/1.44K/2.0Cu and 100Fe/4.6Si/5.0K/2.0Cu catalysts were found to be 0.92 and 0.94, respectively. Since the amount of Cu was essentially identical in both catalysts and the steady CO conversion rate of the 100Fe/4.6Si/5.0K/2.0Cu catalyst was similar to that of the 100Fe/5.1Si/5.0K catalyst, we postulate that the difference in CO conversion level between the two Cu-promoted catalysts is due primarily to a difference in the K (and perhaps Si) content. In general, selectivities in FTS are compared at a similar CO conversion level for each catalyst. However, wide variations in the CO conversion level in the current study make it difficult to evaluate the effects of different promoters on product selectivity. In spite of the differences in CO conversion rates, an effort was made to qualitatively compare the promotional effect of Cu and K on product selectivity at similar, or nearly similar, CO conversion levels.

Comparison of CH_4 selectivities among the catalysts (Figure 7.13b) revealed that unpromoted α-Fe_2O_3, 100Fe/4.6Si, and 100Fe/4.6Si/1.5K catalysts displayed

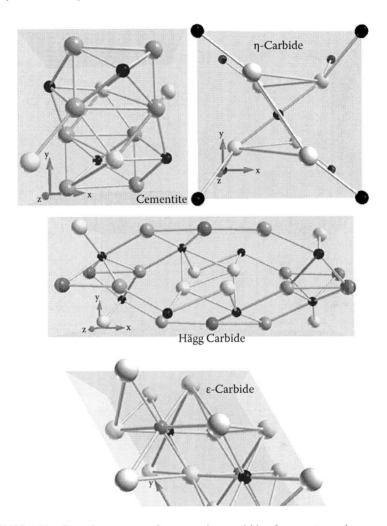

FIGURE 7.11 Crystal structures of common iron carbides. Larger atoms, iron; smaller atoms, carbon.

higher CH_4 selectivity relative to the 100Fe/5.1Si/5.0K, 100Fe/4.6Si/1.44K/2.0Cu, and 100Fe/4.6Si/5.0K/2.0Cu catalysts. The CO conversion levels for both 100Fe/5.1Si/5.0K and 100Fe/4.6Si/5.0K/2.0Cu catalysts were similar (about 45%, Figure 7.13a), and the steady CO conversion rates for both unpromoted α-Fe_2O_3 and 100Fe/4.6Si/1.44K/2.0Cu catalysts were nearly similar (about 15 to 20% after 250 h of time on stream; Figure 7.13a). Hence, a direct comparison of CH_4 selectivity could be made for each catalyst pair mentioned above. Addition of Cu as a promoter was found to slightly increase CH_4 selectivity for catalysts containing higher amounts of K (~2% for 100Fe/4.6Si/5.0K/2.0Cu while ~1% for 100Fe/5.1Si/5.0K; Figure 7.13b). Comparing CH_4 selectivity for unpromoted

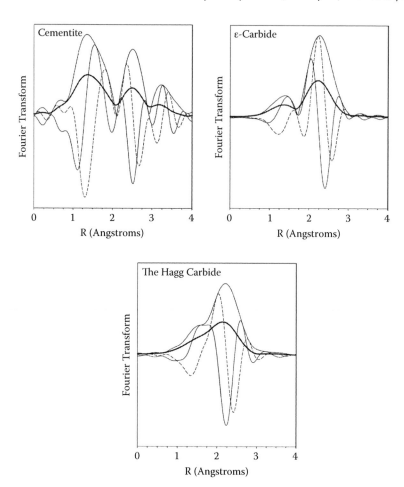

FIGURE 7.12 Fourier tranform magnitude of $k^1\chi(k)$ versus k theoretical EXAFS spectra generated from FEFFIT software by setting coordination numbers to their nominal values, and other EXAFS parameters as described in the experimental section. (Light solid line) Debye-Waller factor of 0.000 and (heavy solid line) Debye-Waller factor of 0.010. k-range of 2.5 to 10 Å^{-1}.

α-Fe$_2$O$_3$ and 100Fe/4.6Si/1.44K/2.0Cu catalysts revealed that addition of Si, Cu, and K significantly inhibited the CH$_4$ selectivity (from 9% for unpromoted α-Fe$_2$O$_3$ to about 2.5% for the 100Fe/4.6Si/1.44K/2.0Cu catalyst; Figure 7.13b).

CO can be converted into either hydrocarbon products and water (via FTS) or CO$_2$ and H$_2$ via the water-gas shift (WGS) reaction. The reversible WGS reaction accompanies FTS over the iron-based catalyst only at high temperature conditions. The individual rates of FTS (r_{FTS}) and the WGS reaction (r_{WGS}) can be calculated from experimental results as $r_{WGS} = r_{CO_2}$ and $r_{FTS} = r_{CO} - r_{CO_2}$, where r_{CO_2} is the rate of CO$_2$ formation and r_{CO} is the rate of CO conversion.

TABLE 7.4

Summary of Catalyst Composition and Reaction Conditions Selected for Comparing Fischer-Tropsch Synthesis Activity in CSTR

Catalyst (Atomic Basis)	Reaction T, °C	Reaction P, psig	Initial Solvent	SV sl/h/g Fe	α (Calculated from C_{5+} Onward)
α-Fe$_2$O$_3$	270	175	PW3000	3.0	0.85
100Fe/4.6Si	270	175	C-30 oil	5.0	0.89
100Fe/4.6Si/1.5K	270	175	PW3000	5.0	0.91
100Fe/5.1Si/5.0K	270	175	PW3000	5.0	0.93
100Fe/4.6Si/1.44K/2.0Cu	270	175	C-30 oil	3.0	0.92
100Fe/4.6Si/5.0K/2.0Cu	230	175	C-30 oil	3.0	0.94

Note: Prior to Fischer-Tropsch synthesis, all catalysts were activated with CO at 270°C and atmospheric pressure for 24 h with a CO SV of 3.0 sl/h/g Fe.

As presented in Figure 7.13c, it can be seen that among the six catalysts, CO_2 selectivity (calculated as % = $r_{CO_2} / r_{CO_2} + r_{FTS}$) for 100Fe/4.6Si/1.5K catalyst was found to be at the highest level (about 45%), and for 100Fe/5.1Si/5.0K catalyst it was found to be lowest (about 6%). It should be mentioned that the CO conversion rates for the above two catalysts varied greatly (Figure 7.13c). Since the CO conversion levels for both 100Fe/5.1Si/5.0K and 100Fe/4.6Si/5.0K/2.0Cu catalysts are about 45% (Figure 7.13a), a comparison of CO_2 selectivity is deemed justified. It can be observed from Figure 7.13c that addition of Cu as a promoter was found to significantly increase CO_2 selectivity (about 40% for the catalyst 100Fe/4.6Si/5.0K/2.0Cu, compared to about 5% for the 100Fe/5.1Si/5.0K catalyst). Similar results were reported by several researchers for Group I alkali metal promoters, although the mechanism by which K or similar alkali metal dopants promote high CO_2 selectivity or WGS activity over the iron catalyst is still not well understood [18–21].

C_{5+} selectivities of different catalysts are presented in Figure 7.13d. Comparison of C_{5+} selectivities for unpromoted α-Fe$_2$O$_3$ and 100Fe/4.6Si/1.44K/2.0Cu catalysts (as the CO conversions of these two catalysts are nearly similar; Figure 7.13a) indicates that the promoting combined effect of Si, Cu, and K results in a significant increase in C_{5+} selectivity (about 82% to about 92%; Figure 7.13d). It is evident from Figure 7.13d that the C_{5+} selectivity of 100Fe/5.1Si/5.0K catalyst is slightly higher than that of the Cu-promoted analog (100Fe/4.6Si/5.0K/2.0Cu), suggesting that promotion by Cu in high K-promoted iron catalyst may slightly decrease the C_{5+} selectivity, in agreement with the increase in CH_4 selectivity observed upon Cu promotion (see Figure 7.13b).

The 1-alkene/2-alkene ratio for the C_4 hydrocarbon fraction is presented in Figure 7.13e, while the alkene/(alkene + alkane) fractions for the C_2 hydrocarbon products from FTS for unpromoted α-Fe$_2$O$_3$ and promoted catalysts are

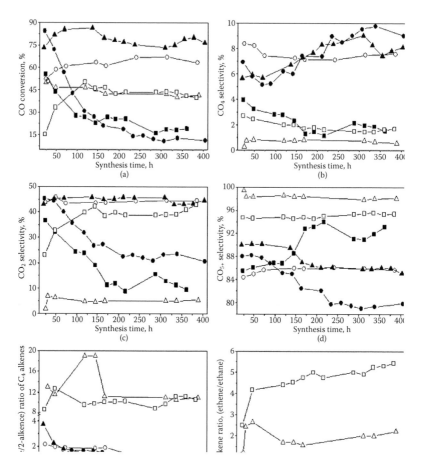

FIGURE 7.13 Comparison of catalytic activity and selectivities of different catalysts in Fischer-Tropsch synthesis (FTS). Plot of (a) CO conversion, %; (b) CH_4 selectivity, %; (c) CO_2 selectivity, %; (d) C_{5+} selectivity, %; (e) 1-alkene/2-alkene ratio of C_4 alkenes; and (f) C_2 alkene ratio (ethene/ethane) against synthesis time (h). Details of FTS conditions are summarized in Table 7.1. Graph legends for catalyst composition: ●, α–Fe_2O_3; ○, 100Fe/4.6Si; ▲, 100Fe/4.6Si/1.5K; △, 100Fe/5.1Si/5.0K; ■, 100Fe/4.6Si/1.44K/2.0Cu; □, 100Fe/4.6Si/5.0K/2.0Cu.

displayed in Figure 7.13f. Both the C_2 alkene ratio and 1-alkene/2-alkene ratios for the 100Fe/5.1Si/5.0K and 100Fe/4.6Si/5.0K/2.0Cu catalysts were measured to be significantly higher than those for the other catalysts studied. At a similar CO conversion level (about 45%; Figure 7.13a), a comparison between the 100Fe/5.1Si/5.0K and 100Fe/4.6Si/5.0K/2.0Cu catalysts reveals that addition of Cu to high K-promoted iron catalysts results in higher alkene selectivity, while only marginally affecting the 1-alkene/2-alkene ratio (i.e., isomerization of 1-alkene).

The effects of potassium on the FTS and WGS activity and product selectivities have been investigated over a variety of iron-based catalysts [17–19,21–27]. Luo et al. [17] reported that in the case of K-promoted iron-based FTS catalyst, the induction period and the peak CO conversion were dependent on the K loading, with lower potassium loadings producing a higher maximum conversion rate and a shorter induction period. It has been suggested that K promotes the activation of CO and the rate of carburization of Fe_3O_4 when the catalyst is activated by CO [22,23]. The significant promoting effects of potassium can be summarized as follows: (1) favors the dissociative chemisorption of CO, strengthens the Fe–C bond, and weakens the Fe–H bond; (2) enhances the selectivity of long-chain products and olefin selectivity, lowers methane selectivity; (3) increases WGS activity; and (4) influences FTS activity. Promotion by higher amounts of K (Fe/K atomic ratio of 100:5) was also found to increase the selectivity of long-chain products (higher α-values), and for higher alkene, while lowering methane selectivity. Li et al. [22–24] concluded that potassium promotes the formation of an increased number of active sites during reduction and carburization of iron oxides (i.e., by facilitating the rapid formation of nucleation sites resulting in the formation of smaller iron carbide crystallites), thus playing a role as a structural promoter [24]. Although potassium enhances the FTS activity and heavy product fraction, high potassium loading may cover too large of a fraction of the surface of the iron catalyst, resulting in a limited promotion effect or even a decrease in CO conversion [25]. Addition of small amounts of K (e.g., 100Fe/4.6Si/1.5K) was found to increase the CO conversion rate significantly. However, higher amounts (e.g., 100Fe/5.1Si/5.0K) resulted in a decrease in the CO conversion rate under identical reaction conditions (Figure 7.13a).

Traditionally Cu has been added to precipitated iron FTS catalysts to facilitate reduction of Fe_2O_3 to zero valent iron during activation [27] by lowering the reduction temperature when activating with H_2, CO, or syngas. The effect of copper on activity and selectivity has not been studied as thoroughly as its effect on catalyst activation. Kölbel and Ralek [28] reported that Cu loadings less than 0.1 wt% (relative to Fe) were sufficient to produce an active catalyst, and that increased copper loading had no effect on FTS activity. However, in a separate investigation, promotion by Cu was reported to increase the FTS and WGS activities, average molecular weight of the product, hydrogenation of alkenes, and isomerization of 1-alkenes [29,30]. O'Brien and Davis [31] reported the effects of Cu (0–2 atomic ratio per 100Fe) incorporation in low K-promoted (Fe/K ratio of 100/1.4) catalyst on the FTS and WGS activities and product selectivities over a wide range of syngas conversions. It was found that addition of Cu decreased CH_4 selectivity and enhanced the selectivity for higher hydrocarbon products; however, Cu did not significantly affect alkene selectivity or the 1-alkene/2-alkene ratio.

It is found that addition of Cu in high K content iron catalyst (100Fe/4.6Si/5.0K/2.0Cu) results in an increase in CO_2 selectivity compared to catalysts without Cu (100Fe/5.1Si/5.0K) at a similar CO conversion level. Promotion by Cu in high K content catalysts also enhances the olefin selectivity (higher olefin ratio of 100Fe/4.6Si/5.0K/2.0Cu in comparison with 100Fe/5.1Si/5.0K;

Figure 7.13f). Although addition of Cu initially promotes isomerization of 1-alkene (initially higher 1-alkene/2-alkene ratio for 100Fe/4.6Si/5.0K/2.0Cu in comparison with 100Fe/5.1Si/5.0K catalyst; Figure 7.13e), at steady conditions the effect is marginal.

It has been suggested [21,22] that the presence of Cu and K increases the rates and extent of Fe_3O_4 carburization during reaction and the FTS rates, by providing multiple nucleation sites that lead to the ultimate formation of smaller carbide crystallites with higher active surface area. In the present investigation, Cu- and K-promoted iron catalysts performed better than the unpromoted catalysts in terms of (1) a lower CH_4 selectivity, (2) higher C_{5+} and alkene product selectivities, and (3) an enhanced isomerization rate of 1-alkene.

7.4 CONCLUSIONS

The TPR-XAFS technique confirmed that doping Fischer-Tropsch synthesis catalysts with Cu and alkali (e.g., K) remarkably promotes the carburization rate relative to the undoped catalyst. The EXAFS results suggest that either the Hägg or ε-carbides were formed during the reduction process over the cementite form. A correlation is observed between the α-value of the product distribution and the carburization rate.

7.5 SUPPLEMENTARY INFORMATION

Tables 7.3A to 7.3D provide the path parameters generated by FEFF for the various carbide forms investigated in this work.

ACKNOWLEDGMENTS

The work carried out at the CAER was supported in part by funding from a seed grant from the Kentucky Governor's Office of Energy Policy (Solicitation 08-GOEP-02), as well as the Commonwealth of Kentucky. Argonne's research was supported in part by the U.S. Department of Energy (DOE), Office of Fossil Energy, National Energy Technology Laboratory (NETL). The use of the Advanced Photon Source was supported by the U.S. Department of Energy, Office of Science, Office of Basic Energy Sciences, under Contract DE-AC02-06CH11357. MRCAT operations are supported by the Department of Energy and the MRCAT member institutions.

REFERENCES

1. Espinoza, R.L., Shingles, T., Duvenhage, D.J., and Langenhoven, P.L., Method of modifying and controlling catalyst selectivity in a Fischer-Tropsch process. U.S. patent 6,653,357, Sasol Technology, Nov. 25, 2003.
2. O'Brien, R.J., Xu, L., Spicer, R.L., Bao, S., Milburn, D.R., and Davis, B.H. 1997. Activity and selectivity of precipitated iron Fischer-Tropsch catalysts. *Catal. Today* 36:325–34.

3. Iglesia, E., Reyes, S.C., Madon, R.J., and Soled, S.L. 1993. Selectivity control and catalyst design in the Fischer-Tropsch synthesis: Sites, pellets, and reactors. *Adv. Catal.* 39:221–302.

4. Eliason, S.A., and Bartholomew, C.H. 1997. Temperature-programmed reaction study of carbon transformations on iron Fischer-Tropsch catalysts during steady-state synthesis. *Stud. Surf. Sci. Catal.* 111:517–26.

5. O'Brien, R.J., Xu, L., Spicer, R.L., and Davis, B.H. 1996. Activation study of precipitated iron Fischer-Tropsch catalysts. *Energy & Fuels* 10:921–26.

6. Ravel, B. 2001. ATOMS: Crystallography for the x-ray absorption spectroscopist. *J. Synchrotron Rad.* 8:314–16.

7. Faraoun, H.I., Zhang, Y.D., Esling, C., and Aourag, H. 2006. Crystalline, electronic, and magnetic structures of 2-Fe_3C, P-Fe_5C_2, and 0-Fe_2C from first principle calculation. *J. Appl. Phys.* 99:093508/1–8/8.

8. Nagakura, S. 1959. Study of metallic carbides by electron diffraction. III. Iron carbides. *J. Phys. Soc. Jpn.* 14:186–95.

9. Hägg, G. 1931. Regularity in crystal structure in hydrides, borides, carbides and nitrides of transition elements. *Z. Physik. Chem.* 12B:33–56.

10. Rehr, J.J., and Albers, R.C. 2000. Theoretical approaches to x-ray absorption fine structure. *Rev. Mod. Phys.* 72:621–54.

11. Schoeni, N., and Chapuis, G. 2004. CrystalOgraph: An interactive applet for drawing crystal structures. Ecole Polytechnique Fédérale de Lausanne, Switzerland.

12. Newville, M. 2001. IFEFFIT: Interactive XAFS analysis and FEFF fitting. *J. Synchrotron Rad.* 8:322–24.

13. Jacoby, M. 2001. ACS award in colloid or surface chemistry. *Chem. Eng. News* 79:36–37.

14. Ressler, T. 1998. WinXAS: A program for x-ray absorption spectroscopy as an in-situ tool in materials science. *J. Synchrotron Rad.* 5:118–22.

15. Li, S., O'Brien, R.J., Meitzner, G.D., Hamdeh, H., Davis, B.H., and Iglesia, E. 2001. Structural analysis of unpromoted Fe-based Fischer–Tropsch catalysts using x-ray absorption spectroscopy. *Appl. Catal. A Gen.* 219:215–22.

16. Miller, J.T., Marshall, C.L., and Kropf, A.J. 2001. (Co)MoS_2/alumina hydrotreating catalysts: An EXAFS study of the chemisorption and partial oxidation with O_2. *J. Catal.* 202:89–99.

17. Luo, M., O'Brien, R.J., Bao, S., and Davis, B.H. 2003. Fischer–Tropsch synthesis: Induction and steady-state activity of high-alpha potassium promoted iron catalysts. *Appl. Catal. A Gen.* 239:111–20.

18. Arakawa, H., and Bell, A.T. 1983. Effects of potassium promotion on the activity and selectivity of iron Fischer-Trospch catalysts. *Ind. Eng. Chem. Res. Process. Des. Dev.* 22:97–103.

19. Dictor, R.A., and Bell, A.T. 1986. Fischer-Tropsch synthesis over reduced and unre-duced iron oxide catalysts. *J. Catal.* 97:121–36.

20. Ngantsoue-Hoc, W., Zhang, Y., O'Brien, R.J., Luo, M., and Davis, B.H. 2002. Fischer–Tropsch synthesis: Activity and selectivity for Group I alkali promoted iron-based catalysts. *Appl. Catal.* 236:77–89.

21. Bukur, D.B., Mukesh, D.S., and Patal, A. 1990. Promoter effects on precipitated iron catalysts for Fischer-Tropsch synthesis. *Ind. Eng. Chem. Res.* 29:194–204.

22. Li, S., Li, A., Krishnamoorthy, S., and Iglesia, E. 2001. Effects of Zn, Cu, and K promoters on the structure and on the reduction, carburization, and catalytic behavior of iron-based Fischer-Tropsch synthesis catalysts. *Catal. Lett.* 77:197–205.

23. Li, S., Krishnamoorthy, S., Li, A., Meitzner, G.D., and Iglesia, E. 2002. Promoted iron-based catalysts for the Fischer–Tropsch synthesis: Design, synthesis, site densities, and catalytic properties. *J. Catal.* 206:202–17.

24. Li, S., Ding, W., Meitzner, G.D., and Iglesia, E. 2002. Spectroscopic and transient kinetic studies of site requirements in iron-catalyzed Fischer-Tropsch synthesis. *J. Phys. Chem. B* 106:85–91.

25. Raje, A.P., O'Brien, R.J., and Davis, B.H. 1998. Effect of potassium promotion on iron-based catalysts for Fischer–Tropsch synthesis. *J. Catal.* 180:36–43.

26. Miller, D.G., and Moskovits, M. 1988. A study of the effects of potassium addition to supported iron catalysts in the Fischer-Tropsch reaction. *J. Phys. Chem.* 92:6081–85.

27. Dry, M.E. 1981. In *Catalysis science and technology*, ed. J.R. Anderson and M. Boudart, 179. Vol. 1. New York: Springer-Verlag.

28. Kölbel, H., and Ralek, M. 1980. The Fischer-Tropsch synthesis in the liquid phase. *Catal. Rev. Sci. Eng.* 21:225–74.

29. Huff, Jr., G.A., and Satterfield, C.N. 1984. Intrinsic kinetics of the Fischer-Tropsch synthesis on a reduced fused-magnetite catalyst. *Ind. Eng. Chem. Process Des. Dev.* 23:696–705.

30. Bukur, D.B., Mukesh, D., and Patel, S.A. 1990. Promoter effects on precipitated iron catalysts for Fischer-Tropsch synthesis. *Ind. Eng. Chem. Res.* 29:194–204.

31. O'Brien, R.J., and Davis, B.H. 2004. Impact of copper on an alkali promoted iron Fischer-Tropsch catalyst. *Catal. Lett.* 94:1–6.

8 Characterization of Co/Silica Catalysts Prepared by a Novel NO Calcination Method

Gary Jacobs, Wenping Ma, Yaying Ji,
Syed Khalid, and Burtron H. Davis

CONTENTS

A novel conversion of cobalt nitrate to cobalt oxide using NO (J. R. A. Sietsma et al., patent applications WO 2008029177 and WO 2007071899) was utilized to prepare silica-supported cobalt research catalysts, in order to test the materials for their sensitivity to Fischer-Tropsch synthesis process parameters. In the current contribution, extensive characterization of activated air calcined and nitric oxide calcined 15 and 25% cobalt-loaded silica catalysts by temperature-programmed reduction (TPR), hydrogen chemisorption/pulse reoxidation, extended x-ray absorption fine structure (EXAFS), and x-ray absorption near-edge spectroscopy (XANES) is described. For catalysts activated at a standard condition of 350°C for 10 h in hydrogen, despite a lower percentage of reduction observed with the nitric oxide calcined catalysts, the smaller average metal cobalt crystallite size more than compensates for this effect, leading to a higher cobalt surface metal active site density on a per gram of catalyst basis. The H_2-activated nitric oxide calcined catalysts were found to result in higher CO conversion rates on a per gram of catalyst basis relative to their activated air calcined counterparts.

8.1 INTRODUCTION

Supported cobalt catalysts are important for the slurry phase Fischer-Tropsch synthesis of hydrocarbons, which can be subsequently processed to produce an ultra-clean, virtually sulfur-free diesel. Owing to their low selectivity for the water-gas shift reaction, cobalt catalysts are well suited for the conversion of synthesis gas mixtures with a high H_2/CO ratio, such as methane-derived syngas. Iron-based catalysts, on the other hand, exhibit higher intrinsic water-gas shift selectivity and, although much lower in cost, are more suitable for converting syngas with a lower H_2/CO ratio. Due to the high cost of cobalt relative to iron, catalyst activity and stability are important considerations. Iglesia[1] reported that the turnover frequency is relatively constant over a range of dispersion for supported cobalt catalysts, indicating that the catalysts are, relatively speaking, structurally insensitive.

Regarding the use of Co/Al_2O_3 catalysts, from the standpoint of activity, it is often reported that the metal oxide–support interaction is a major problem (Figure 8.1). Certainly, it is true that after a standard reduction at 350°C, only a fraction of the cobalt is reduced. However, the number of active sites depends not only on the degree of reduction, but also on the size of the cobalt crystallites. In comparing hydrogen chemisorption/pulse reoxidation results shown in Table 8.1 for 12.4% Co/SiO_2 with those of 15% Co/Al_2O_3 and 10% Co/TiO_2 catalysts, all of which were prepared by standard impregnation and calcination, while cobalt silica offered a much higher degree of reduction, the number of active sites was found to be low in comparison with cobalt alumina, owing to the larger average diameter of the Co crystallites.[2] This was true even considering that the Brunauer-Emmett-Teller (BET) surface area of the silica support was much higher than the alumina support. In the past, researchers have considered alternate preparation and reduction procedures,[3–5] which have resulted in the formation of smaller Co crystallites on SiO_2, and consequently, interactions between cobalt species and the support were observed in temperature-programmed reduction (TPR) studies (Figure 8.1 and Table 8.1). Further inroads in active site density are made when adding small amounts of metal promoters, such as Pt, Ru, and Re (Figure 8.1 and Table 8.1).[1,2,6–14]

Recently, it has been reported that a novel calcination procedure relying on nitric oxide gas in lieu of air also results in smaller cobalt crystallites over silica supports.[15–17] The idea is to use a less oxidative gas to prevent rapid decomposition of the nitrate precursor during thermal nitrate decomposition, which has been observed when O_2 is present.[17] As a result, the mobility of the precursor on the oxide carrier surface is hindered, resulting in a smaller average Co oxide cluster

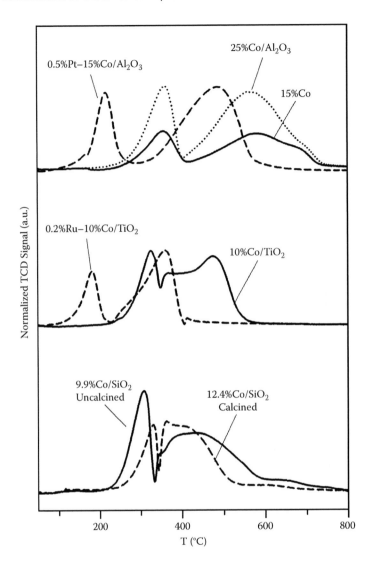

FIGURE 8.1 Temperature-programmed reduction profiles of supported cobalt catalysts, including (top) Co/Al$_2$O$_3$ catalysts, (middle) Co/TiO$_2$ catalysts, and (bottom) Co/SiO$_2$ catalysts.

size.[17] NO was reported to act as an oxygen scavenger during decomposition of nickel nitrate, such that no O$_2$ was observed in the product effluent during the decomposition of the precursor.[17] As such, upon reduction, the active site densities are reported to be considerably higher.[17] With the aim of testing these catalysts for stability and sensitivity to Fischer-Tropsch synthesis process parameters (see Chapter 3), we first prepared and characterized supported cobalt catalysts based

TABLE 8.1
H$_2$ Chemisorption (TPD) and Pulse Reoxidation

Catalyst Description	Reduction Temperature (°C)	µmol H$_2$ Desorbed per g cat.	Uncorrected % Dispersion	µmol O$_2$ Consumed per g cat.	% Reduction	Corrected % Dispersion	Corrected Diameter (nm)
12.4% Co/SiO$_2$ Air calcined	350	18.2	1.7	902	64	1.7	38.2
Co/SiO$_2$ Nitrate reduction route	350	25.6	3.0	433	39	3.0	13.2
10.0% Co/TiO$_2$	300	42.8	5.1	593	52	9.7	10.6
0.2% Ru–10% Co/TiO$_2$	300	66.6	7.8	722	64	12.2	8.5
15% Co/Al$_2$O$_3$	350	66.9	5.3	509	30	17.5	5.9
0.5% Pt–15% Co/Al$_2$O$_3$	350	140.6	11.0	1,024	71	18.4	5.6
25% Co/Al$_2$O$_3$	350	77.7	77.7	1,174	42	8.7	11.8

on this new procedure, extensively characterizing the materials by TPR, hydrogen chemisorption/pulse reoxidation, and extended x-ray absorption fine structure (EXAFS)/x-ray absorption near-edge spectroscopy (XANES) methods.

8.2 EXPERIMENTAL

8.2.1 CATALYST PREPARATION

PQ silica CS-2133 was used as the support for the cobalt FTS catalysts. A slurry phase method was used to load cobalt nitrate to the support, such that the loading solution volume was 2.5 times that of the measured pore volume. To obtain a cobalt loading of 15 or 25% cobalt, multiple steps were used, due to the limited solubility of the cobalt nitrate salt. Following cobalt addition, the catalyst was dried at 80 and 100°C in a rotary evaporator following each slurry impregnation. Catalysts were calcined in either flowing air or flowing 5% nitric oxide in nitrogen at a rate of 1 L/min for 4 h at 350°C.

8.2.2 BET MEASUREMENTS

BET and Barrett-Joyner-Halenda (BJH) measurements for the catalysts were conducted to determine the loss of surface area with loading of the metal and changes in pore size distributions. These measurements were conducted using a Micromeritics Tri-Star system. Prior to the measurement, samples were slowly ramped to 160°C and evacuated for 24 h to approximately 50 mTorr.

8.2.3 HYDROGEN CHEMISORPTION WITH PULSE REOXIDATION

Hydrogen chemisorption measurements were performed using a Zeton Altamira AMI-200 unit, which incorporates a thermal conductivity detector (TCD). The sample weight was always 0.220 g. The catalyst was activated at 350°C for 10 h using a flow of pure hydrogen and then cooled under flowing hydrogen to 100°C. The sample was held at 100°C under flowing argon to prevent physisorption of weakly bound species prior to increasing the temperature slowly to the activation temperature. At that temperature, the catalyst was held under flowing argon to desorb the remaining chemisorbed hydrogen so that the TCD signal returned to the baseline. The TPD spectrum was integrated and the number of moles of desorbed hydrogen determined by comparing to the areas of calibrated hydrogen pulses. Prior to experiments, the sample loop was calibrated with pulses of nitrogen in helium flow and compared against a calibration line produced from gas-tight syringe injections of nitrogen under helium flow.

After TPD of hydrogen, the sample was reoxidized at the activation temperature by injecting pulses of pure oxygen in helium referenced to helium gas. After oxidation of the cobalt metal clusters, the number of moles of oxygen consumed was determined and the percentage reduction calculated, assuming that the Co^0 reoxidized to Co_3O_4. While the uncorrected dispersions are based on the assumption of complete reduction, the corrected dispersions reported include the percentage of reduced cobalt as follows:

$$\%D = (\text{no. of } Co^0 \text{ atoms on surface} \times 100\%)/(\text{total no. of } Co^0 \text{ atoms})$$

$$\%D = (\text{no. of } Co^0 \text{ atoms on surface} \times 100\%)/$$
$$[(\text{total no. of Co atoms})(\text{fraction reduced})]$$

8.2.4 TEMPERATURE-PROGRAMMED REDUCTION

Temperature-programmed reduction (TPR) profiles of fresh catalyst samples were obtained using a Zeton Altamira AMI-200 unit. Calcined fresh samples were first heated and purged in flowing argon to remove traces of water. TPR was performed using 30 cc/min 10% H_2/Ar mixture referenced to argon. The ramp was 5°C/min from 50 to 1,100°C, and the sample was held at 1,100°C for 30 min.

8.2.5 EXTENDED X-RAY ABSORPTION FINE-STRUCTURE/X-RAY ABSORPTION NEAR-EDGE SPECTROSCOPY

Catalysts and Co references were evaluated by XANES and EXAFS spectroscopy at Brookhaven National Laboratory. Catalysts were first reduced in-house at 350°C for 10 h using 30 ccm of 33% H_2 (balance He) and cooled in hydrogen to room temperature, prior to a helium purge and passivation with 1% O_2/He. Catalysts were re-reduced in an *in situ* flow cell in flowing H_2 (100 ccm) and He (300 ccm) at 350°C and held for 30 min, prior to cooling to liquid nitrogen temperatures in flowing H_2. EXAFS/XANES spectra were recorded in transmission mode at the National Synchrotron Light Source (NSLS) at Brookhaven National Laboratory, Upton, New York, Beamline X18-b. The beamline was equipped with a Si (111) channel-cut monochromator. A crystal detuning procedure was used to help remove harmonic content from the beam and make the relative response of the incident and transmission detectors more linear. The x-ray flux for the beamline was on the order of 1 E 10 photons per second at 100 mA and 2.8 GeV, and the usable energy range at X-18b is from 5.8 to 40 keV. EXAFS/XANES spectra were recorded near the Co K edge. The spectra were recorded near the boiling temperature of liquid nitrogen to minimize contributions to the dynamic Debye-Waller factor. A sample thickness was determined by calculating the amount in grams per square centimeter of sample, w_D, by utilizing the thickness equation

$$w_D = \ln(I_0/I_t)/\Sigma\{(m/r)_j w_j\}$$

where m/r is the total cross section (absorption coefficient/density) of element j in the sample at the absorption edge of the EXAFS element under study in cm^2/g, w_j is the weight fraction of element j in the sample, and $\ln(I_0/I_t)$ was taken over a typical range of 1 to 2.5. An average value of w_D from inputting both values was employed. Based on the calculation for w_D, and the cross-sectional area of the pellet, the grams were calculated. Boron nitride was utilized to dilute the sample, such that the wafer could be self-supported. Smooth wafers, free of pinholes, were pressed and loaded into the *in situ* x-ray adsorption spectroscopy (XAS) flow cell, and the treatment gas was directed to the sample area. The cell was purged for a long duration of time with a high flow rate of inert gas to ensure removal of air, prior to the re-reduction treatment.

EXAFS data reduction and fitting were carried out using the WinXAS,[18] Atoms,[19] FEFF,[20] and FEFFIT[20] programs. The k- and r-ranges were chosen to be 3–15 $Å^{-1}$ and 1.0–3.0 Å, respectively. XANES spectra were compared qualitatively after normalization.

8.3 RESULTS AND DISCUSSION

BET surface area and porosity data are tabulated in Table 8.2. There is a decrease in the BET surface area with loading of cobalt onto the silica support. However, in each case where the nitric oxide calcination is used, the drop in surface area is lower. For the case of 15% Co/SiO_2, the difference is $\Delta 20$ m^2/g, while in the case of the 25% Co/SiO_2 catalysts, the difference is $\Delta 12.5$ m^2/g. The results suggest that there is less blocking of the narrower pores by large crystallites in the case of the nitric oxide calcined catalysts. In agreement with this, the average pore diameter was found to be slightly lower for the case of the nitric oxide catalysts. A sample comparison of pore size distributions from application of the BJH method is shown in Figure 8.2 for the 15% Co air calcined (solid) and NO calcined (dashed) catalysts. It is evident that while the wider pores do appear to be blocked for the NO

TABLE 8.2
BET Surface Area and Porosity Measurements

Catalyst Description	Calcination	BET SA (m²/g)	Single-Point Pore Volume (cm³/g)	BJH Pore Volume (cm³/g)	Single-Point Pore Diameter (nm)	BJH Pore Diameter (nm)
PQ silica CS-2123	Air	352	2.362	2.373	28.0	25.8
15% Co/SiO₂	Air	278	1.047	1.057	15.1	14.3
15% Co/SiO₂	5% NO	298	1.061	1.072	14.3	13.5
25% Co/SiO₂	Air	226	0.760	0.767	13.4	12.8
25% Co/SiO₂	5% NO	238	0.728	0.740	12.2	11.5

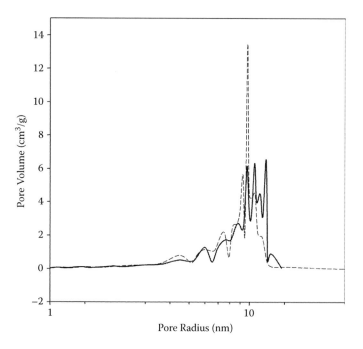

FIGURE 8.2   Comparison of pore size distributions for the (solid) air calcined and (dashed) NO calcined 15% Co/SiO$_2$ catalysts.

calcined catalyst relative to the air calcined catalyst, the narrower pores remain more accessible.

Temperature-programmed reduction profiles, provided in Figure 8.3, reveal that the air calcined samples reduce over a narrow temperature range in the region near 350°C, and reduction resembles the two-step process of bulk Co$_3$O$_4$ crystallites, where the second step of the process (3CoO + 3H$_2$ = 3Co + 3H$_2$O) consumes three times as much hydrogen as the first step (Co$_3$O$_4$ + H$_2$ = 3CoO + H$_2$O). Remarkably, when nitric oxide is used to calcine the catalysts, the reduction steps are spread out over a much wider range, with a shoulder extending up to 600°C. These species are likely more strongly interacting CoO species due to the presence of smaller particles (i.e., smaller particles equal a stronger surface interaction with the support), as has been observed even with air calcined Co/Al$_2$O$_3$ catalysts. Continuing reduction of more strongly interacting species occurs up to 900°C, which may be attributed to the reduction of cobalt silicates. It is worth pointing out that there is clearly an important difference between the surface free energies of Co/silica and Co/alumina catalysts, since the latter yield small crystallites even with air calcination procedures, while special methods must be used to form small crystallites on Co/silica.

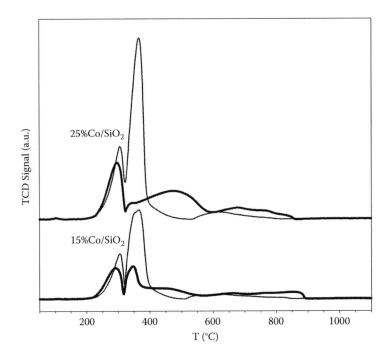

FIGURE 8.3 Temperature-programmed reduction profiles of (top) 25% Co/SiO$_2$ and (bottom) 15% Co/SiO$_2$, including catalysts calcined using (bold) 5% NO in N$_2$ and (light) standard airflow.

In agreement with the TPR results, the hydrogen chemisorption/pulse reoxidation data provided in Table 8.3 indicate that, indeed, the extents of reduction for the air calcined samples are ~20% higher upon standard reduction at 350°C (compare O$_2$ uptake values). Yet in spite of the higher extent of reduction, the H$_2$ desorption amounts, which probe the active site densities (assume H:Co = 1:1), indicate that the activated nitric oxide calcined samples have higher site densities on a per gram of catalyst basis. This is due to the much smaller crystallite that is formed. The estimated diameters of the activated air calcined samples are between 27 and 40 nm, while the H$_2$-reduced nitric oxide calcined catalysts result in clusters between 10 and 20 nm, as measured by chemisorption/pulse reoxidation.

Turning to the XANES results (Figure 8.4), upon reduction at 350°C, the extent of reduction is found to be higher for the H$_2$-activated air calcined catalysts. This is evident in the shoulder at the edge (~7,709 eV), which is a measure of metallic content, as well as the lower white line intensity for the activated air calcined catalyst at ~7,725 eV. The catalysts appear to contain a combination of mainly Co metal and CoO, in agreement with the interpretation of TPR profiles previously discussed.

TABLE 8.3

Results of Hydrogen Chemisorption/Pulse Reoxidation Measurements over Activated Silica-Supported Cobalt Catalysts Calcined at 350°C Using either Flowing Air or 5% Nitric Oxide in Nitrogen

Catalyst Description	Calcination Type	μmoles H$_2$ Desorbed per g$_{cat}$	Uncorrected Diameter (nm)	Uncorrected % Dispersion	μmoles O$_2$ Consumed per g$_{cat}$	% Reduced	Corrected Diameter (nm)	Corrected % Dispersion
15% Co/SiO$_2$	Air	26.3	49.9	2.07	1,309	77.1%	38.5	2.68
15% Co/SiO$_2$	Air	28.6	46.0	2.24	1,408	83.0%	38.2	2.70
15% Co/SiO$_2$	5% NO	38.1	34.4	3.00	944	55.6%	19.2	5.39
15% Co/SiO$_2$	5% NO	37.7	34.9	2.96	965	55.9%	19.8	5.21
25% Co/SiO$_2$	Air	50.4	43.5	2.37	1,859	65.7%	28.6	3.61
25% Co/SiO$_2$	Air	57.2	38.3	2.70	1,988	70.3%	26.9	3.84
25% Co/SiO$_2$	Air	54.6	40.1	2.57	1,915	67.7%	27.1	3.80
25% Co/SiO$_2$	5% NO	72.1	30.4	3.40	1,442	51.0%	15.5	6.66
25% Co/SiO$_2$	5% NO	84.6	25.9	3.99	1,476	52.2%	13.1	7.85

Note: Catalysts reduced in hydrogen at 350°C for 10 h. Multiple measurements are reported for each sample.

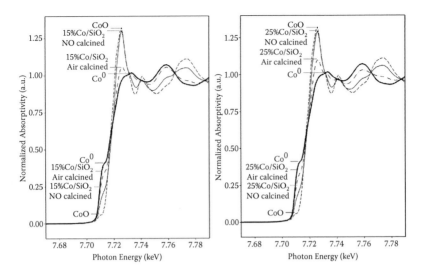

FIGURE 8.4   XANES spectra of normalized absorptivity versus photon energy depicting spectra of H_2-activated Co/SiO$_2$ catalysts calcined using (dash-dotted) air or (thin solid line) nitric oxide. Also, spectra of (dashed line) CoO and (thick solid line) Co metal reference compounds are provided.

As shown in Figure 8.5, the $\chi(k)$ and FT magnitude EXAFS spectra of the reference compounds for Co metal and CoO are very different. While the first coordination shell of cobalt metal has ideally twelve atoms of Co surrounding the absorber, CoO has six atoms of O and twelve atoms of Co, albeit at a farther distance away from the absorber. To fit the EXAFS data of the catalysts, provided in Figure 8.6, a mixed model was constructed relying on not only Co–Co metal coordination, but also the Co–O and Co–Co contributions due to the presence of strongly interacting CoO species that do not reduce at 350°C. As summarized in Table 8.4, the average Co–Co coordination in the metal was significantly higher in the case of the H_2-activated air calcined catalysts than in that of the activated catalysts that were calcined using nitric oxide. The H_2-activated nitric oxide calcined catalysts, on the other hand, exhibited greater coordination to Co–O and Co–Co in the oxide. All of the r-factors were well below 0.02, indicating that the fittings were excellent.

Figure 8.7 demonstrates that the H_2-activated catalysts that were calcined using nitric oxide resulted in higher initial CO conversion rates on a per gram of catalyst basis in a CSTR reactor at 220°C and 280 psig, and using a H_2/CO ratio of 2.5.

Most recently, we have attempted to use this procedure to alter the dispersion of cobalt particles over the more strongly interacting 25% Co/Al$_2$O$_3$ catalyst. However, as shown in Table 8.5, the cluster size was not found to change significantly, and the TPR profiles (not shown for the sake of brevity) were observed

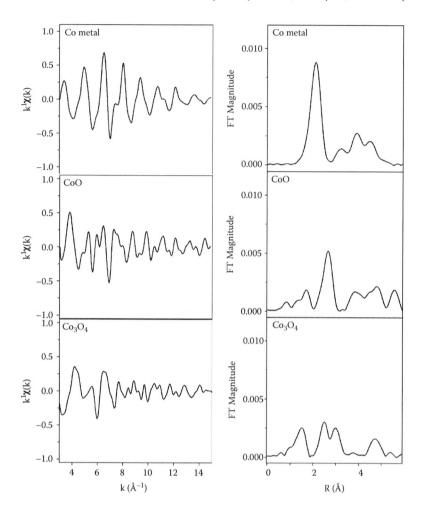

FIGURE 8.5   EXAFS spectra of reference compounds (top) Co metal, (middle) CoO, and (bottom) Co_3O_4, including (left) $k^1 \cdot \chi(k)$ vs. k spectra and (right) corresponding Fourier transform magnitude spectra.

to be quite similar. As shown in Table 8.5, only a slight decrease in the cluster size was observed, and the extents of reduction are quite similar between the activated air calcined and NO calcined catalysts. The findings suggest that, for strongly interacting supports (e.g., Al_2O_3 and TiO_2), the support interaction plays a more important role in governing the resulting Co oxide average cluster size, and therefore, upon reduction, the Co metal dispersion. In the case of Al_2O_3 in particular, the cluster sizes reported for heavily loaded Co/Al_2O_3 catalysts (e.g., 25% Co)[21–23] are found to offer good resistance to certain deactivation phenomena (e.g., H_2O-induced reoxidation and/or sintering), which are problematic especially with noble metal– loaded lower loaded Co/Al_2O_3 catalysts

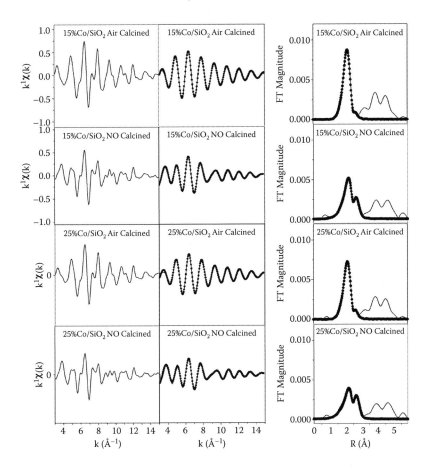

FIGURE 8.6 EXAFS spectra of H_2-activated catalysts, including (top) 15% Co/SiO$_2$ air calcined, (upper middle) 15% Co/SiO$_2$ calcined using 5% nitric oxide, (lower middle) 25% Co/SiO$_2$ air calcined, and (bottom) 25% Co/SiO$_2$ calcined using 5% nitric oxide. Results include (left) $k^1 \times \chi(k)$ vs. k spectra of raw data, (middle column) $k^1 \times \chi(k)$ vs. k spectra of (solid line) filtered data and (circles) EXAFS fitting, and (right) corresponding Fourier transform magnitude spectra of (solid light line) raw data and (circles) EXAFS fitting.

(e.g., Pt-promoted 15% Co/Al$_2$O$_3$), where the noble metal assists in reducing strongly interacting, but smaller, Co oxide species.[24–26] Since smaller Co species are reported to be less resistant to these deactivation processes, it is suggested that, generally speaking, the NO calcination procedure may not be effective in improving catalysts prepared using the more strongly interacting supports. In those cases, adequate Co loading and the addition of reduction promoters (e.g., Pt, Re, and Ru) appear to be effective for increasing active site densities, while ensuring catalyst stability. On the other hand, the NO calcination method appears to be more effective with Co catalysts prepared using more weakly

TABLE 8.4

Results of EXAFS Fitting[a] of H_2-activated Catalysts for Data Acquired Near the Co K Edge

Co Edge	N Co–O	R Co–O (Å)	N Co–Co Metal	R Co–Co Metal (Å)	N Co–Co Oxide	R Co–Co Oxide (Å)	e_0 (eV)	σ^2 (Å²)	r-factor
15% Co/SiO$_2$ Air calcined	**0.62** (0.097)	2.037 (0.0215)	**9.07** (0.301)	2.496 (0.0025)	**1.25** (0.195)	3.019 (0.0143)	0.619 (0.421)	0.00249 (0.000254)	0.00237
15% Co/SiO$_2$ NO calcined	**2.33** (0.176)	2.085 (0.0098)	**5.74** (0.475)	2.509 (0.0056)	**4.66** (0.352)	3.025 (0.0090)	2.142 (0.807)	0.00321 (0.000620)	0.0111
25% Co/SiO$_2$ Air calcined	**0.84** (0.112)	2.043 (0.0180)	**7.47** (0.348)	2.499 (0.0034)	**1.68** (0.224)	3.017 (0.0127)	0.231 (0.576)	0.00246 (0.000356)	0.00457
25% Co/SiO$_2$ NO calcined	**2.44** (0.206)	2.096 (0.0113)	**4.15** (0.466)	2.508 (0.0074)	**4.88** (0.412)	3.021 (0.0098)	2.024 (0.987)	0.00306 (0.000771)	0.0188

[a] Note that $S_0^2 = 0.764$ and $\Delta S_0^2 = 0.0229$.

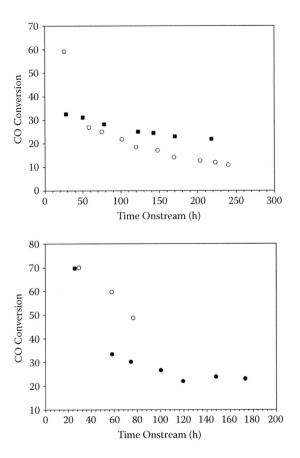

FIGURE 8.7 CO conversion vs. time on stream in the CSTR (220°C, 280 psig, H_2/CO = 2.5) at (squares) 10 $NL/g_{cat}/h$ and (circles) 20 $NL/g_{cat}/h$ for H_2-activated (filled circles) air calcined and (unfilled circles) NO calcined catalysts, including (top) 15% Co/SiO_2 and (bottom) 25% Co/SiO_2. The NO calcined catalysts clearly exhibit higher CO conversion rates on a per gram of catalyst basis.

interacting supports, where air calcination would otherwise result in a large Co cluster size (e.g., some silica and carbon supports).

8.4 CONCLUSIONS

The results confirm that the novel metal nitrate conversion method using nitric oxide in place of air advocated by Sietsma et al. in patent applications WO 2008029177 and WO 2007071899 leads to, after activation in H_2, catalysts with smaller cobalt crystallites, as measured by EXAFS and hydrogen chemisorption/ pulse reoxidation. In spite of the lower extent of cobalt reduction for H_2-activated nitric oxide calcined catalysts, which was recorded by TPR, XANES, EXAFS,

TABLE 8.5

Results of Hydrogen Chemisorption/Pulse Reoxidation Measurements over H_2-activated Alumina-Supported Cobalt Catalysts Calcined at 350°C Using either Flowing Air or 5% Nitric Oxide in Nitrogen

Catalyst Description	Calcination Type	μmoles H_2 Desorbed per g_{cat}	Uncorrected Diameter (nm)	Uncorrected % Dispersion	μmoles O_2 Consumed per g_{cat}	% Reduced	Corrected Diameter (nm)	Corrected % Dispersion
25% Co/Al$_2$O$_3$	Air	98.4	22.2	4.64	1,695	60	13.3	7.7
25% Co/Al$_2$O$_3$	5% NO	97.4	22.5	4.59	1,527	54	11.4	9.0

Note: Catalysts reduced in hydrogen at 350°C for 10 h.

and pulse reoxidation, cobalt surface metal active site densities were found to be considerably higher than those of H_2-activated catalysts prepared using the air calcination procedure. The activated nitric oxide calcined catalysts were found to result in higher CO conversion rates on a per gram of catalyst basis relative to their activated air calcined counterparts. The utility of the method appears to be particularly effective with weakly interacting supports (e.g., SiO_2) and less so with more strongly interacting supports (e.g., Al_2O_3).

ACKNOWLEDGMENTS

This effort was funded by the National Aeronautics and Space Administration (NASA) Grant NNX07AB93A under a project entitled "Basic Studies for the Production and Upgrading of Fischer-Tropsch Synthesis Products to Fuels" and the Commonwealth of Kentucky. This research was carried out, in part, at the National Synchrotron Light Source, Brookhaven National Laboratory, which is supported by the U.S. DOE, Divisions of Materials Science and Chemical Sciences. Special thanks to Dr. Nebojsa Marinkovic (Beamline X18b, NSLS, Brookhaven) for help with XAFS studies and Joel Young (University of Oklahoma, Department of Physics) for XAFS cell construction.

REFERENCES

1. Iglesia, E. 1997. Design, synthesis, and use of cobalt-based Fischer-Tropsch synthesis catalysts. *Appl. Catal.* 161:59–78.
2. Jacobs, G., Das, T.K., Zhang, Y.-Q., Li, J., Racoillet, G., and Davis, B.H. 2002. Fischer–Tropsch synthesis: Support, loading, and promoter effects on the reducibility of cobalt catalysts. *Appl. Catal. A Gen.* 233:263–81.
3. Davis, B.H., and Iglesia, E. 2000. *DOE Quarterly Report*, No. 8, July September.
4. Li, J., Jacobs, G., Das, T.K., Zhang, Y.-Q., and Davis, B.H. 2002. Fischer–Tropsch synthesis: Effect of water on the catalytic properties of a Co/SiO_2 catalyst. *Appl. Catal. A Gen.* 236:67–76.
5. Barbier, A., Hanif, A., Dalmon, J.-A., and Martin, G.A. 1998. Preparation and characterization of well-dispersed and stable Co/SiO_2 catalysts using the ammonia method. *Appl. Catal. A Gen.* 168:333–43.
6. Vada, S., Hoff, A., Adnanes, E., Schanke, D., and Holmen, A. 1995. Fischer-Tropsch synthesis on supported cobalt catalysts promoted by platinum and rhenium. *Topics Catal.* 2:155–62.
7. Schanke, D., Vada, S., Blekkan, E.A., Hilmen, A.M., Hoff, A., and Holmen, A. 1995. Study of Pt-promoted cobalt CO hydrogenation catalysts. *J. Catal.* 156:85–95.
8. Hilmen, A.M., Schanke, D., and Holmen, A. 1996. TPR study of the mechanism of rhenium promotion of alumina-supported cobalt Fischer-Tropsch catalysts. *Catal. Lett.* 38:143–47.
9. Ronning, M., Nicholson, D.G., and Holmen, A. 2001. In situ EXAFS study of the bimetallic interaction in a rhenium-promoted alumina-supported cobalt Fischer–Tropsch catalyst. *Catal. Lett.* 72:141–46.
10. Bazin, D., Borko, L., Koppany, Zs., Kovacs, I., Stefler, G., Sajo, L.I., Schay, Z., and Guczi, L. 2002. Re-Co/NaY and Re-Co/Al_2O_3 bimetallic catalysts: In situ EXAFS study and catalytic activity. *Catal. Lett.* 84:169–82.

11. Kogelbauer, A., Goodwin, Jr., J.G., and Oukaci, R. 1996. Ruthenium promotion of Co/Al$_2$O$_3$ Fischer–Tropsch catalysts. *J. Catal.* 160:125–33.

12. Tsubaki, N., Sun, S., and Fujimoto, K. 2001. Different functions of the noble metals added to cobalt catalysts for Fischer–Tropsch synthesis. *J. Catal.* 199:236–46.

13. Iglesia, E., Soled, S.L., and Fiato, R.A. 1992. Fischer-Tropsch synthesis on cobalt and ruthenium. Metal dispersion and support effects on reaction rate and selectivity. *J. Catal.* 137:212–24.

14. Rygh, L.E.S., and Nielsen, C.J. 2000. Infrared study of CO adsorbed on a Co/Re/-Al$_2$O$_3$-based Fischer–Tropsch catalyst. *J. Catal.* 194:401–9.

15. Sietsma, J.R.A., van Dillen, A.J., de Jongh, P.E., and de Jong, K.P. 2008. Metal nitrate conversion method. PCT International Application WO 2008029177.

16. Sietsma, J.R.A., van Dillen, A.J., de Jongh, P.E., and de Jong, K.P. 2007. Metal nitrate conversion method. PCT International Application WO 2007071899.

17. Sietsma, J.R.A., Meeldijk, J.D., den Breejen, J.P., Versluijs-Helder, M., van Dillen, A.J., de Jongh, P.E., and de Jong, K.P. 2007. The preparation of supported NiO and Co$_3$O$_4$ nanoparticles by the nitric oxide controlled thermal decomposition of nitrates. *Angew. Chem. Int. Ed.* 46:4547–49.

18. Ressler, T. 1997. *WinXAS 97.* Version 1.0.

19. Ravel, B. 2001. EXAFS analysis using FEFF and FEFFIT workshop, June 27.

20. Newville, M., Ravel, B., Haskel, D., Stern, E.A., and Yacoby, Y. 1995. Analysis of multiple-scattering XAFS data using theoretical standards. *Physica B* 154:208–9.

21. van Berge, P.J., van de Loosdrecht, J., Barradas, S., van der Kraan, A.M. 2000. Oxidation of cobalt based Fischer–Tropsch catalysts as a deactivation mechanism. *Catal. Today* 58:321–34.

22. Jacobs, G., Patterson, P. M., Das, T.K., Luo, M.-S., and Davis, B.H. 2004. Fischer–Tropsch synthesis: Effect of water on Co/Al$_2$O$_3$ catalysts and XAFS characterization of reoxidation phenomena. *Appl. Catal. A Gen.* 270:65–76.

23. van de Loosdrecht, J., Balzhinimaev, B., Dalmon, J.-A., Niemantsverdriet, J.W., Tsybulya, S.V., Saib, A.M., van Berge, P.J., and Visagie, J.L. 2007. Cobalt Fischer-Tropsch synthesis: Deactivation by oxidation? *Catal. Today* 123:293–302.

24. Jacobs, G., Das, T.K., Patterson, P.M., Li, J., Sanchez, L., and Davis, B.H. 2003. Fischer–Tropsch synthesis: XAFS studies of the effect of water on a Pt-promoted Co/Al$_2$O$_3$ catalyst. *Appl. Catal. A Gen.* 247:335–43.

25. van Steen, E., Claeys, M., Dry, M.E., van de Loosdrecht, J., Viljoen, E.L., and Visagie, J.L. 2005. Stability of nanocrystals: Thermodynamic analysis of oxidation and re-reduction of cobalt in water/hydrogen mixtures. *J. Phys. Chem. B* 109:3575–577.

26. Storsæter, S., Borg, Ø., Blekkan, E.A., and Holmen, A. 2005. Study of the effect of water on Fischer–Tropsch synthesis over supported cobalt catalysts. *J. Catal.* 231:405–19.

9 Fischer-Tropsch Synthesis and Hydroformylation on Cobalt Catalysts
The Thermodynamic Control

Hans Schulz

CONTENTS

Heterogeneous Fischer-Tropsch (FT) synthesis and homogeneous hydroformylation with cobalt catalysts are investigated for common principles. Selectivity and mechanism of FT synthesis are evaluated and compared with hydroformylation. For FT on cobalt, the concepts of self-organization of the kinetic regime and catalyst surface segregation to attain a thermodynamically controlled state with different sites (on-top sites for chain growth, in-pit sites for CH$_2$ monomer formation from CO, and on-plane sites for methanation and olefin reactions) are presented. It is shown how the regime of FT synthesis shifts toward the regime of hydroformylation.

9.1 INTRODUCTION AND EXPERIMENTAL

During the early work on Fischer-Tropsch synthesis, when experimenting with (heterogeneous) cobalt catalysts and the olefinic FT products, Otto Roelen observed the famous reaction of olefins with CO and H_2 to produce aldehydes.[1] Originally seen as a heterogeneous reaction, it turned out that high performance is achieved in a homogeneous regime, with cobalt dissolved as a carbonyl complex, at higher pressure and lower temperature than for FT synthesis.[2,3] It could be thought that there is a gradual shift from Fischer-Tropsch synthesis to hydroformylation.[4] The mechanism of hydroformylation appears understood today, with its individual steps of the catalytic cycle.[2,3] However, the mechanism of FT synthesis is a matter of ongoing debate.[5]

It appears like a miracle how aliphatic chains (mainly olefins and paraffins) are formed from a mixture of CO and H_2. But *miracle* means only high complexity of unknown order (Figure 9.1). Problems in FT synthesis research include the visualization of a multistep reaction scheme where adsorbed intermediates are not easily identified. Kinetic constants of the elemental reactions are not directly accessible. Models and assumptions are needed. The steady state develops slowly. The *true catalyst* is assembled under reaction conditions. Difficulties with product analysis result from the presence of hundreds of compounds (gases, liquids, solids) and from changes of composition with time.

From high-resolution wide-range gas chromatograms[6,7] basic conclusions are possible (Figure 9.2). There are straight-chain hydrocarbons with carbon numbers C_1 to C_{20} and higher, indicating a polymerization-type reaction. This raises a question concerning the identity of the monomer.

The detailed composition, referring to classes of compounds, is shown for C_6 in Figure 9.3 with and without precolumn hydrogenation. In addition to paraffins, there are olefins—mainly with terminal double bond—and small amounts of alcohols (and aldehydes). The low detection limit of gas chromatography (GC) analysis allows precise determination even of minor compounds and provides exhaustive composition data also for use in kinetic modeling. Because of the short sampling duration of ca. 0.1 s,[8] time-resolved selectivity data are obtained.

An essential method used in this work is quick ampoule sampling of volatiles.[8,9] Small samples of the gaseous reactor effluent (e.g., 1 ml) are recovered in glass ampoules for later analysis. The capillary end of the evacuated ampoule is inserted into the product flow. The capillary tip is broken and the ampoule filled

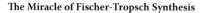

The Miracle of Fischer-Tropsch Synthesis

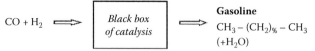

$CO + H_2 \implies$ Black box of catalysis $\implies$ Gasoline $CH_3 - (CH_2)_{96} - CH_3$ $(+H_2O)$

High complexity of unknown order

FIGURE 9.1   General view on Fischer-Tropsch catalysis.

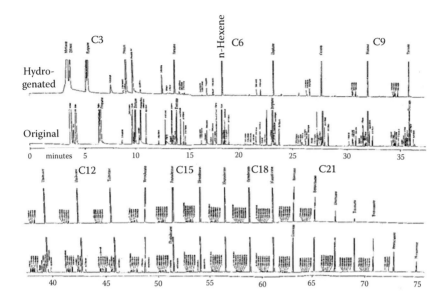

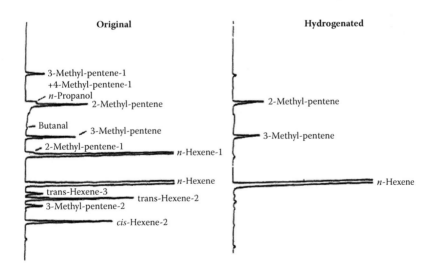

FIGURE 9.2 Gas chromatograms of an FT product, original and precolumn hydrogenated. Ampoule samples taken from the gaseous product flow at ca. 200°C. GC conditions: Capillary, 100 m; d_i, 0.25 mm; d_f, 0.5 μm; methyl silicone cross-linked temperature programm −80 to 270°C; carrier gas H_2; introducing gas N_2; FID.

FIGURE 9.3 C_6 section of the chromatograms in Figure 9.2.

immediately. The ampoule is sealed by fusion of the capillary with a small flame. Figure 9.4 shows the sampling duration in dependence of the capillary diameter. This sampling is the key method for obtaining FT selectivity data—accurately time resolved—as needed for understanding self-organization in FT synthesis. To underline the impact of ampoule sampling, Figure 9.5 demonstrates conventional FT product recovery, as laborious and inaccurate, and temporal resolution would be only in the order of hours. For studying self-organization of the FT regime, an apparatus has been developed, as schematically presented in Figure 9.6.[6,7,9,10]

From the abundant FT literature an important conclusion by Pichler[4,11] is reported in Figure 9.7. It contains the basic understanding from all the famous work of Franz Fischer and coworkers at the Kaiser-Wilhelm-Institute in Muehlheim, today an institute of the Max Plank society. Each of the catalysts—Ni, Co, Fe, and Ru—has its own operating range of pressure and temperature. With Ni, Co,

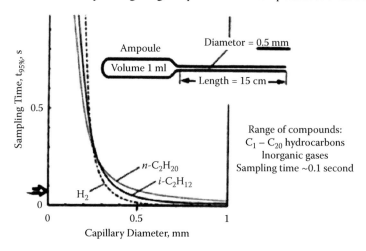

FIGURE 9.4 Quick ampoule sampling of volatiles. Ninety-five percent ampoule filling time as a function of capillary diameter for 3 compounds. Calculation for filling through consecutive Knudsen diffusion into a vacuum, super sonic flow, and laminar flow.

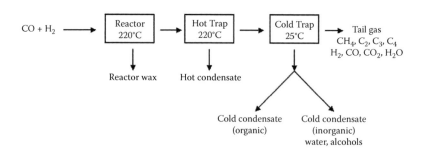

FIGURE 9.5 Conventional product recovery in FT synthesis.

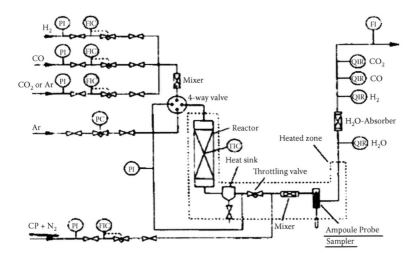

FIGURE 9.6   Apparatus for time resolution kinetic in FT-measurements. Quick ampoule sampling, internal references, accurate control of flows and pressure, precise zero time, accurate mixing of flows, and catalyst powder on inert particles.

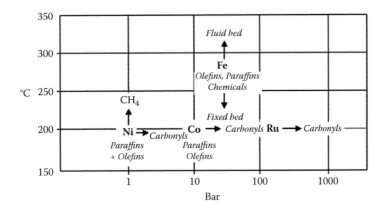

FIGURE 9.7   Operating ranges of the catalysts nickel, cobalt, iron, and ruthenium in FT synthesis as indicated by Pichler.

and Ru, the maximum temperature is 200 to 220°C, because at higher temperature methane selectivity is excessive. With increasing pressure another limit is noticed. This is of interest in relation to the topic of this article: the catalyst metals begin to react with CO to form metal carbonyls—with nickel above ca. 1 bar, cobalt above ca. 30 bar, and ruthenium above more than 100 bar. Because of the strong interaction of CO with the metal surface, Pichler proposed "surface complexes" as intermediates of FT synthesis.[12] It can be concluded that the metals are finally converted to carbonyl complexes, which are optimal catalysts for hydroformylation.

It is also indicated in Figure 9.7 that iron behaves differently. With iron as catalyst the temperature can be increased up to 350°C without excessive methane formation (Sasol fluid bed process for low molecular weight olefin and gasoline production[13] and former U.S. Hydrocol process[14]). Here, promoting with alkali is essential.[15] Increasing the pressure, and specifically the partial pressure of CO, leads to carbon formation on the catalyst—and not to iron-carbonyl formation. Much understanding had been achieved already by the pioneers of this area of catalysis—Franz Fischer, Otto Roelen, and Helmut Pichler (Figure 9.8).

9.2 ACTIVITY IN DEPENDENCE OF TIME ON STREAM

As a general phenomenon, observed already by Fischer and coworkers, activity and FT synthesis selectivity develop in the initial time of a run in a process of *Formierung* (formation)[16]—in modern terms *self-organization* and *catalyst restructuring*. In order to achieve high performance of synthesis with cobalt as catalyst, the temperature had to be raised slowly up to the temperature of steady-state conversion. A distinct thermodynamically controlled state of the Co surface, populated with reactants and intermediates, can be assumed. This state depends on temperature and particularly on CO partial pressure, and its catalytic nature changes with changing conditions.

An example of activity developing with a Co catalyst is shown in Figure 9.9 (right). CO-conversion (respectively the yield of products) increases with time by a factor of about 10, from ca. 4% to ca. 55%.[7,17] Figure 9.9 (left) shows the time dependence of FT with an iron catalyst. There are a strong initial carbon deposition (referring to iron carbide formation) and fast water gas shift reaction, and FT

Franz Fischer
(1877–1947)
Famous professor in Fuel Chemistry
Director Coal-Research Institute at Muehlheim (1914–1943)
Invention of **"gasoline synthesis"** (at normal pressure)
together with **Hans Tropsch** in 1925

Otto Roelen
(1897–1993)
Olefin-hydroformylation on Cobalt-FT-catalysts in 1938
Head of Research at the Ruhrchemie company until 1962
Early co-worker of Franz Fischer

Helmut Pichler
(1904–1974)
FT-synthesis at medium pressure on Co and Fe
Polymethylene on ruthenium
Co-worker of Franz Fischer, FT-research at KWI until 1946
In the USA (Bureau of Mines and HRI) 1946–1956
Professor in Fuel Chemistry, University of Karlsruhe 1956–1974

FIGURE 9.8   Pioneers of Fischer-Tropsch and Oxo synthesis.

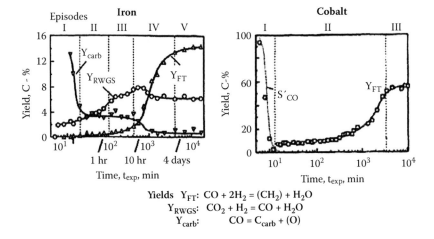

FIGURE 9.9   Self-organization of Fischer-Tropsch synthesis with iron and cobalt as catalysts. Yields of carbon containing compounds as a function of time on stream t_{exp}. Catalysts and conditions: *With iron:* 100Fe-13Al-11Cu-9K (by weight), prepared by precipitation, 250°C, 10 bar, H_2/CO_2 = 3:1. The feed gas composition allows slow catalyst carbiding and fair discrimination of the episodes of reconstruction. *With cobalt:* 100Co-13Zr-0.66Ru-100SiO_2 (Aerosil), prepared by precipitation, 190°C, 10 bar, H_2/CO = 2:1.

activity develops as a function of time, when sufficient carbon has accumulated on the catalyst and reacted with iron to form iron carbide.[18]

9.3   CONCEPT OF COBALT SURFACE SEGREGATION CAUSING FT ACTIVE SITES GENERATION

In recent literature, images with Angstrom-scale resolution of cobalt metal surfaces before and after FT synthesis (obtained by surface tunneling electron microscopy) have been reported. These show surface restructuring through FT synthesis,[19] being addressed as formation of the "true catalyst."[20] The observed features of this process are the following: Crystallite planes (0001) are segregated to small islands of fairly uniform diameter (1.75 nm), the depth of segregation being one atomic step (0.205 nm). It can be assumed that this restructuring is caused by CO adsorption. It has been concluded that restructuring approaches the thermodynamic equilibrium.[20] For catalysis on the segregated surface this would mean:

*   Increase of the number of surface atoms
*   Disproportionation of on-plane sites to such of lower coordination (on-top sites) and higher coordination (in-pit sites)

These sites must be considered to have different catalytic properties. The sites on top (Co atoms of low coordination) would be similar to that of the central atom of cobalt carbonyl complexes, and reactions on these should be similar to those in hydroformylation. Specifically, insertion reactions between π- and σ-ligands (CO

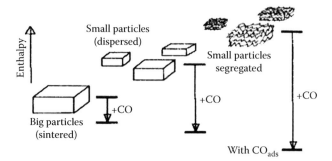

FIGURE 9.10 Thermodynamic view of cobalt surface segregation in the presence of CO (and H_2). Activating catalyst restructuring under Fischer-Tropsch conditions.

or C_2H_4 or CH_2 with R or H) are expected. On the in-pit sites with near-neighbor atoms, CO dissociation would be favored (see Ponec et al.[21] for dissociative adsorption of CO on nickel). Co-surface segregation, as a principle of formation of the true Co FT catalyst,[20] is pictured in Figure 9.10. The new structure then represents the equilibrium between more CO molecules adsorbed (this producing additional free enthalpy of adsorption) against enlarging the specific Co surface (consuming free enthalpy).

It can be expected that, if a highly dispersed, poorly ordered active catalyst has been prepared, and if the reaction conditions (T, p_{CO}) refer to low-equilibrium segregation, a deactivation will also proceed initially through *reverse segregation*.

Excessive segregation—at high pressure and low temperature—then leads to carbonyl complexes.

In commercial Fischer-Tropsch synthesis, catalyst performance is modified by support and promoter interaction[22] to control segregation and stabilize dispersion.

9.4 BASIC KINETIC MODEL OF FT SYNTHESIS

The basic approach of kinetic modeling of FT synthesis is to assume ideal polymerization (Figure 9.11). Such product composition can be described by only one number, the value of probability of chain prolongation p_g. The assumptions would be that p_g is independent of carbon number (Nc) of the growing species (Sp), and that there is only one kind of product compound. With this model the curve of p_g over Nc is just a horizontal line. Calculating such curves from experimental product compositions can be used to characterize the deviations from ideal behavior[17] (Figure 9.12.). In the present example, the deviations are a low value at Nc = 1, referring to high methane selectivity; a high value at Nc = 2, referring to the high reactivity of ethene for readsorption for growth[23–25]; and increasing values in the range of Nc = 5 to Nc = 9, indicating an increasing probability of growth, this being caused by increasing readsorption resulting from increasing reactor residence time.[7,24,25]

Typical results of growth probability as a function of carbon number for FT synthesis on iron and cobalt[7,26]—both for different times on stream during catalyst

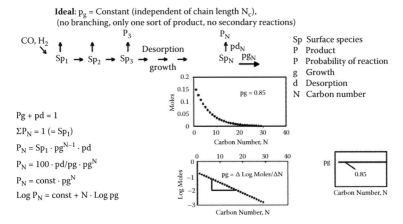

FIGURE 9.11 Ideal polymerization model of Fischer-Tropsch synthesis and ideal Fischer-Tropsch product composition.

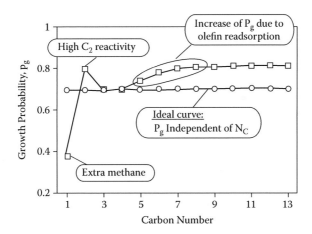

FIGURE 9.12 Ideal and real chain growth. Growth probability as a function of carbon number.

formation—are shown in Figure 9.13. With the cobalt catalyst, the shapes of the curve shows deviations from ideal polymerization, as discussed with Figure 9.12. Depending on time, growth probability is initially low (first hour of experiment, p_g ca. 0.6). It increases with time (during, e.g., 100 h) to p_g ca. 0.8. This means the polymerization nature develops under reaction conditions, when the cobalt catalyst is being reconstructed.

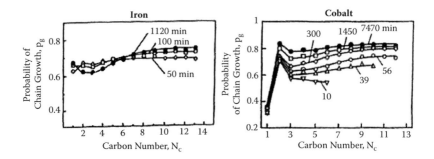

FIGURE 9.13   Chain prolongation probability in dependence of the length of the growing surface species N_C for different times on stream. Left: Iron catalyst. Right: Cobalt catalyst: 100Co-11Zr-0.45Pt-100SiO$_2$ (Aerosil), 5 bar.

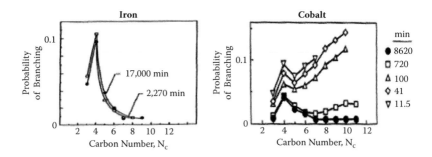

FIGURE 9.14   Chain branching probability. Co-catalyst with 0.15% wt. Pt.

9.5   FT CHAIN BRANCHING

In the extended kinetic model of nontrivial surface polymerization, a branching reaction to form a methyl side group at the aliphatic chain has been proposed.[27] Typical results[10,18] are presented in Figure 9.14. With cobalt,[10] the curve of branching probability over carbon number at steady state shows a strong decline with the size of the chain. This is interpreted as a steric effect, assuming the branching reaction to be demanding in space. The first value at Nc = 3 is exceptionally low, a result attributed (Wojciechowski[28]) to particular spatial demands: hindrance of desorption after branching. Because a tertiary C atom is involved, a bulky transition state of desorption has to be assumed.

The values of branching probability with cobalt are in general more than two times higher in the beginning of an experiment than at steady state. It is concluded that initially the spatial constraints on the growth sites are lower than at steady state. This also indicates a change in the nature of the growth sites.

A further phenomenon is noticed. Initially (during about the first hour), the shapes of the curve shows a different pattern, as exhibiting a minimum. A strong increase of branching probability in the range Nc = 5 to Nc =10 is observed. It follows that initially the spatial constraints at the growth sites are not fully established. The readsorption of alpha-olefins with their second C atom will then be possible, and growth leads to a methyl-branched chain. Olefin readsorption in general is favored by the chain length of the olefin (along with the increased reactor residence time).

Branching reactions appear to be a unique indication of the existence of spatial constraints at growth sites. Analogies between homogeneous and heterogeneous catalysis are pertinent.

Iron as catalyst for FT synthesis does not exhibit the dynamic characteristics observed with cobalt[26] (Figure 9.14, left).

9.6   FT OLEFINS

By extending the FT model to the formation of two kinds of products—olefins and paraffins—and including secondary olefin reactions, the kinetic schemes shown in Figure 9.15 are obtained. In parallel primary reactions (from the growth sites), paraffins and alpha-olefins are desorbed—by irreversible associative desorption (the paraffins) and by dissociative desorption (the olefins) (upper scheme in Figure 9.15).

Olefins react secondarily for isomerization and hydrogenation (on cobalt sites that are not active for chain growth; lower scheme in Figure 9.15). There is a *first reversible H-addition* (at the alpha- or beta-C-atom of the double bond) to form an alkyl species, and a *slow irreversible second H-addition* to form the paraffin (lower scheme in Figure 9.15). Thus, double-bond shift and double-bond hydrogenation are interrelated by a common intermediate to produce olefins with internal double bonds or paraffins from the primary FT alpha-olefins. Experimental results[10,18] are presented in Figures 9.16 and 9.17.

On cobalt, the typical shape of the curve of olefin content vs. carbon number fractions in Figure 9.16 shows a low value at Nc = 2, indicating the high reactivity of ethene for hydrogenation and growth. The olefin content declines from Nc = 3

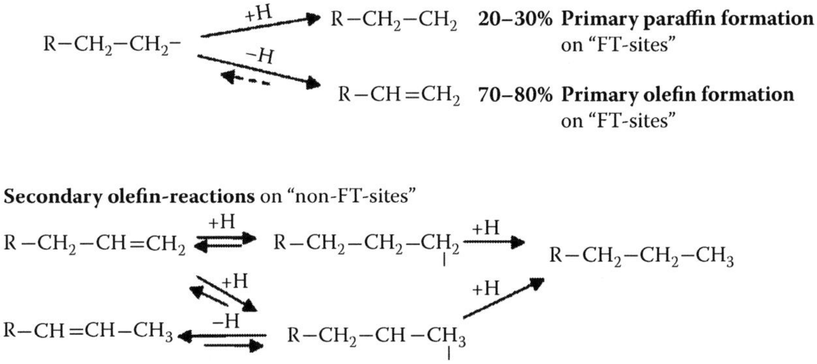

$R-CH_2-CH_2-$ $\xrightarrow{+H}$ $R-CH_2-CH_2$ **20–30% Primary paraffin formation**
on "FT-sites"

$\xrightarrow{-H}$ $R-CH=CH_2$ **70–80% Primary olefin formation**
on "FT-sites"

Secondary olefin-reactions on "non-FT-sites"

$R-CH_2-CH=CH_2 \xrightleftharpoons{+H} R-CH_2-CH_2-CH_2 \xrightarrow{+H} R-CH_2-CH_2-CH_3$

$R-CH=CH-CH_3 \xleftarrow{-H} R-CH_2-CH-CH_3 \xrightarrow{+H}$

FIGURE 9.15   Kinetic schemes of olefin formation and reaction.

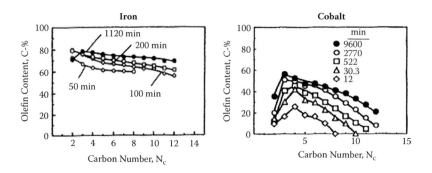

FIGURE 9.16 Olefin contents in carbon number fractions: primary and secondary olefin selectivity.

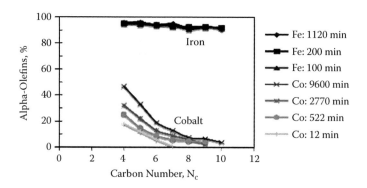

FIGURE 9.17 Contents of α-olefins in carbon number fractions as referring to secondary double-bond shift.

with increasing carbon number of the olefins because of increasing reactivity and longer reactor residence time of the olefins. In the early time of the run (when the FT regime is being established), the olefin content is particularly low. In view of ongoing surface segregation, this reflects more on-plane sites.

The shapes of the curves for short time on stream (12–30 min; Figure 9.16, right) differ principally from the common shape by exhibiting a lower value at Nc = 3 than at Nc = 4. The plain explanation is that at the actual high degree of secondary hydrogenation, the relative concentration of olefins with internal double bonds is high (see Figure 9.17). This means that at C_4 and higher, the concentrations of the more reactive alpha-olefins are low. But at C_3 there are only two identical alpha-olefin isomers possible. Thus, olefin hydrogenation at C_3 is more extensive than at olefins with higher carbon numbers.

With iron, almost no secondary olefin double-bond shift and double-bond hydrogenation are possible (see Figures 9.16, left, and 9.17). Consequently, in systems, where iron carbide is the true FT catalyst, no sites for olefin secondary reactions are present.

9.7   SHIFT OF THE FT REGIME AT INCREASING PRESSURE

Increasing pressure strongly affects performance of FT synthesis[17,29] (Figure 9.18). In the present case, the increase from 1.2 to 30 bar causes the reaction rate to increase by a factor of 5. The methane selectivity declines from 15% to 7%, and the olefin content in the C_3 fraction increases from 20% to 65% C. The concept of activation through surface segregation is hereby strongly supported, the segregation being more extensive at higher pressure. More FT sites (on-top for growth, in-pit for CH_2 formation) are generated, and more on-plane sites (active in methanation and secondary olefin reactions) are consumed. The decline of olefin hydrogenation and olefin double-bond shift with increasing pressure[29] is shown in Figure 9.19.

Increased spatial constraint effects on growth sites are evident from Figure 9.20. This figure shows the contents of individual monomethyl-branched compounds in carbon number product fractions. Note that their values decline by a factor of 5 to

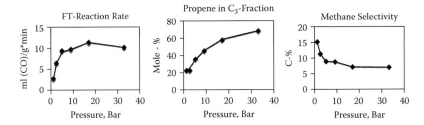

FIGURE 9.18   Influence of pressure on reaction rate, olefin content in the C_3 fraction, and methane selectivity with cobalt as the catalyst for FT synthesis. Catalyst: $100Co$-$18ThO_2$-100 SiO_2 (Kieselguhr), $H_2/CO = 1.8$, $175°C$.

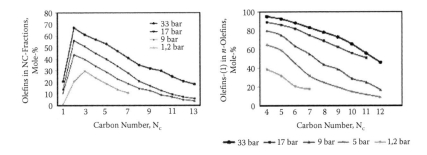

FIGURE 9.19   Influence of pressure on the secondary olefin reactions of hydrogenation (left) and isomerization (right). Further legend as in Figure 9.18.

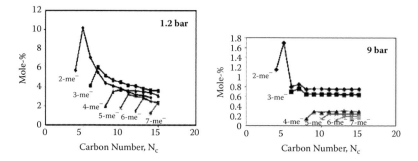

FIGURE 9.20 Methyl-branched isomers in carbon number fractions at 1.2 bar (left) and 9 bar (right) during FT synthesis on cobalt as indications for spatial constraints on the growth sites. Further legend as in Figure 9.18.

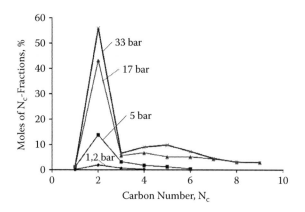

FIGURE 9.21 Alcohol distributions at different pressures of FT synthesis on cobalt. Further legend as in Figure 9.18.

10 when the pressure increases from 1.2 to 9 bar.[29] The selectivity of the isomers with methyl groups more inside the chain is even more reduced. With $N_c = 9$, the value of 2-methyl octane decreases from 5% C at 1.2 bar to 0.7% C at 9 bar (a factor of 0.14), but the value of 4-methyl octane decreases from 4% C to 0.2% C (a factor of 0.05).

9.8 FT ALCOHOLS

Increasing pressure affects alcohols selectivity (Figure 9.21). The contents of alcohols in product carbon number fractions generally increase, and the shape of the distribution curve changes, exhibiting a second maximum at higher reaction pressure.[29] At $N_c = 2$ the alcohol content increases drastically from ca. 3 to ca. 55 mole% when the pressure is raised from 1.2 to 33 bar.

It is of interest to relate the formation of alcohols to the basic reactions of the kinetic scheme. For this reason, co-feeding of 1 vol% of propene was performed[29] (Figure 9.22). An increase of butanol selectivity was observed. The interpretation is that propene has reacted with CO via CO insertion, as known in hydroformylation. The oxygen remains in the species, and the species does not grow. Generalizing this finding, it is concluded that with cobalt as FT catalyst, the oxygenates (aldehydes, alcohols) are formed by CO insertion. However, this is not the common FT reaction of chain growth, because the obtained acyl species does not grow but must desorb.

To explain the specifically high selectivity of ethanol (Figure 9.22), it can be assumed that CO insertion with the small CH_3 species is preferred if compared with bigger alkyl species. The reaction is favored at high pressure, because of stronger surface segregation, shifting the regime from Fischer-Tropsch synthesis toward hydroformylation. Hydroformylation with FT olefins causes the maximum in the alcohol distribution at Nc = 4. This refers to the high olefin content in the FT C_3 fraction.

The kinetic scheme in Figure 9.23 pictures the alternative reaction possibilities of the alkyl species (here the CH_3 species) on a growth site to react either with CO for hydroformylation or with CH_2 for FT synthesis.

9.9   CH₂ AS MONOMER IN FT SYNTHESIS

Among the indications for CH_2 to be the main monomer in FT synthesis, a result from co-feeding ^{14}C-labeled n-hexadecene-(1)-(1-^{14}C) during FT synthesis on cobalt[17] is being recalled. Figure 9.24 presents the relative molar radioactivity of the product compounds as a function of their carbon number. Evidently, the labeled olefin has reacted for growth, as indicated by the constant molar radioactivity of compounds bigger than C_{16}. This means the expected reaction of

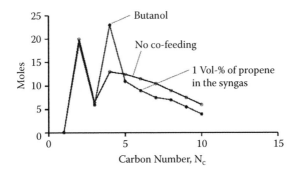

FIGURE 9.22   Influence of co-feeding propene with the synthesis gas on the alcohol distribution during FT synthesis on cobalt. T = 178°C, P = 30 bar, H_2/CO = 2, CO conversion = 40%.

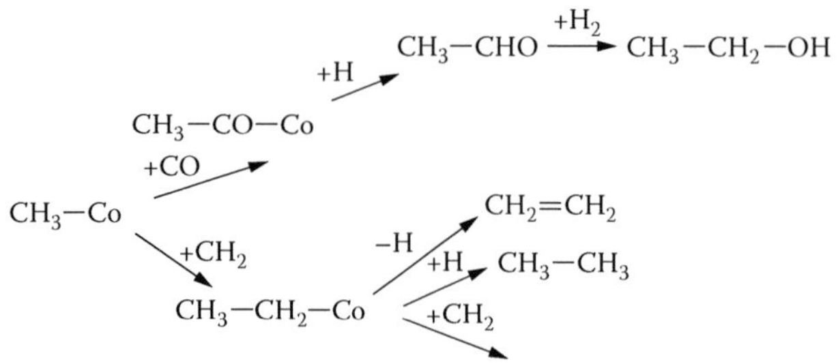

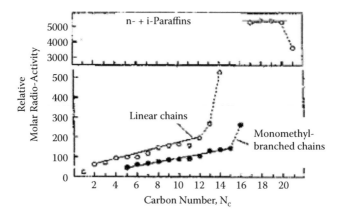

FIGURE 9.23 Alternative chain propagation with CO or CH_2.

FIGURE 9.24 Molar radioactivity of the reaction products when co-feeding n-hexa-decene-(1)-(1-^{14}C) together with the synthesis gas to FT synthesis on cobalt. Catalyst: 100Co-18ThO$_2$-100Kieselguhr (by weight), precipitated, 190°C, 1 bar, H$_2$/CO = 2, X$_{CO}$ = 70%, 0.1 vol% of hexadecene-^{14}C in the synthesis gas.

adsorption of the olefin on a growth site and indicates reversibility of FT olefin desorption from growth sites.

Surprisingly, the compounds smaller than C_{16} are also radioactive. Their molar radioactivity increases linearly with carbon number. This indicates a carbon number constant probability to incorporate the ^{14}C-labeled C atom. The radioactive alpha-C-atom is split from the C_{16}-olefin and reacts as monomer for FT chain prolongation. Because the probability to react with the ^{14}CH$_2$ is the same for each growth step, the molar radioactivity of the product compounds increases proportionally with their carbon number. The general mechanistic conclusion for FT synthesis is that the observed splitting of the C_1 atom from the alkyl chain is the reverse of chain growth with a CH_2 species. This indicates reversibility of chain growth by CH_2 insertion.

A sketch of the FT mechanism, involving a surface intermediate with carbon number N, is presented in Figure 9.25. There are:

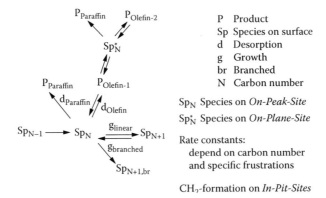

Kinetic Scheme/Alternative Reactions

P   Product
Sp   Species on surface
d   Desorption
g   Growth
br   Branched
N   Carbon number

Sp_N Species on *On-Peak-Site*
Sp_N^* Species on *On-Plane-Site*

Rate constants:
   depend on carbon number
   and specific frustrations

CH_2-formation on *In-Pit-Sites*

FIGURE 9.25   Basic kinetic scheme of FT hydrocarbon formation on cobalt.

1. Sites for growth (on-top sites) and the alkyl intermediate (Sp_N) bound on them
2. Sites for secondary reactions (on-plane sites) and alkyl species (Sp_N^*) bound on them
3. Two desorption reactions, for paraffins and for olefins, the latter one being reversible
4. Reversibility of the growth reaction

The sites for secondary reactions are not capable of chain growth.

9.10   CONCLUSIONS

Comparing heterogeneous Fischer-Tropsch synthesis with homogeneous olefin hydroformylation can be seen as a source for understanding catalytic principles, particularly because the selectivity is complex and therefore highly informative. Reliable analytical techniques must be readily available.

It is concluded that with cobalt the regime of FT synthesis is established through self-organization and catalyst restructuring. Surface segregation is visualized to form on-top sites for growth and in-pit sites for CO dissociation. The steady state of the FT regime with cobalt is dynamic and thought to be thermodynamically controlled, thus depending on experimental parameters as temperature and CO partial pressure.

Advanced insight is obtained from temporal resolution of activity and selectivity during formation of the catalytic regime. Applying a polymerization model, it is deduced how the ratio of chain growth to chain desorption, or of linear to branched growth, or of desorption as olefin to that as paraffin, varies with time and reaction conditions. Specifically, reaction pressure and reaction temperature are relevant, shifting the catalytic regime from heterogeneous to homogeneous catalysis. The trends of this shift are characterized in Figure 9.26 as follows:

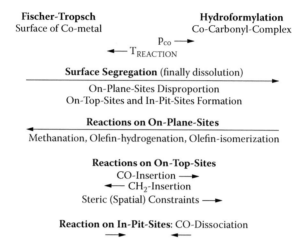

FIGURE 9.26   Fischer-Tropsch synthesis vs. Oxo synthesis on cobalt catalysts. The thermodynamically controlled shift from heterogeneous to homogeneous catalysis.

1. Active sites generation
2. CO or CH_2 insertion
3. Desorption after CO insertion
4. Spatial constraints on the growth sites
5. Methanation and olefin hydrogenation or isomerization (secondary and side reactions)

Progress in understanding FT reaction mechanisms shall be useful for theoretical calculations, catalyst design, and reaction engineering.

REFERENCES

1. O. Roelen. 1938. DRP: 849.548.
2. A. Behr. 2003. Organometallic complexes and homogeneous catalysis. In *Ullmann's encyclopaedia of industrial chemistry*, 429. 6th ed., Vol. 24. Weinheim: Wiley-VHC.
3. R.F. Heck, D.S. Breslow. 1961. *J. Am. Chem. Soc.* 83:4023.
4. H. Pichler. 1952. Twenty-five years of synthesis of gasoline by catalytic conversion of carbon monoxide and hydrogen. In *Advances in catalysis*, ed. W. Frankenburg, E. Rideal, V. Komarewsky. Vol. 4:271. New York: Academic Press.
5. H. Schulz. 2007. *Stud. Surf. Sci. Cat.* 163:127.
6. E. van Steen. 1993. Dissertation, University of Karlsruhe, Karlsruhe.
7. M. Claeys. 1997. Dissertation, University of Karlsruhe, Karlsruhe.
8. H. Schulz, K. Beck. Unpublished results.
9. H. Schulz, W. Böhringer, C. Kohl, N. Rahman, A. Will. 1984. Entwicklung und Anwendung der Kapillar-GC-Gesamtprobentechnik fuer Gas/Dampf-Vielstoffgemische, DGMK-Forschung: Hamburg.

10. Zh. Nie. 1996. Dissertation, University of Karlsruhe, Karlsruhe.
11. H. Pichler. 1952. *Brennstoff-Chemie* 33.
12. H. Pichler, H. Buffleb. 1940. *Brennstoff-Chemie* 21:273.
13. A. Steynberg, R. Espinoza, B. Jager, A. Vosloo. 1999. *Appl. Catal. A Gen.* 186:41.
14. A.W. Weitkamp et al. 1953. *Ind. Eng. Chem.* 45:343, 350, 539, 363.
15. M. Dry. 1981. The Fischer-Tropsch synthesis. In *Catalysis*, ed. J. Anderson and M. Boudart, 159. Berlin: Springer-Verlag.
16. H. Pichler. Personal communication.
17. H. Schulz. 2003. *Topics Catal.* 26:73.
18. T. Riedel, H. Schulz, G. Schaub, K. Jun, J. Hwang, K. Lee. 2003. *Topics Catal.* 26:41.
19. J. Wilson, G. de Groot. 1995. *J. Phys. Chem.* 99:7860.
20. H. Schulz, Zh. Nie, F. Ousmanov. 2002. *Catal. Today* 71:351.
21. V. Ponec, W.L. van Dijk, J.A. Groenewegen. 1976. *J. Catal.* 45:277.
22. S.L. Soled, E. Iglesia, R.A. Fiato, J.E. Baumgartner, H. Vroman, S. Miseo. 2003. *Topics Catal.* 26:101.
23. H. Schulz, B.R. Rao, M. Elstner. 1970. Erdoel und Kohle: Erdgas. *Petrochemie* 23:65.
24. H. Schulz, M. Claeys. 1999. *Appl. Catal. A Gen.* 186:71.
25. H. Schulz, M. Claeys. 1999. *Appl. Catal. A Gen.* 186:91.
26. H. Schulz, G. Schaub, M. Claeys, T. Riedel. 1999. *Appl. Catal. A Gen.* 186:215.
27. H. Schulz, K. Beck, E. Erich. 1988. Fischer-Tropsch CO-hydrogenation, a non trivial surface polymerization: Selectivity of branching. In *Proceedings of the 9th International Congress on Catalysis*, Calgary, Vol. 2, p. 829.
28. B.W. Wojciechowski. 1988. *Catal. Rev. Sci. Eng.* 30:629.
29. S. Roesch. 1980. Dissertation, University of Karlsruhe, Karlsruhe.

10 The Value of a Two Alpha Model in the Elucidation of a Full Product Spectrum for Fe-LTFT

Johan Huyser, Matthys Janse van Vuuren, and Godfrey Kupi

CONTENTS

In this chapter a two α selectivity model is proposed that is based on the premise that the total product distribution from an Fe-low-temperature Fischer-Tropsch (LTFT) process is a combination of two separate product spectrums that are produced on two different surfaces of the catalyst. A carbide surface is proposed for the production of hydrocarbons (including n- and iso-paraffins and internal olefins), and an oxide surface is proposed for the production of light hydrocarbons (including n-paraffins, 1-olefins, and oxygenates) and the water-gas shift (WGS) reaction. This model was tested against a number of Fe-catalyzed FT runs with full selectivity data available and with catalyst age up to 1,000 h. In all cases the experimental observations could be justified in terms of the model proposed.

10.1 INTRODUCTION

The determination of an accurate product spectrum for the Fischer-Tropsch (FT) reaction is not trivial because there is not an analytical technique available to accurately quantify a product sample consisting of hydrocarbons and oxygenates with carbon numbers ranging from C1 to C100+. In the case of a continuously stirred tank reactor (CSTR), the product exits the reactor in either the gas or liquid phase, with the heavier products condensed in two separate catch pots. For a full product spectrum, the four different samples (tail gas, water, oil, and wax) are analyzed separately and combined for the construction of a full product spectrum to represent the catalyst produced during the sample period. Based on this product spectrum, a selectivity model was developed and used to gain an understanding of the factors influencing selectivity changes in a low-temperature Fischer-Tropsch (LTFT) process.

10.2 METHOD

There are several different reactor types available for generating kinetic and selectivity data. A properly mixed CSTR is used because of the accurate control of reaction conditions (temperature and reagent partial pressures) to which the catalyst is exposed. To account for all the products that leave the reactor, the contents of both knockout pots and the tail gas need to be analyzed and combined to give the total product spectrum. Because the product spectrum changes with catalyst age, it is important to isolate the products the catalyst produced within the sample period. The wax sample that leaves the reactor as a liquid, due to dilution in the reactor slurry, will represent the average of the heavy product made during the previous few days and will not represent the product made during the sampling period. The lighter products that leave the reactor as a vapor and condense in a cold pot that is drained once during the sampling period will represent the products made during the sampling period, and this difference of sampling periods to represent the full product spectrum signifies a real challenge. For the same reason, the wax analysis of a reaction in the first few days will not be representative of the products made by the catalyst but will be contaminated with the medium used to start up the reaction.

To avoid these negative effects, results were recently reported[1] from reactions done where an FT run, at any selected age, was stopped and the catalyst placed under inert conditions. This was followed by the replacement of the reactor contents by flushing out with degassed and dried poly-alpha-olefin oil (Durasyn P164) and restarting the reaction again in order to determine a full product spectrum. The freshly produced FT products that are free of accumulation can then be considered as real products produced at that particular period with a catalyst of defined age. These analyses can then be used to construct a true product spectrum of the catalyst with a specific age. The complete product spectrum showing the contribution of the four different gas chromatography (GC) analyses is shown in Figure 10.1.

The description of the product distribution for an FT reaction can be simplified and described by the use of a single parameter (α value) determined from the Anderson-Schulz-Flory (ASF) plots. The α value (also called the chain growth probability factor) is then used to describe the total product spectrum in terms of carbon number weight fractions during the FT synthesis. In the case

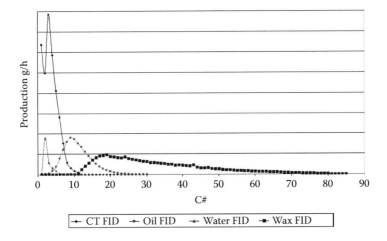

FIGURE 10.1   Complete product spectrum showing contributions from the four separate analyses.

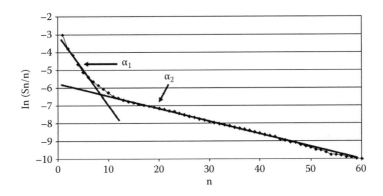

FIGURE 10.2   ASF plot for a typical iron-catalyzed LTFT process.

of a precipitated iron catalyst in an LTFT process, the ASF plot does not give a straight line, as shown in Figure 10.2.

As indicated in Figure 10.2, there is a distinct change in the slope of the line at carbon numbers 8 to 12, and this has also been observed by other researchers.[2,3] This change in the slope cannot be explained by the ASF model, which is based on the premise that the chain growth probability factor (α) is independent from the carbon number. Some further developments of the ASF model by Wojciechowski et al.[3] made use of a number of abstract kinetic parameters for the calculation of a product spectrum. Although it still predicts a straight line for the α plot, they suggested that the break in the line is due to different mechanisms of chain termination and could be explained by the superposition of two ideal distributions. This bimodal distribution explained by two different mechanisms has also been proposed by other researchers.[4] Other authors[5] have suspected the existence of two different growth sites, with each yielding ideal distributions with different chain growth probabilities. Another explanation was recently proposed by Botes[6] in which the main assumption was that the rates of chain growth and hydrogenation (termination step for paraffins) are independent of carbon chain length, but that the rate of desorption (the termination step for olefins) is a function of carbon number. This model allows us to explain the exponential decrease in the olefin/paraffin ratio and the gradual increase of the α value with increasing carbon number until a limiting value is reached.

Another explanation for the changing slope has been proposed by Schulz and Claeys,[7] who suggest that the product olefins undergo secondary reactions and, because of changing product olefin solubility, result in chain length dependence on the chain growth probability (α).

The approach proposed in this report is similar to those proposed independently by Satterfield and Gaube[5] in a sense that the total product spectrum is a combination of two distinct sets of products produced as a result of either two different mechanisms, for instance, two different reactive intermediates,[4] or two different catalytic surfaces,[5] each producing a different product spectrum.

If we assume that α1 and α2 are the results of two distinct sets of products, then it is imperative to first separate the contribution of each in terms of mass or molar production before determining the α1 value. This is because the total product spectrum for C1 to C10 (typically used for the calculation of α1) will be a combination of the two product spectra and will give an erroneous α1 value. Since we are interested in not only the paraffin and olefin production but also the oxygenates, we use molar production. A typical molar production of a standard LTFT run is shown in Figure 10.3.

If the product distribution follows a perfect ASF product distribution, then the graph in Figure 10.3 should fit an exponential function. However, this is not the case and can only be done if the production is separated in two product spectra, one being described by α1 and consisting mostly of <C15, and the other one C1–C100+, with the products >C15 being mostly from the spectrum described by α2.

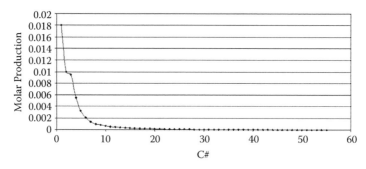

FIGURE 10.3   Molar production of standard LTFT reaction.

Separation of the two product spectra is done by fitting an exponential function to the total molar production graph for the carbon number higher than 15 and then extrapolating down to C1. This exponential function would then describe the molar production of the α2 contribution and could just be subtracted from the total production to give the contribution described by α1. Figure 10.4 shows the exponential function fitted to the >C15 part of the total molar production shown in Figure 10.3. Subtracting the α2 contribution from the total production will give the α1 production, and the individual molar production is shown in Figure 10.5.

This technique allows us to separate the total product spectrum into an α1 and an α2 contribution. This also highlights the point that for this technique to give accurate results, the product spectrum needs to be determined very accurately, especially in the region that is used to calculate α2. Small variations in the products measured will skew the α2 determination and result in the inaccurate product separation. This only highlights the fact that care should be taken with

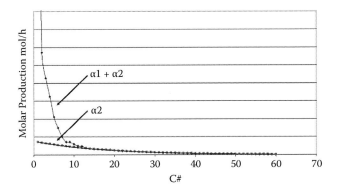

FIGURE 10.4   Molar production with an exponential function fitted to the C >15 part of the graph and extrapolated down to C1.

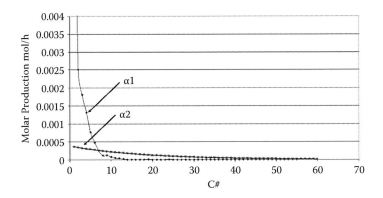

FIGURE 10.5   Molar production of α1 and α2.

the interpretation of analysis and the calculation of α values. Before we can specify which products represent the $\alpha1$ and $\alpha2$ product spectrum, it is first necessary to have a closer look at the catalyst surface.

10.3 A CLOSER LOOK AT THE CATALYST SURFACE

For the development of a selectivity model it is helpful to have a picture of the surface of the catalyst to fit the explanation of how the product spectrum is formed. The fundamental question regarding the nature of the active phase for the FT and water-gas shift (WGS) reactions is still a controversial and complex topic that has not been resolved.[8] Two very popular models to describe the correlations between carbide phase and activity are the carbide[9] and competition models.[10] There are also proposals that magnetite and metallic iron are both active for the FT reaction and carbides are not active[11]. These proposals will not be discussed in detail and are only mentioned to highlight the uncertainty that is still present on the exact phase or active site responsible for the FT and WGS reactions.

For a precipitated iron catalyst, several authors propose that the WGS reaction occurs on an iron oxide (magnetite) surface,[12,13] and there are also some reports that the FT reaction occurs on a carbide surface.[14] There seems to be a general consensus that the FT and WGS reactions occur on different active sites,[13] and some strong evidence indicates that iron carbide is active for the FT reaction and that an iron oxide is active for the WGS reaction,[15] and this is the process we propose in this report. The most widely accepted mechanism for the FT reaction is surface polymerization on a carbide surface by $*CH_2$ insertion.[16] The most widely accepted mechanism for the WGS reaction is the direct oxidation of CO with surface $*O$ (from water dissociation).[17] Analysis done on a precipitated iron catalyst using bulk characterization techniques always shows iron oxides and iron carbides, and the question of whether there can be a sensible correlation made between the bulk composition and activity or selectivity is still a contentious issue.[18]

Because of all these uncertainties and the questionable relationships found between bulk composition and activity, or even selectivities, we followed a different approach in order to gain an understanding and formulate a theory on surface composition responsible for FT and WGS reactivity. A logical sequence of catalyst surface transformations based on scientific principles was proposed and adapted to form a general model that can be used to explain our experimental observations. These proposed catalyst transformations will be discussed in a little more detail below.

After calcinations, the precipitated iron catalyst is composed of a mixture of iron oxide phases before activation. The exact nature of this phase is not critical for the discussion and will be referred to in general as an oxide phase. During activation the catalyst is subjected to a reducing environment that will lead to the formation of either metallic iron if pure hydrogen is used or some iron carbide if the reduction is done with either CO or syngas. During reduction with a gas containing CO there is a transformation of an iron oxide phase to an iron carbide phase.[13] The density of these types of phases is given in Table 10.1.

During the reduction with the conversion of an iron oxide to iron carbide there is a substantial increase in density (see values in Table 10.1). The formation of a smooth carbide layer over an oxide bulk can therefore not proceed without the creation of a void or shattering of the carbide layer because of the density mismatch between the two phases. This shrinking core type of reduction is therefore very unlikely. Because of the Cu and K promoters present and evenly distributed in the bulk, the formation of a large number of small iron carbide regions will lead not only to more complete carburization (lower the diffusion path of oxygen and limit the negative effects of the density mismatch between the two phases) but also to a high concentration of small carbide islands on a bulk oxide phase.[19] Considering only the surface of the catalyst containing iron, it will be very difficult to predict the percentage oxide and carbide, but what is sure is that both will be present on the surface of the catalyst, most probably small carbide islands covering most of the oxide surface.

To better understand the difference in reactivity behavior of these carbide and oxide surfaces, it is helpful to look at the charge distribution among iron, carbon, and oxygen. The difference in the electronegativity of the atoms in a complex can be used to classify it as being either covalent or ionic. The electronegativity of some atoms is given in Table 10.2. The carbide phase, consisting of Fe–C bonds, will have a nonpolar, covalent character ($\Delta X_{Fe-C} = 0.72$), and the oxide phase, consisting of Fe–O bonds, will have a polar, ionic character ($\Delta X_{Fe-O} = 1.61$) with a more prominent positive charge on the iron than in the case of the carbide phase.

TABLE 10.1
Density of Selected Iron Carbide and Oxide Phases[20]

Iron Oxide	Iron Carbide
Fe_2O_3 density, 5.25 g/cm^3	Fe_3C density, 7.7 g/cm^3
Fe_3O_4 density, 5.17 g/cm^3	Fe_5C_2 density, 7.57 g/cm^3
$Fe(O)OH$ density, 4.26 g/cm^3	Fe_2C density, 7.19 g/cm^3
	Fe_7C_3 density, 7.61 g/cm^3

TABLE 10.2
Electronegativity Values of Selected Elements[20]

Elements	Electronegativity (X)
Fe	1.83
O	3.44
C	2.55

10.4   CORRELATION OF SURFACE PROPERTIES
WITH PRODUCT SELECTIVITIES

As discussed above, the product spectrum of an iron-catalyzed LTFT reaction can best be described by using a two α model with two separate product spectra added together to give the total product distribution. For the production of paraffins (see discussions above), it is reasonable to assume it is produced on the carbide surface since this is where the CH_2* monomer responsible for chain growth is most abundant. For the oxygenates it is a little more challenging, since they can be explained as being formed on either the carbide or oxide surface. There are, however, a few clues that we can use to favor one of the surfaces. If the oxygenates were formed on the carbide surface by reacting of the growing chains with the very active surface O* species, we would expect the product spectrum of the types of oxygenates (alcohols and acids) to follow the paraffin distribution in terms of α values. Even though the oxygenate production is a lot lower than that of the paraffins, both termination steps would be expected to be similarly affected by the chain length, and so the α values of paraffin and oxygenate should be similar. However, this is not the case, and a much more realistic explanation is that they are actually formed on two different surfaces. Since the oxide surface, active for WGS, mostly consists of O*/OH*, it is a logical conclusion that this is where oxygenates are formed.

Following the discussions above, the product spectra were separated using the following assumptions. The carbide (nonpolar) surface is responsible for the production of paraffins and olefins, while the oxide (polar) surface is responsible for the production of light hydrocarbons, olefins, and oxygenates. A proposed mechanism of olefin formation is shown in Figure 10.6. The consequence of this is that the formation of branched paraffins and internal olefins must be linked due to the common intermediate. Considering this reaction (Figure 10.6) on a nonpolar (carbide) surface, it is realistic to assume that reaction A will be far closer to equilibrium than the same reaction on a polar (oxide) surface. The reason is the much stronger coordination of a nonpolar molecule to a nonpolar surface than of the same nonpolar molecule to a polar surface. The more electronegative oxygen atom will lower the electron density on the iron in an oxide environment, lowering the back donation from the iron d-orbital to the antibonding π molecular orbital, resulting in a weaker bond that will leave the surface as a 1-olefin. This means that on a nonpolar surface (carbide surface) the formation of internal olefins and branched paraffins will be favored compared to a polar surface, which will prevent equilibrium A from being established and the alpha olefin will leave the surface as soon as it is formed.

The approach described above to calculate a true α2 value as well as to separate the α1 and α2 product distributions was followed to test if this description could be used to account for the selectivity results observed in the iron-catalyzed LTFT reaction. The assumptions for the calculations are given below:

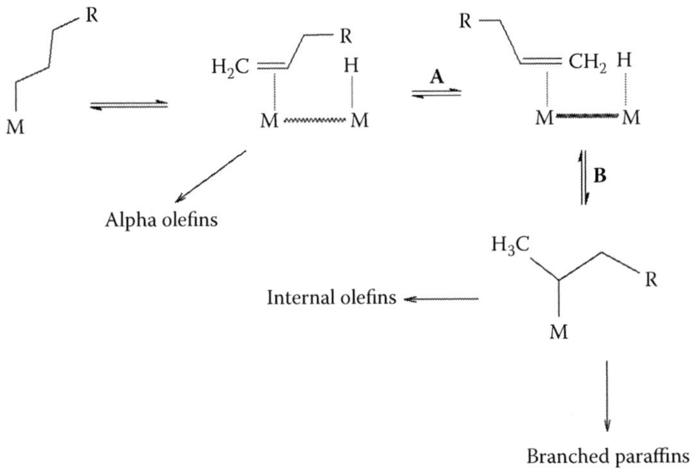

FIGURE 10.6 Mechanism for the formation of branched paraffins and internal olefins.

- In the total product spectrum the contribution of $\alpha 1$ is negligible in the carbon range higher than C15.
- There are two different surfaces (polar or oxide and nonpolar or carbide) responsible for the formation of FT products.
- One of these surfaces is also responsible for the WGS reaction (polar or oxide surface).
- The WGS reaction as well as n-paraffins, 1-olefins, and oxygenates are formed on the polar surface.
- Paraffins (normal and branched) and internal olefins are formed on the nonpolar surface.

These assumptions might not be 100% correct, and a more correct statement will, for instance, be that the bulk of the 1-olefins are formed on the polar surface but will make separation of products impossible, so the assumption was made that all the 1-olefins are formed on the polar surface. These assumptions should not change the conclusions reached.

Following the reasoning above, the product spectrum was separated with the stated assumptions. For the carbide surface the molar productions of the n- and iso-paraffins and internal olefins were added together and a growth function fitted for the carbon number range higher than C15 (C30+). This growth function was extrapolated to include C1 up to C60. However, there was still a small deviation from the function in the lower carbon range, and since paraffins are formed on both the polar and nonpolar surfaces, it was assumed that this deviation is due to the paraffins formed on the polar surface. This deviation was subtracted from the n-paraffin production that left the molar production of internal olefins and n- and iso-paraffins. The molar production of the remaining products is then the production on the polar (oxide) surface. The total molar production of the products formed on the two surfaces can then just be added together and expressed as a percentage. From this, a few interesting observations can be made from the product distribution and the WGS reaction. Because the WGS reaction also takes place on the polar (oxide) surface, there should be a correlation between the molar production corresponding to the $\alpha 1$ surface and the CO_2 selectivity. This correlation is shown in Figure 10.7 from the plot of $\alpha 1$ mole% vs. CO_2 mole %. It is also expected that since most of the CH_4 is produced on the oxide (polar) surface (same as the WGS), there should be a correlation between the CH_4% selectivity and the WGS or CO_2% selectivity. This is shown in Figure 10.8 for a catalyst at different ages.

Another way of comparing the CH_4 and CO_2 selectivities that will reflect changes in the surface on which the reactions take place is to look at the CH_4 formation rate expressed as mole CO converted to CH_4/g catalyst/second, and the WGS reaction rate expressed as mole CO_2 formed/g catalyst/second. Because the units only differ in specifying the products made, the number of active sites must be the same for CH_4 and CO_2 formation *only if* there is a direct correlation between the two rates. The two rates are plotted in Figure 10.9 and clearly show that there is a very strong correlation. It must be kept in mind, however, that strictly speaking the parameters that can influence the CH_4 and CO_2 formation rates to a different degree were not constant for the duration of the run; specifically the H_2 and CO partial pressures are known to influence the CH_4 and CO_2 reactions differently, so the correlation is expected to fluctuate a bit. Another interesting observation is in the selectivity of the products formed on the oxide surface. This product selectivity corresponds to the $\alpha 1$ production and is shown in Figure 10.10.

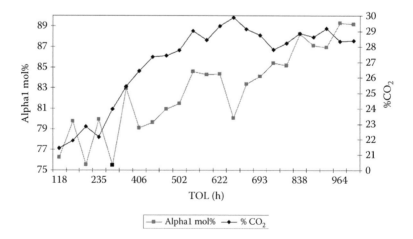

FIGURE 10.7 $\alpha 1$ mole% vs. CO_2 mole% selectivity.

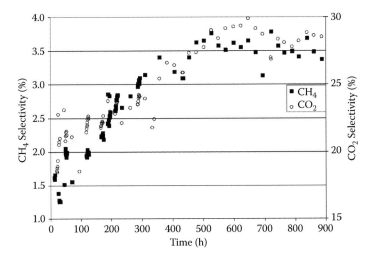

FIGURE 10.8 CO_2 vs. CH_4 selectivity.

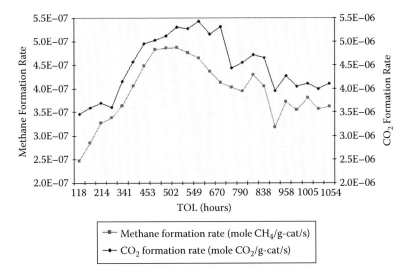

FIGURE 10.9   Comparison of the CH_4 and CO_2 formation rates.

Keeping in mind that the molar production in Figure 10.10 is a combination of three FID GC (Flame Ionization Detector) results and dependent on six parameters, it is remarkable that a good exponential fit is possible and a consistent α1 value can be calculated. What is also immediately apparent is that the region from C1 to C4 deviates only slightly from the exponential fit. Since reincorporation of C2 products (mostly ethylene and ethanol) is known to occur, the C2 negative deviation corresponds remarkably to the sum of the C3 and C4 positive

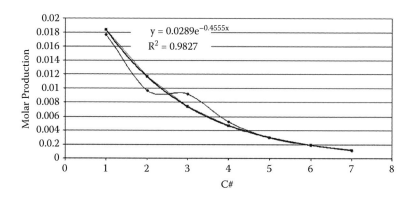

FIGURE 10.10   Exponential function fitted for C1–C7 to the molar production.

deviations, indicating a possibility of C2 being transformed into C3 and C4 products explaining these deviations. This is in accordance with similar work with co-feeding olefins.[4] The C1 deviation (although small) can be related to an error in the combination of at least two FID GC traces and a thermal conductivity detector (TCD) GC trace, making the C1 fraction difficult to determine accurately. (Some methanol reincorporation might also play a role.)

10.5   CONCLUSIONS

A selectivity model was proposed based on a theory that the product spectrum consists of two separate products formed on two different catalytic surfaces. Chemical reactivity arguments were used to show that the two surfaces (carbide and oxide) are responsible for the production of different groups of products. Separation of the product spectrum in this way enabled us to explain and sometimes predict changes in selectivity and gave us a handle on tracking the full selectivity changes by following the CO_2 and CH_4 selectivity.

REFERENCES

1. G. Kupi, M. Janse van Vuuren, E. Jordaan, K. Wilken, and J. Huyser. 2007. Poster presentation at CATSA conference, Richardsbay South Africa, November 11.
2. R.B. Anderson. 1956. In *Catalysis*, ed. P. Emmet, 1. Vol IV. New York: Reinhold. A. Rautavuoma and H. van der Baan. 1981. *Appl. Catal.* 1:247. A. Takuchi and J. Katzer en G. Schuit. 1984. *J. Catal.* 82:3271. J. Baker and A.T. Bell. 1982. *J. Catal.* 78:165. R. Dictor and A.T. Bell. 1986. *J. Catal.* 97:121.
3. R. Taylor and B.W. Wojciechowski. 1984. *Fuel Proc. Technol.* 8:135. R. Taylor and B.W. Wojciechowski. 1983. *Can. J. Chem. Eng.* 61:98. B.W. Wojciechowski. 1986. *Can. J. Chem. Eng.* 64:149. B.W. Wojciechowski. 1988. *Catal. Rev.-Sci. Eng.* 30:629. B. Sarup and B.W. Wojciechowski. 1984. *Can. J. Chem. Eng.* 62:249. R. Taylor and B.W. Wojciechowski. 1988. *Can. J. Chem. Eng.* 66:831. N.O. Egiebor, W.C. Cooper, and B.W. Wojciechowski. 1985. *Can. J. Chem. Eng.* 63:826.

4. J. Patzlaff, Y. Liu, C. Graffmann, and J. Gaube. 1999. *Appl. Catal. A Gen.* 186:109. J. Patzlaff, Y. Liu, C. Graffmann, and J. Gaube. 2002. *Catal. Today* 71:381. P. Biloen and W.M.H. Sachtler. 1981. *Adv. Catal.* 30:165. M.E. Dry. 1990. *Catal. Today* 6:183.
5. L. König and J. Gaube. 1983. *Chem.-Ing. Tech.* 55:14. G. Huff Jr. and C.N. Satterfield. 1984. *J. Catal.* 85:370.
6. F. G. Botes. 2007. *Energy & Fuels* 21:1379.
7. H. Schulz and M. Claeys. 1999. *Appl. Catal. A Gen.* 186:91.
8. D.B. Bukur, X. Lang, and Y. Ding. 1999. *Appl. Catal. A Gen.* 186:255–75.
9. J.A. Amelse, J.B. Butt, and L.J. Schwartz. 1978. *J. Phys. Chem.* 82:558. G.B. Raupp and W.N. Delgass. 1979. *J. Catal.* 58:361.
10. J.W. Niemandsverdriet and A.M. van der Kraan. 1981. *J. Catal.* 72:385.
11. J.P. Reymond, P. Meriadeau, and S.J. Teichner. 1982. *J. Catal.* 75:39.
12. D.S. Newsome. 1980. *Catal. Rev.-Sci. Eng.* 21:275. H.-B. Zhang and G.L. Schrader. 1985. *J. Catal.* 95:325. D.G. Rethwisch and J.A. Dumesic. 1986. *J. Catal.* 101:35.
13. E.S. Lox and G.F. Froment. 1993. *Ind. Eng. Chem. Res.* 32:71. K.R.P.M Rao, F.E. Huggins, V. Mahajan, G.P. Huffman, V.U.S. Rao, B.L. Bhatt, D.B. Bukur, B.H. Davis, and R.J. O'Brien. 1995. *Top. Catal.* 2:71.
14. R.J. O'Brien, L. Xu, R.L. Spicer, and B.H. Davis. 1996. *Energy & Fuels* 10:921–26.
15. G.P. van der Laan and A.C.M. Beenackers. 2000. *Appl. Catal. A Gen.* 193:39–53.
16. B.W. Wojciechowski. 1988. *Catal. Rev.-Sci. Eng.* 30:4629. M.E. Dry and J.C. Hoogendoorn. 1981. *Catal. Rev.-Sci. Eng.* 23:265. H. Schulz, E. van Steen, and M. Claeys. 1993. *Selective hydrogenation and dehydrogenation.* Kassel, Germany: DGMK.
17. S. Krishnamoorthy, A. Li, and E. Iglisia. 2002. *Catal. Lett.* 80:77–86.
18. R.B. Anderson. 1984. *The Fischer-Tropsch synthesis.* Orlando, FL: Academic Press. M.E. Dry. 1981. In *Catalysis—Science and technology,* ed. J.R. Anderson and M. Boudart, 159–255. Vol. 1. New York: Springer-Verlag. R.J. O'Brien, L. Xu, D.R. Milburn, Y.-X. Li, K.J. Klabunde, and B.H. Davis. 1995. *Top. Catal.* 2:1.
19. S. Li, G.D. Meitzer, and E. Iglisia. 2001. *J. Phys. Chem. B* 105:5743–50.
20. D. R. Lide, ed. CRC *handbook of chemistry and physics.* 2007. Boca Raton, FL: Taylor & Francis. 4–43; 9–77.

11 Studies on the Reaction Mechanism of the Fischer-Tropsch Synthesis
Co-Feeding Experiments and the Promoter Effect of Alkali

Johann Gaube and Hans-Friedrich Klein

CONTENTS

The readsorption and incorporation of reaction products such as 1-alkenes, alcohols, and aldehydes followed by subsequent chain growth is a remarkable property of Fischer-Tropsch (FT) synthesis. Therefore, a large number of co-feeding experiments are discussed in detail in order to contribute to the elucidation of the reaction mechanism. Great interest was focused on co-feeding CH_2N_2, which on the catalyst surface dissociates to CH_2 and dinitrogen. Furthermore, interest was focused on the selectivity of branched hydrocarbons and on the promoter effect of alkali on product distribution. All these effects are discussed in detail on the basis of a recently proposed reaction mechanism characterized by the superposition of two incompatible mechanisms resting exclusively on $-CH_2-$ and on CO insertion.

11.1 INTRODUCTION

Since the invention of the Fischer-Tropsch (FT) synthesis numerous studies have been concerned with its mechanism. However, this subject still remains controversial. Only rare case studies were focused on co-feeding of compounds such as alcohols, aldehydes, 1-alkenes, or diazomethane. But most of these investigations were undertaken in order to support one of the specific hypotheses that were favored at that time. Unfortunately, these informative co-feeding experiments were hardly taken into consideration for discussion of more recent mechanistic hypotheses.

An outstanding example is the excellent and comprehensive work of Emmett and coworkers in the 1950s,[1-4] where ^{14}C-labeled alcohols, aldehydes, and ethene were co-fed to the syngas in Fischer-Tropsch synthesis over iron and cobalt catalysts in order to study the buildup of hydrocarbons of higher carbon numbers. These authors have already shown that for iron catalysts, the incorporation of alcohols, and their activity to start chains, is much higher than that of ethene. They found that in the case of co-feeding of ethane, the ^{14}C activity of formed hydrocarbons decreases with increasing carbon number, while for alcohol co-feeding, the ^{14}C activity increases and finally approaches a constant value at high carbon number hydrocarbons in the fraction of wax. Furthermore, this series of articles presents a wealth of experimental results that are useful for current discussion of the mechanism of FT synthesis.

As a refinement of the well-known studies of Brady and Petit,[5] Quyoum and coworkers were able to show, by co-feeding of [13]CH_2N_2 to syngas [12]CO/H_2, that the [13]CH_2 intermediates react in the same way as surface [12]CH_2 groups formed from [12]CO/H_2, leading to random incorporation into the formed hydrocarbons.[6] If hydrocarbons were formed exclusively via CH_2 insertion, a constant fraction of incorporated ^{13}C would be expected in all hydrocarbons. However, the experiments revealed a strong decrease of ^{13}C incorporation with increasing carbon number, an observation that was not commented on by the authors. This result shows without a doubt that the mechanism characterized by CH_2 insertion is not the only one operating in Fischer-Tropsch synthesis.

The carbon number distribution of Fischer-Tropsch products on both cobalt and iron catalysts can be clearly represented by superposition of two Anderson-Schulz-Flory (ASF) distributions characterized by two chain growth probabilities and the mass or molar fraction of products assigned to one of these distributions.[7-10] In particular, this bimodal-type distribution is pronounced for iron catalysts promoted with alkali (e.g., K_2CO_3). Comparing product distributions obtained on alkali-promoted and -unpromoted iron catalysts has shown that the distribution characterized by the lower growth probability α_1 is not affected by the promoter, while the growth probability α_2 and the mass fraction f_2 are considerably increased by addition of alkali.[9] This is a further indication of two different and incompatible mechanisms governing the Fischer-Tropsch synthesis.

Based on these observations and several other experimental results with co-feeding of ethene and 1-alkene,[9] the selectivity of branched hydrocarbons,[11] and the different promoter effects of Li-, Na-, K-, and Cs-carbonate/oxide,[12,13] a novel mechanism has been proposed that is consistent with these various experimental results.[14] The formulation of this mechanism follows the knowledge of analogous reactions in homogeneous catalysis and gives a detailed insight in the crucial step of C–C linkage formation. The aim of this work is to discuss in detail these experiments and their relationship to the proposed mechanism.

11.2   THE BIMODAL CARBON NUMBER DISTRIBUTION

With the exception of methane and the C_2 fraction, the carbon number distribution of Fischer-Tropsch products can be well represented by superposition of two ASF distributions:

$$S_{Ci,exp.}/i = S_{Ci,1}/i + S_{Ci,2}/i = A \cdot \alpha_1^{i-1} + B \cdot \alpha_2^{i-1}$$

The carbon selectivity S_{Ci} of hydrocarbons of carbon number i is defined by the mass of carbon in the components related to the mass of carbon of all hydrocarbons in the 3 to 40 carbon number range. The slopes of the straight lines in the diagram $\log(S_{Ci}/i)$ vs. the carbon number i give α_1 and α_2. In Figures 11.1 and 11.2 the open symbols represent $S_{Ci,1}/i$ of distribution 1, which is the difference of

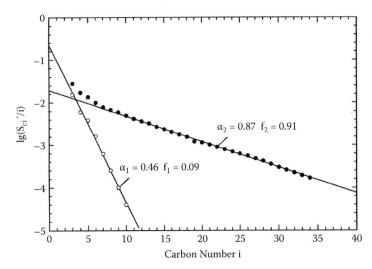

FIGURE 11.1   Bimodal ASF distribution obtained on Co catalyst. T = 493 K, p_{H2} = 3 bar, p_{CO} = 1.5 bar, pure unsupported cobalt.[15] (From J. Gaube, H.-F. Klein, *J. Mol. Catal. A. Chemical* 283, 2008. pp. 60–88. Elsevier. With permission.)

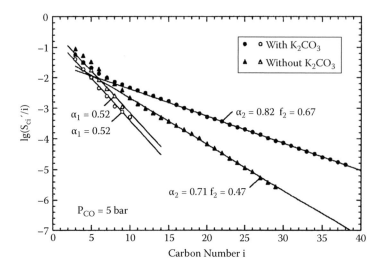

FIGURE 11.2 Promoter effect of K_2O/K_2CO_3 on ASF distribution. Iron catalysts Fe(0) and Fe(2 wt% K_2CO_3), T = 493 K, P_{H2} = 5 bar, p_{CO} = 5 bar.[15]

the experimentally obtained value $S_{Ci,exp}/i$ and the extrapolated value for distribution 2, $S_{Ci,2}/i$. A more detailed description is given by Patzlaff et al.[9] and Huff and Satterfield.[10]

For the experiment of Figure 11.1 the degree of syngas conversion was kept at a low level, 0.2–0.3, so that the partial pressures of 1-alkenes also remained low. Under these conditions incorporation of 1-alkenes is negligible. As discussed in Section 11.2, for a high degree of conversion, a deviation from this strict bimodal ASF distribution is observed due to incorporation of these compounds and their subsequent chain growth.

For iron catalysts in general, the incorporation of 1-alkenes is negligible, and that of ethene is much lower than that of cobalt.[16,17] Therefore, for all published carbon number distributions for iron catalysts, a strict representation by two superimposed ASF distributions is obtained. Examples are given by Schliebs and Gaube,[7] Dictor and Bell,[8] and Huff and Satterfield.[10] Also, the old experiments of the Schwarzheide tests are well represented by this model.[7]

The conclusion of this strict representation is that if two reaction mechanisms are responsible for this bimodal distribution, these mechanisms have to be incompatible.[9]

11.3   CO-FEEDING EXPERIMENTS

11.3.1   CO-FEEDING OF DIAZOMETHANE

It has been assumed that CH_2N_2 dissociates on the catalyst surface to N_2 and CH_2, producing methylene units as active surface intermediate for chain growth of hydrocarbons. Quyoum and coworkers have shown that these CH_2 intermediates react in the same way as surface CH_2 groups formed from CO/H_2, leading to random incorporation into the formed hydrocarbons. A very interesting result of these experiments is the strong decrease of ^{13}C incorporation with increasing carbon number.[6]

Figure 11.3 shows the fraction of hydrocarbons with incorporated ^{13}C and the ratio of hydrocarbons formed according to distribution 1 and the total of hydrocarbons formed. The decrease of both curves is the same. The conclusion is drawn that hydrocarbons assigned to distribution 1 are formed via insertion of CH_2.

The same conclusion has already been drawn on the basis of co-feeding experiments with ethene and 1-alkene, as discussed in detail in the following section.

11.3.2   CO-FEEDING OF ETHENE AND 1-ALKENES

Detailed studies by Patzlaff et al.[9,18] have shown that addition of ethene causes an increased fraction f_1 of the distribution characterized by α_1 and a small increase of α_1. This indicates that ethene mainly acts as a chain initiator of hydrocarbons formed according to distribution 1, and to a very small extent as a surface intermediate for insertion into a growing chain. Concurrent experimental results were obtained by Schulz and Claeys.[19] Distribution 2 and also α_2 are not affected by co-feeding of ethene. Figure 11.4 shows that ethene changes the ASF plot only in the range of low carbon numbers.

The carbon number distribution of the synthesis run with co-fed 1-hexene shows an increased C selectivity of the C_7 fraction (Figure 11.5). However, the increase of the following fractions strongly declines with increasing carbon number, so that the distribution approaches, within a few carbon numbers, the one obtained without co-feeding 1-hexene. This result suggests that readsorbed and

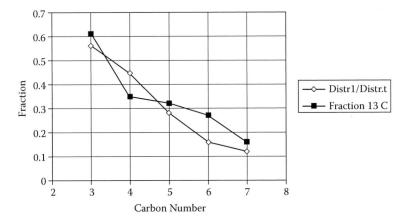

FIGURE 11.3   Cobalt catalyst (for reaction conditions, see Figure 11.1) fraction distribution 1/distribution total fraction ^{13}C incorporation.

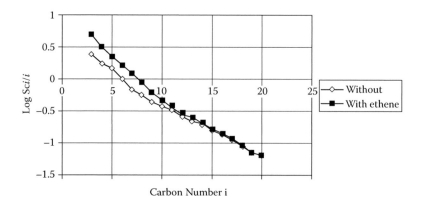

FIGURE 11.4 Effect of ethene addition. Co catalyst, T= 493 K, p_{CO} = 0.1 MPa, p_{H2} = 0.1 MPa.[18] (From J. Patzlaff, Y. Liu, C. Graffmann, and J. Gaube, *Catal. Today* 71, 2002. pp. 381–394. Elsevier. With permission.)

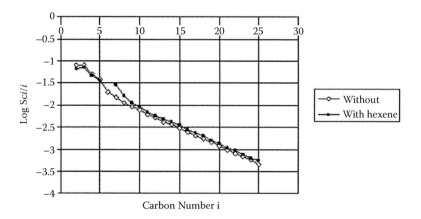

FIGURE 11.5 Effect of 1-hexene addition. Co catalyst, T= 493 K, p_{CO} = 0.1 MPa, p_{H2} = 0.1 MPa.[18]

finally incorporated 1-alkenes grow according to the mechanism characterized by the lower growth probability α_1. The experiments with co-feeding of 1-hexene at low p_{CO}, as presented in Figure 11.5, have revealed a low fraction of incorporated 1-hexene (about 0.1), while most co-fed 1-hexene is hydrogenated. At elevated p_{CO} the degree of incorporation reaches relatively high values up to 0.5.[19,20] This result can be interpreted by the decrease of secondary 1-alkene hydrogenation with increasing p_{CO}, as demonstrated in Figure 11.6.

The kinetics of secondary hydrogenation and isomerization of 1-alkenes as represented by the reaction scheme is characterized by a negative reaction order with respect to carbon monoxide.[13,15]

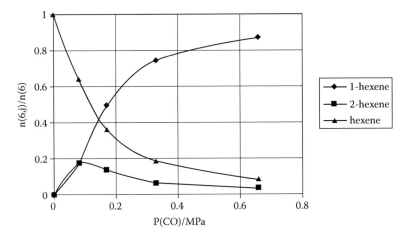

FIGURE 11.6   Dependence of hydrogenation and isomerization of 1-hexene on p_{CO} pure cobalt catalyst. $T = 493$ K, $p_{H2} = 0.33$ MPa, $m_{Cat} = 7.42$ g Fe, $j_{1\text{-hexene}} = 0.0024$ mol/h.[15] (From J. Gaube and H.-F. Klein, *Catal. A. General* 350, 2008. pp. 126–132. Elsevier. With permission.)

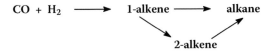

Hydrogenation of 1-alkene: $r_1 = k_1 \cdot p_{H2} \cdot c_{1-en} \cdot p_{CO}^{-1.5}$

Isomerization of 1-alkene: $r_2 = k_2 \cdot c_{1-en} \cdot p_{CO}^{-1.5}$

Hydrogenation of 2-alkene: $r_3 = k_3 \cdot p_{H2} \cdot c_{2-en} \cdot p_{CO}^{-1.5}$

Therefore, for elevated p_{CO} it is expected that with increasing residence time in a fixed bed reactor, a marked incorporation of 1-alkenes followed by subsequent chain initiation occurs. The consequence is an increased formal $\alpha_{1,\text{eff}}$ and an increased fraction f_1, respectively; f_2 is decreased. This effect can be demonstrated by experiments of Madon et al.[21] While at low residence times and at a respectively low degree of syngas conversion the carbon number distribution can be strictly represented by superposition of two ASF distributions at increased residence times, a positive bulging in the middle range of carbon numbers is observed. However, at elevated carbon numbers the straight line, and consequently α_2, remains unchanged. Later Iglesia[22] showed that for an experiment with a high degree of syngas conversion, in comparison to one with low conversion, a decreased termination probability is observed only up to carbon number 15. In our view this corresponds to an increased $\alpha_{1,\text{eff}}$ and f_1.

For an accurate representation of the carbon number distribution in the case of elevated p_{CO} and a high degree of syngas conversion, the bimodal model must be completed by including 1-alkene incorporation and subsequent chain growth, a task still to be carried out.

11.3.3   Co-Feeding of Alcohols and Aldehydes

Co-feeding of alcohols effects an increased rate of hydrocarbon formation, as shown in early experiments of Emmett and coworkers[1–4] using ^{14}C-labeled alcohols. These experiments were carried out in order to support the hydroxyl-carbene mechanism favored at that time. Their experiments were confirmed by Shi and Davis[23] for Co catalysts and co-feeding of ethanol. Furthermore, in their study, the argument that ethanol may be dehydrated to ethene, incorporated, and followed by subsequent chain growth via CH_2 insertion could be excluded, as co-fed ethanol incorporated much faster than ethene.

For co-feeding of ^{14}C-labeled ethanol and 1-propanol, the molar ^{14}C activity of hydrocarbons formed on iron catalysts (unpromoted and promoted with K_2O/K_2CO_3) shows in the range up to C_6 a decelerating increase, and then a constant value up to the wax fraction, as shown for ethanol in Figure 11.7.[1,3] This characteristic behavior of alcohol incorporation is complementary to the incorporation of ethene and 1-alkenes. Therefore, incorporation of alcohol and subsequent chain growth can only occur in the course of mechanism 2, with the high growth probability α_2 dominating at elevated carbon numbers.

For both ethanol and 1-propanol, co-feeding of methylene- and methyl-labeled alcohols shows nearly the same activity distribution of formed hydrocarbons, indicating that C–C bonds of incorporated alcohols are not cleaved. Accordingly, the formation of methane is very small.[1,3]

Propionaldehyde is incorporated to an extent similar to that of 1-propanol. For formation of hydrocarbons the hydroxoalkyl group of incorporated alcohol

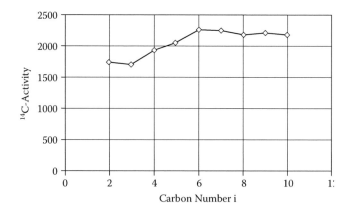

FIGURE 11.7   Molar ^{14}C activity of hydrocarbons vs. carbon number of co-feeding experiments with ethanol, labeled in 1-position, ^{14}C activity of wax (average carbon number 26) = 2,200.[1]

is inevitably hydrogenated to a CH_2 group. Besides incorporation and subsequent chain growth, hydrogenation of incorporated alcohol toward hydrocarbons of the same carbon number also occurs.[1-4]

11.4   SYNTHESIS OF BRANCHED HYDROCARBONS

Van Steen[11] and Schulz et al.[24,25] have presented a detailed analysis of FT products obtained on iron and cobalt catalysts that revealed an exponential decrease of branching with increasing carbon number, as demonstrated in Figure 11.8. At elevated carbon numbers the fractions of branched hydrocarbons approach a constant value.

The trend in branching reflects the superposition of the two mechanisms. The decrease of branching corresponds to the decrease of the fraction of hydrocarbons that are formed via mechanism 1 as presented in Figure 11.3, while the constant fraction of branching corresponds to mechanism 2, which dominates at elevated

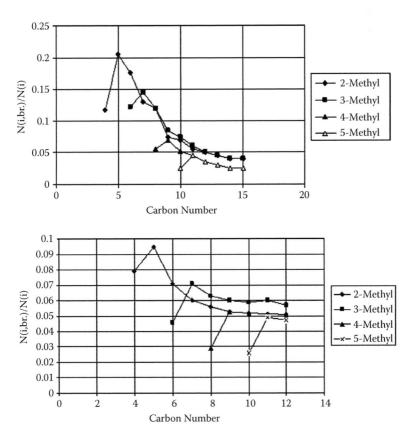

FIGURE 11.8   Ratio of branched and linear hydrocarbons vs. carbon number.

carbon numbers. A detailed analysis and modeling of the course of branching is in preparation.

11.5   DEVELOPMENT OF A NOVEL MECHANISM

Without doubt, the superimposed ASF distributions are the result of two incompatible mechanisms. In particular, the experiment with co-fed $^{13}CH_2N_2$ suggests that mechanism 1 assigned to distribution 1 and characterized by the growth probability α_1 is based on the insertion of CH_2. It is assumed that mechanism 2 assigned to distribution 2 and characterized by the higher growth probability α_2 is based on the insertion of CO.

This view is supported by the formation of alcohols and aldehydes, which is not possible via mechanism 1. For the formation of these oxygenates, insertion of CO is necessary. Therefore, several authors and first of all M. E. Dry[26] proposed a combined mechanism where hydrocarbons are mainly formed via CH_2 insertion and oxygenates via CO insertion. We extend this proposal by the assumption that hydrocarbons are also formed via CO insertion in the same way as oxygenates.

Further support comes from the alcohol co-feeding experiments of Emmett and coworkers, who have shown that co-fed alcohol is hydrogenated to a considerable extent to hydrocarbons. Claeys and Schulz[27] have shown that the yield of alcohols strongly decreases with increasing particle size of the catalyst due to the increased residence time of products favoring consecutive dehydration.

Based on these manifold experimental data and the concept of two incompatible mechanisms, we have proposed a novel mechanism[14] that in a slightly revised version is presented in the following sections.

11.5.1   FORMATION OF CH_2

Current opinion is that CO dissociatively adsorbs on the catalyst surface, and that both C and O are subsequently hydrogenated, yielding CH_2 and H_2O.

However, recently Inderwildi et al.[28] showed by density functional theory (DFT) calculations that hydrogenation of CO leading to formyl (oxomethylidyne) and subsequent conversion toward CH_2 show lower activation barriers than CO dissociation.

Therefore, we prefer the formal reduction of CO toward a "formyl" species. Subsequently, C–O bond cleavage gives coordinated methylene and an oxide ion that is transformed to OH^- and H_2O.

$$CO + 2\ H + 2\ e^- \circledR CH_2O^{2-} \circledR CH_2 + O^{2-}$$

$$O^{2-} + 2\ H \circledR H_2O + 2\ e^-$$

The sequence of steps requires at least two metal atoms: One of them accepts the oxide ion and the adjacent one carries the carbon chain.

11.5.2 Mechanism 1 (CH$_2$ Insertion)

In order to formally insert CH$_2$, the growing alkyl chain must attain a situation of metal-to-carbon bonding that favors CH$_2$ insertion over coupling of two CHR groups (R=H, alkyl).[29] A (C, H)-chelating coordination mode characterized by agostic M-H-C interaction[30] would meet this requirement.

Valence-bond representations of the (C, H)-chelating ("agostic") bonding mode in a two-metal-atom site

When compared with α-alkylmetal bonding, the (C–H)-chelating coordination mode is characterized by a longer C–H bond and a smaller HCM angle and requires a decreased activation energy for alkylidene-methylene coupling. Valence bond representation (b) is chosen to demonstrate the particular reactivity. This novel mechanism is in accordance with the experiments of Brady and Pettit, who could show that chain growth can only occur in the presence of hydrogen. In the absence of hydrogen, dimerization toward ethene occurs.

The mechanism presented in Scheme 11.1 substantially differs from the formal CH$_2$ insertion mechanism assumed in most preceding studies. Each growth cycle consists of:

1. Transformation of an alkyl metal species via α-H elimination to the (C, H)-chelating ground state (written as (b) of the scheme)
2. Coordination of a migrating CH$_2$ unit to the adjacent metal center
3. Alkylidene-methylene coupling affording a coordinated olefin
4. Reductive coupling leading to the alkyl metal state

Chain propagation is started from a methylene group and terminated by desorption of 1-alkenes or alkanes. Propeller-type mobility of the olefin ligand renders possible CH$_3$ branching of the growing chain, as demonstrated by the scheme. The growth probability is determined by the ratio of rates of formation of the alkyl intermediate and of the desorption of 1-alkenes, and to a minor extent of alkanes.

11.5.3 Mechanism 2 (CO Insertion)

The presented CO insertion mechanism (Scheme 11.2) partly follows the one proposed by Pichler and coworkers,[31,32] but differs in several important points. In

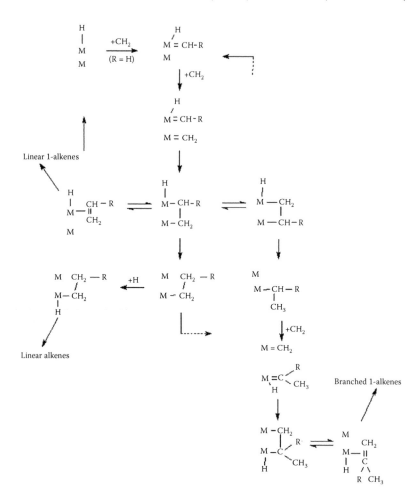

SCHEME 11.1   Mechanism 1 (CH$_2$ insertion) M = Fe, Co.

both proposals insertion of CO proceeds as in alkyl(carbonyl)-metal complexes by alkyl migration. Reductive C–O cleavage and elimination of water lead back to the starting situation, the alkyl metal. In the present version it is assumed that termination occurs by conversion of the C,O-bridging aldehyde intermediate to alcoholate, or by release of aldehyde. This formulation has been inspired by a study of Tau et al.,[33] who have shown that alcohol and aldehyde are reversibly converted on iron. Hutchings et al.[34] interpreted this redox process by the adsorption of alcohol via oxygen.

The alcoholate is either dehydrogenated via β-CH activation to afford an M–OH function and a hydrido (olefin) metal species or, to a minor extent, converted to alcohol. The hydrido (olefin) metal species can release 1-alkenes. The route toward alcohol is regarded as reversible, and thus allows the incorporation of co-fed alcohol, causing initiation of the FT synthesis via the alcoholate and the C,O-bridging aldehyde

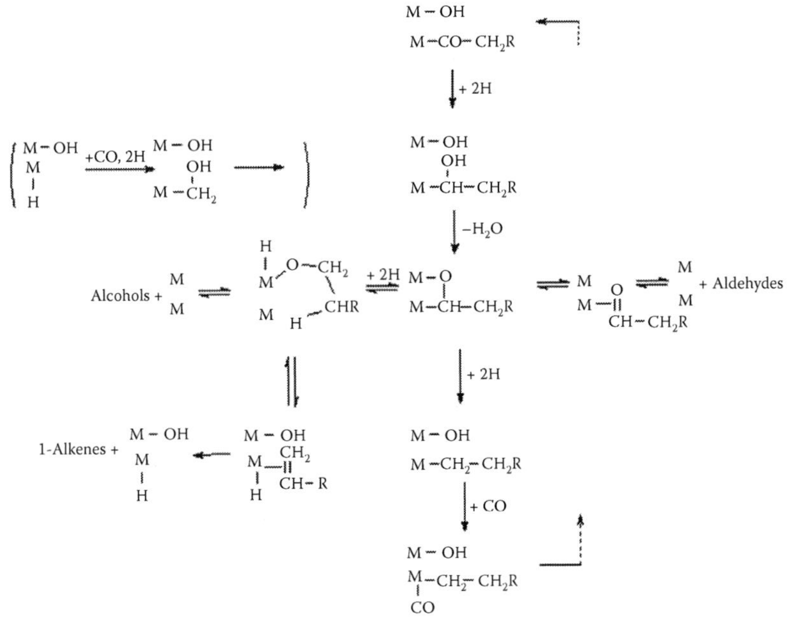

SCHEME 11.2   Mechanism 2 (CO insertion), M = Fe, Co.

intermediate. But co-fed alcohol can also be dehydrated to olefin, as expressed by the novel mechanism and in line with the experiments of Emmett and coworkers.

As shown by van Steen,[11] the fractions of monomethyl-branched hydrocarbons decrease with increasing carbon numbers and change into a constant level. Since mechanism 2 dominates the carbon number distribution in the range of elevated carbon numbers, the constant molar fraction of branched hydrocarbons must be assigned to mechanism 2. Since there is hardly a possibility of branching in the cycle of chain prolongation, we suspect that the origin of branching is in the reversible conversion of alcoholate intermediate/olefine metal species. Support comes from the experiments of Kummer and Emmett[3] with co-feeding of ^{14}C-labeled 1-propanol in which a marked formation of ^{14}C-labeled isobutene was observed.

11.5.4   INTERPRETATION OF THE PROMOTER EFFECT OF ALKALI

Alkalization of iron catalysts causes two different effects. The selectivities of 1-alkenes are raised and both the growth probability α_2 and the fraction f_2 are markedly increased, as already shown in Figure 11.2. Detailed studies on the promoter effect of alkali have revealed the effect on 1-alkene selectivity to saturates at 1 mass% of K_2CO_3, while the effect on f_2 already begins at 0.2 mass% of K_2CO_3.[12,13] This difference points to specific active sites in Fischer-Tropsch syn-

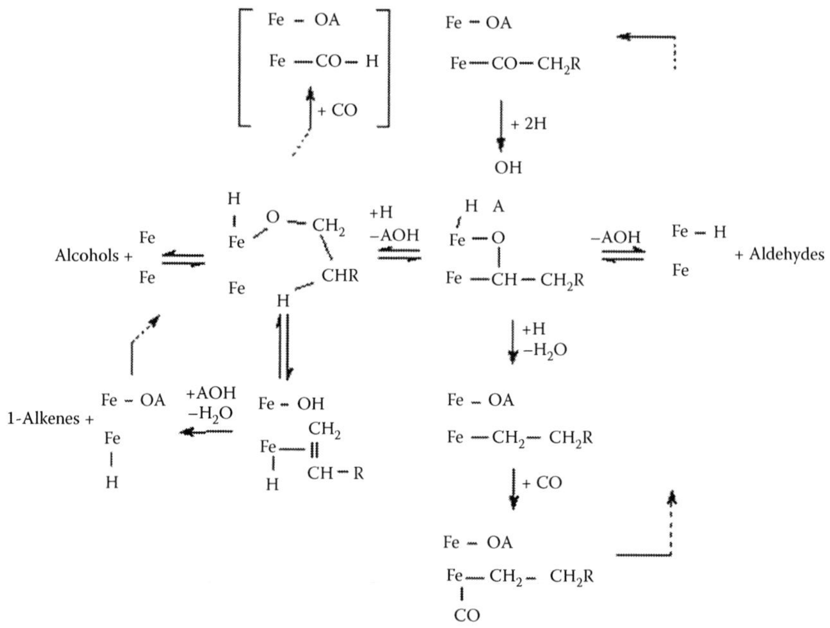

$$
\left[
\begin{array}{c}
Fe - OA \\
Fe -\!\!-\!\!CO-H \\
\uparrow + CO
\end{array}
\right]
\qquad
\begin{array}{c}
Fe - OA \\
Fe-\!\!-CO-CH_2R \\
\downarrow + 2H \\
OH
\end{array}
$$

Alcohols +
$$
\begin{array}{c}
H \\
| \\
Fe \\
Fe
\end{array}
\begin{array}{c}
O \\
Fe \nearrow \;\; CH_2 \\
\diagdown \\
CHR \\
H
\end{array}
\begin{array}{c}
\xrightarrow{+H} \\
\xleftarrow{-AOH}
\end{array}
\begin{array}{c}
H \;\; A \\
Fe - O \\
| \\
Fe -\!\!-CH - CH_2R
\end{array}
\xrightarrow{-AOH}
\begin{array}{c}
Fe - H \\
Fe
\end{array}
+ \text{Aldehydes}
$$

$$
\downarrow \begin{array}{c}+H\\-H_2O\end{array}
$$

1-Alkenes +
$$
\begin{array}{c}
Fe - OA \\
Fe \\
| \\
H
\end{array}
\begin{array}{c}
+AOH \\
\xleftarrow{-H_2O}
\end{array}
\begin{array}{c}
Fe - OH \\
CH_2 \\
Fe-\!\!-|| \\
H \quad CH-R
\end{array}
\qquad
\begin{array}{c}
Fe - OA \\
Fe -\!\!-CH_2 - CH_2R \\
\downarrow + CO
\end{array}
$$

$$
\begin{array}{c}
Fe - OA \\
Fe-\!\!-CH_2 - CH_2R \\
| \\
CO
\end{array}
$$

SCHEME 11.3   Modified mechanism 2 on alkalized iron surfaces, A = Li, Na, K, Cs.

thesis, while subsequent isomerization and hydrogenation of 1-alkenes may occur at various other sites on the catalyst.

The mass fraction f_2 increases strongly in the order Li-, Na-, K-, and Cs-oxide/carbonate. However, the increase of the growth probability α_2 is the same for all alkali promoters. The growth probability α_1 and, consequently, mechanism 1 are not affected. Therefore, it is necessary to modify only mechanism 2 of the novel hypothesis for alkali ions to take part in the catalytic cycle.[13]

Formally, the proton of a surface Fe-OH group is replaced by an alkali cation A^+ (Scheme 11.3).

One of the main consequences of this change concerns the branching toward the formation of the alcoholate or the release of aldehyde and hydrogenation toward the alkyl intermediate formation. Coordination of the alkali cation weakens the O–C bond, favoring the hydrogenolysis of this bond. Correspondingly, it reduces the chance of alcoholate formation and aldehyde release. This assumption may explain why α_2 is only marginally affected by the nature of A (A = Li, Na, K, Cs).

For interpretation of the dependence of f_2 on the nature of the alkali cation, we must consider the effect of hydrolysis expressed by the following equilibrium:

$$
Fe-OA+H_2O \rightleftharpoons Fe-OH+AOH
$$

With increasing size of the alkali cation, the left-hand side is favored, lowering the chance of Fe-OH hydrogenation. Thereby the number of active centers with mechanism 2, and thus f_2, is increased. This effect may explain the increasing promoter effect on f_2 in the order: H $\lll$ Li $\ll$ Na $<$ K $\sim$ Cs.

11.6 CONCLUSION

In Fischer-Tropsch synthesis the readsorption and incorporation of 1-alkenes, alcohols, and aldehydes and their subsequent chain growth play an important role on product distribution. Therefore, it is very useful to study these reactions in the presence of co-fed ^{13}C- or ^{14}C-labeled compounds in an effort to obtain data helpful to elucidate the reaction mechanism. It has been shown that co-feeding of CH_2N_2, which dissociates toward CH_2 and N_2 on the catalyst surface, has led to the sound interpretation that the bimodal carbon number distribution is caused by superposition of two incompatible mechanisms. The distribution characterized by the lower growth probability is assigned to the CH_2 insertion mechanism.

Detailed studies of co-feeding experiments with alcohol and aldehyde, first undertaken by Emmett and coworkers, have led to the conclusion that the distribution with the higher growth probability is with high probability due to a mechanism based on CO insertion.

The discussion of all these results and other experimental results on the basis of our recently proposed novel mechanism gives a consistent interpretation of this highly complicated synthesis.

REFERENCES

1. J.T. Kummer, H.H. Podgurski, W.B. Spencer, P.H. Emmett. 1951. *J. Am. Chem. Soc.* 73:564–69.
2. R.J. Kokes, W.K. Hall, P.H. Emmett. 1957. *J. Am. Chem. Soc.* 79:2989–96.
3. J.T. Kummer, P.H. Emmett. 1953. *J. Am. Chem. Soc.* 75:5177–84.
4. W.K. Hall, R.J. Kokes, P.H. Emmett. 1960. *J. Am. Chem. Soc.* 82:1027–37.
5. R.C. Brady, R. Pettit. 1980. J. Am. Chem. Soc. 102:6181–82.
6. R. Quyoum, V. Berdini, M.L. Turner, H.C. Long, P.M. Maitlis. 1998. *J. Catal.* 173:355–65.
7. B. Schliebs, J. Gaube. 1985. *Ber. Bunsenges. Phys. Chem.* 89:68–73.
8. R. A. Dictor, A.T. Bell. 1986. *J. Catal.* 97:121–36.
9. J. Patzlaff, Y. Liu, C. Graffmann, J. Gaube. 1999. *Appl. Catal. A* 186:109–19.
10. G.A. Huff, Jr., C.N. Satterfield. 1984. *J. Catal.* 85:370–79.
11. E. van Steen. 1993. Thesis, University of Karlsruhe, Karlsruhe, Germany.
12. P. Hnatow, L. König, B. Schliebs, J. Gaube. 1984. In *Proceedings of the International Congress on Catalysis*, Berlin, Vol. II, pp. 113–22.
13. J. Gaube, H.-F. Klein. 2008. *Appl. Catal. A*: General 350:126–132.
14. J. Gaube, H.-F. Klein. 2008. *J. Mol. Catal. A Chem.* 283:60–68.
15. Y. Liu. 1992. Thesis, TH Darmstadt.
16. G. Sudheimer, J. Gaube. 1985. *Ger. Chem. Eng.* 8:195–202.
17. H. Schulz, B.R. Rao, M. Elstner. 1970. *Erdöl Kohle Erdgas Petrochem.* 23:651–55.
18. J. Patzlaff, Y. Liu, C. Graffmann, J. Gaube. 2002. *Catal. Today*, 71:381–94.

19. H. Schulz, M. Claeys. 1999. *Appl. Catal. A*:General 186:71–90.
20. H. Schulz, M. Claeys. 2000. *Prepr. Am. Chem. Soc. Div. Petr. Chem.* 45:206–9.
21. R.J. Madon, E. Iglesia, S.C. Reyes. 1993. *ACS Symp. Ser.* 517:383–96.
22. E. Iglesia. 1997. *Appl. Catal. A* 161:59–78.
23. B. Shi, G. Jacobs, D. Sparks, B.H. Davis. 2005. *Fuel* 84:1093. B. Shi, B.H. Davis. 2003. *Top. Catal.* 26:157–61.
24. H. Schulz. 2003. *Top. Catal.* 26:7–85.
25. H. Schulz, G. Schaub, M. Claeys, T. Riedel. 1999. *Appl. Catal. A Gen.* 186:215–27.
26. M.E. Dry. 1990. *Catal. Today* 6:183.
27. M. Claeys, H. Schulz. 2004. *Prepr. Am. Chem. Soc. Div. Petr. Chem.* 49:195–99.
28. O.R. Inderwildi, S.J. Jenkins, D.A. King. 2008. *J. Phys. Chem. C* 112:1305–7.
29. R.H. Grubbs. 2003. *Handbook of metathesis.* Weinheim: Wiley-VCH.
30. M. Brookhart, M.L.H. Green. 1983. *J. Organomet. Chem.* 250:395.
31. H. Pichler, H. Schulz. 1970. *Chem.-Ing.-Tech.* 42:1162–74.
32. A. Zein El Deen. 1970. Thesis, University of Karlsruhe, Karlsruhe, Germany.
33. L.M. Tau, R. Robinson, R. Dudley Ross, B.H. Davis. 1987. *J. Catal.* 105:335.
34. G.J. Hutchings, M. van der Riet, R. Hunter. 1989. *J. Chem. Soc. Faraday Trans. 1* 85:2875.

12 Modeling of Internal Diffusion Limitations in a Fischer-Tropsch Catalyst

Anke Jung, C. Kern, and Andreas Jess

CONTENTS

Today, Fischer-Tropsch (FT) synthesis is carried out in diverse reactor designs, such as fixed beds, bubble columns, or circulating fluid beds. If a fixed bed mode of operation is favored, the Fischer-Tropsch catalyst will generally consist of particles of a few millimeters in size in order to minimize the pressure drop. Unfortunately, for particle diameters of more than about 1 mm, the effective reaction rate decreases significantly due to pore filling of the catalyst with high molecular weight hydrocarbons formed during synthesis.[1] As a result, a limited diffusion rate of dissolved hydrogen and carbon monoxide is produced. Furthermore, steam is formed that induces strong inhibiting effects on the reaction rate, especially in the rear part of a fixed bed reactor.[2–5]

Taking these effects into account, internal pore diffusion was modeled on the basis of a wax-filled cylindrical single catalyst pore by using experimental data. The modeling was accomplished by a three-dimensional finite element method as well as by a respective differential-algebraic system. Since the Fischer-Tropsch synthesis is a rather complex reaction, an evaluation of pore diffusion limitations by the classical Thiele modulus is problematic, as this approach is in general limited to simple kinetics such as a power law equation. Hence, the numerically obtained solution was compared to data calculated by means of the classical Thiele modulus.

12.1 INTRODUCTION

Production of synthetic fuels via Fischer-Tropsch (FT) synthesis has the potential to produce high-value automotive fuels like gasoline and diesel oil, as well as petrochemicals from fossil and renewable resources. The availability of cheap natural gas and solid raw materials like coal and biomass has given momentum to synthesis technologies already developed in the 1920s. Thus, the worldwide Fischer-Tropsch capacities will increase significantly in the near future, with natural gas as the favored feedstock for today, and coal and biomass as potential feedstocks for the future. Therefore, it is worthwhile to have a closer look at this long known but still important and fascinating technology.

The most difficult problem to solve in the design of a Fischer-Tropsch reactor is its very high exothermicity combined with a high sensitivity of product selectivity to temperature. On an industrial scale, multitubular and bubble column reactors have been widely accepted for this highly exothermic reaction.[6] In case of a fixed bed reactor, it is desirable that the catalyst particles are in the millimeter size range to avoid excessive pressure drops. During Fischer-Tropsch synthesis the catalyst pores are filled with liquid FT products (mainly waxes) that may result in a fundamental decrease of the reaction rate caused by pore diffusion processes. Post et al. showed that for catalyst particle diameters in excess of only about 1 mm, the catalyst activity is seriously limited by intraparticle diffusion in both iron and cobalt catalysts.[1]

Modeling of pore diffusion phenomena can be a helpful tool mainly in terms of catalyst design considerations but also in terms of understanding the effects caused by diffusional restrictions. For example, a modeling study by Wang et al.[7] demonstrated a negative impact on selectivity by particle diffusion limitations.

12.2 FUNDAMENTALS

12.2.1 Reactions in Fischer-Tropsch Synthesis

Fischer-Tropsch synthesis can be regarded as a surface polymerization reaction since monomer units are produced from the reagents hydrogen and carbon monoxide *in situ* on the surface of the catalyst. Hence, a variety of hydrocarbons (mainly n-paraffines) are formed from hydrogen and carbon monoxide by successive addition of C_1 units to hydrocarbon chains on the catalyst surface (Equation 12.1). Additionally, carbon dioxide (Equation 12.3) and steam (Equations 12.1 and 12.2) are produced; CO_2 affects the reaction just a little, whereas H_2O shows a strong inhibiting effect on the reaction rate when iron catalysts are used.

$$CO + 2\,H_2 \rightarrow (-CH_2-) + H_2O \qquad \Delta_R H^0_{298} = -152\ kJ/mol \qquad (12.1)$$

$$CO + 3\,H_2 \rightarrow CH_4 + H_2O \qquad \Delta_R H^0_{298} = -206\ kJ/mol \qquad (12.2)$$

$$CO + H_2O \rightarrow CO_2 + H_2 \qquad \Delta_R H^0_{298} = -41\ kJ/mol \qquad (12.3)$$

The model presented in this paper will display the concentration gradients in a cylindrical, fully wax-filled pore of the catalyst. For simplification reasons, only the reactants CO and H_2 as well as the reaction product H_2O will be considered. CO_2 formation is disregarded in the model due to its comparatively low concentration in FT products and because compared to steam, CO_2 has no impact on the main reaction rate. Methane formation (usually not more than 5% of total CO conversion) was also neglected.

12.2.2 Mass Balance of a Single Cylindrical Pore and Diffusive Effects (Classical Thiele Approach)

For a single cylindrical pore of length L and a reactant A diffusing into the pore, where a first-order reaction takes place at the pore surface, the power law rate expression

$$r_A = -\frac{d\dot{n}_A}{dA_{pore}} = k_A c_A \qquad (12.4)$$

is obtained. The rate constant k_A ($m^3\ m^{-2}\ s^{-1}$) is related to the surface of the pore A_{pore} (m^2). As shown in Figure 12.1, at steady state the mass balance for a small

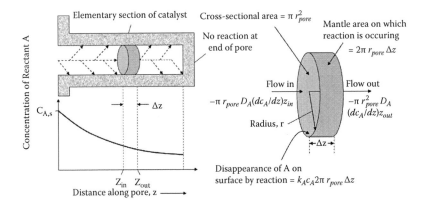

FIGURE 12.1   Representation of a single cylindrical catalyst pore and mass balance for an elementary slice of the pore.

slice with thickness Δz, cross-sectional area $\pi\, r^2_{pore}$, and mantle area $2\pi\, r_{pore}$, Δz is given by

$$\underbrace{-D_A\,\pi r^2_{pore}\left(\frac{dc_A}{dz}\right)_z}_{\text{Input flux of A}} = \underbrace{k_A\,c_A\,2\pi r_{pore}\,\Delta z}_{\text{Dissappearance by reaction}} -\underbrace{D_A\,\pi r^2_{pore}\left(\frac{dc_A}{dz}\right)_{z+\Delta z}}_{\text{Output flux of A}} \tag{12.5}$$

if Fick's first law is used for the flux of A into and out of the slice. In this equation, D_A is the diffusion coefficient in the pore and equals the molecular diffusion coefficient, if the influence of Knudsen diffusion in very narrow pores is negligible (which is the case here, as we have diffusion in liquid wax). Rearrangement of Equation 12.5 leads to

$$\frac{d^2 c_A}{dz^2} = \frac{2k_A}{D_A\,r_{pore}}c_A \tag{12.6}$$

The solution of this frequently met linear differential equation for the boundary conditions

$$c_A = c_{A,s} \text{ for } z = 0 \text{ (pore entrance, external surface to bulk phase)} \tag{12.7}$$

and

$$\frac{dc_A}{dz} = 0 \text{ for } z = L \text{ (interior end of pore)} \tag{12.8}$$

is

$$\frac{c_A(z)}{c_{A,s}(z=0)} = \frac{\cosh(\Phi(1-z/L))}{\cosh\Phi} \tag{12.9}$$

where Φ is the Thiele modulus of a single pore:

$$\Phi = L\sqrt{\frac{k_A\,2}{D_A\,r_{Pore}}} = L\sqrt{\frac{k_A\,A_{int,v}}{D_A}} \tag{12.10}$$

$A_{int,v}$ ($m^2\,m^{-3}$) is the internal surface area per volume ($2/r_{pore}$ for a single cylindrical pore), and k_A ($m^3\,m^{-2}\,s^{-1}$) is the rate constant related to the surface of the pore.

As a measure of how much the effective rate is lowered by the resistance to pore diffusion, the effectiveness factor η_{pore} is used. This factor is defined as the ratio of the actual mean reaction rate within the pore to the maximum rate if not slowed down by pore diffusion. For a first-order reaction, η_{pore} is also equivalent to the ratio of the mean concentration to the one at the pore entrance:

$$\eta_{pore} = \frac{r_{A,eff}}{r_{A,max}} = \frac{r_{A,eff}}{r_A\left(c_A = c_{A,s}\right)} = \frac{k_A\, \overline{c}_A}{k_A\, c_{A,s}} = \frac{\overline{c}_A}{c_{A,s}} \qquad (12.11)$$

In case of Fischer-Tropsch synthesis, we have to consider that the first-order reaction rate constant is related to the concentration in the gas phase (e.g., c_{H2}), and that the diffusive flux in the liquid-filled pores is related to the concentration in the liquid ($c_{H2,l}$). Thus, instead of Equation 12.10, we have to use

$$\Phi = L\sqrt{\frac{k_A\, A_{int,v}\, c_{H_2,g}}{D_{H_2,l}\, c_{H_2,l}}} \qquad (12.12)$$

and if we use Henry's law $\left(c_{H2,l} = \dfrac{RT}{H_{H_2}} c_{H_2,g} \right.$, we get

$$\Phi = L\sqrt{\frac{k_A\, A_{int,v}\, H_{H_2}}{D_{H_2,l}\, RT}} \qquad (12.13)$$

Unfortunately, Equation 12.13 is only a rough approximation, as we do not have a simple first-order reaction, as discussed in the subsequent section.

12.3 PORE DIFFUSION MODELING IN FISCHER-TROPSCH SYNTHESIS

12.3.1 Basis of Calculation

In reality, not only the main reaction (the *Fischer-Tropsch reaction*) leading to the formation of higher hydrocarbons (Equation 12.1), but also methane formation (Equation 12.2) and the water-gas shift reaction (Equation 12.3) have to be considered. The rate equations for these three reactions on a commercial Fe-catalyst were determined by Popp[8] and Raak[2] and summarized by Jess et al.[9] However, to simplify matters, just the Fischer-Tropsch reaction forms the basis of the approach presented here:

$$r_{m,H_2,FT} = -\frac{d\dot{n}_{H_2}}{dm_{cat}} = k_{m,H_2,FT}\, \frac{c_{H_2,g}}{1 + 1.6\, \dfrac{c_{H_2O,g}}{c_{CO,g}}} \qquad (12.14)$$

with

$$k_{m,H_2,FT} = 1.2 \cdot 10^7 \; m^3 \, kg^{-1} \, s^{-1} \, e^{\frac{-109000}{RT}} \tag{12.15}$$

as the intrinsic mass-related rate constant.

In the special case of an ideal single catalyst pore, we have to take into account that diffusion is quicker than in a porous particle, where the tortuous nature of the pores has to be considered. Hence, the tortuosity τ has to be regarded. Furthermore, the mass-related surface area $A_{m,BET}$ is used to calculate the surface-related rate constant based on the experimentally determined mass-related rate constant. Finally, the gas phase concentrations of the kinetic approach (Equation 12.14) were replaced by the liquid phase concentrations via the Henry coefficient. This yields the following differential equation:

$$D_{H_2,l} \frac{d^2 c_{H_2,l}}{dz^2} = \underbrace{\frac{2}{r_{pore}} \frac{\tau}{A_{m,BET}} k_{m,H_2,FT}}_{k_{A,H_2,FT}} \; \frac{\dfrac{H_{H_2}}{RT} c_{H_2,l}}{1 - 1.6 \dfrac{\dfrac{H_{H_2O}}{RT} c_{H_2O,l}}{\dfrac{H_{CO}}{RT} c_{CO,l}}} \tag{12.16}$$

A similar equation was derived for carbon monoxide.

Henry's law was also used to calculate the H_2, CO, and H_2O concentrations in the liquid Fischer-Tropsch products (wax) at the entrance of the pore ($z = 0$):

$$c_{i,l} = \frac{1}{H_i} p_{i,g} = \frac{RT}{H_i} c_{i,g} \tag{12.17}$$

The molecular diffusivity was evaluated by the Wilke-Chang[10] equation:

$$D_{i,l} = 5.88 \cdot 10^{-17} \frac{T \sqrt{\chi_{wax} M_{wax}}}{\mu_{wax} \left(v_{mol,cp,i} \right)^{0.6}} \, , m^2 s^{-1} \tag{12.18}$$

with χ as the association parameter, M in kg mol^{-1}, $v_{mol,cp}$ (molar volume of solute at the condensation point at 1 bar) in m^3 mol^{-1}, and μ in P = Pa s = kg m^{-1} s^{-1}. The applied pore geometry resulted from BET measurements (pore diameter) and an approximation of the pore length according to

$$L_{pore} = \frac{d_{particle,cyl}}{4} \tag{12.19}$$

based on dimensions of a commercially used Fischer-Tropsch catalyst.

12.3.2 MODELING

The modeling of the internal pore diffusion of a wax-filled cylindrical single catalyst pore was accomplished by the software Comsol Multiphysics (from Comsol AB, Stockholm, Sweden) as well as by Presto Kinetics (from CiT, Rastede, Germany). Both are numerical differential equation solvers and are based on a three-dimensional finite element method. Presto Kinetics displays the results in the form of diagrams. Comsol Multiphysics, instead, provides a three-dimensional solution of the problem.

All subsequent considerations are based on the following constant reaction conditions:

- $T = 523$ K
- $p = 2$ MPa
- Molar ratio of H_2:CO = 2:1

The initial gas phase concentrations, the respective concentrations of the components CO, H_2, and H_2O at the gas-wax phase boundary (for a conversion $X = X_{CO} \approx X_{H2}$ of 5%), as well as the molecular diffusion coefficients applied in the model are listed in Table 12.1.

Assuming that there are only axial and no radial concentration gradients in the pore due to the negligible size of the pore diameter, the modeled concentration profiles of CO, H_2, and H_2O in a wax-filled cylindrical pore are given in Figure 12.2 (left, Presto Kinetics). For verification reasons of the underlying model and to obtain a better visual impression of the respective processes in the catalyst pore,

---

TABLE 12.1

Molecular Diffusion Coefficients (at 523 K), Gas Concentrations, and the Phase Interface Concentrations of CO, H_2, and H_2O (assumed reaction conditions: conversion $X = 5\%$, 2 MPa, 523 K, H_2:CO = 2:1)

$D_{H_2,l}$[a]	$D_{CO,l}$[a]	$D_{H_2O,l}$[a]	$c_{H_2,g}$	$c_{CO,g}$	$c_{H_2O,g}$	$c_{H_2,l}$[b]	$c_{CO,l}$[b]	$c_{H_2O,l}$[b]
m²/s	m²/s	m²/s	mol/m³	mol/m³	mol/m³	mol/m³	mol/m³	mol/m³
$4.2 \cdot 10^{-8}$	$2.2 \cdot 10^{-8}$	$2.8 \cdot 10^{-8}$	301.4	150.7	7.9	48.2	32.6	15.7

[a] Calculated by Wilke-Chang (Equation 12.14).
[b] Liquid phase concentration calculated by Henry's law (Equation 12.13).

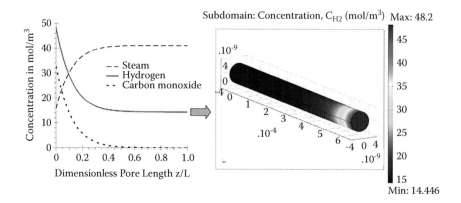

FIGURE 12.2 Concentration profiles of CO, H_2, and H_2O in a wax-filled pore (L_{pore} = $6.75 \cdot 10^{-4}$ m → $d_{particle,cyl}$ = $L_{pore} \cdot 4$ = $2.7 \cdot 10^{-3}$ m) modeled with Presto Kinetics (left) and a three-dimensional concentration profile of H_2 modeled with Comsol Multiphysics (right) (conversion X = 5%).

a second three-dimensional model was calculated by Comsol Multiphysics. As an example, the concentration profile for H_2 is shown in Figure 12.2 (right). As expected, the results of both simulation programs are exactly the same.

With the data obtained it is feasible to determine the actual effectiveness factor of the pore η_{pore}:

$$\eta_{pore} = \frac{r_{eff}}{r_{max}} = \frac{\bar{r}}{r_{max}} = \frac{\dfrac{\bar{c}_{H_2}}{1+1,6\dfrac{\bar{c}_{H_2O}}{\bar{c}_{CO}}}}{\dfrac{c_{H_2,s}}{1+1,6\dfrac{c_{H_2O,s}}{c_{CO,s}}}} \qquad (12.20)$$

where the surface concentrations $c_{i,s}$ at the pore mouth are identical to the phase boundary concentrations (see Table 12.2 and Figure 12.3).

Hence, for a technical particle size in the range of several millimeters (L_{pore} = $6.75 \cdot 10^{-4}$ m → $d_{particle,cyl}$ = $L_{pore} \cdot 4$ = $2.7 \cdot 10^{-3}$ m) and a typical reaction temperature of 523 K (X = 5% at p = 2 MPa), only 13% of the average pore is really used for synthesis, which has to be accepted to limit the pressure loss in fixed bed reactors.

The influence of the increasing conversion degree along the catalyst bed on the effectiveness factor is illustrated in Table 12.3. The calculations were carried out for three different CO conversions X = 5, 40, and 80%.

TABLE 12.2
Gas and Liquid Phase Concentrations at the Phase Change

CO Conversion %	$c_{H_2,g}$ mol/m³	$c_{CO,g}$ mol/m³	$c_{H_2O,g}$ mol/m³	$c_{H_2,l}$ mol/m³	$c_{CO,l}$ mol/m³	$c_{H_2O,l}$ mol/m³
5	301.4	150.7	7.9	48.2	32.6	15.7
40	250.9	125.4	83.6	40.1	27.2	165.0
80	131.4	65.7	262.8	21.0	14.2	518.5

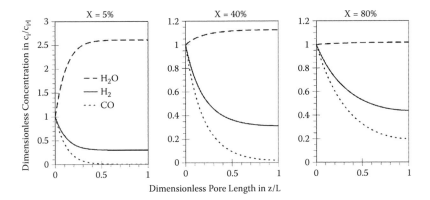

FIGURE 12.3   Concentration profiles of CO, H_2, and H_2O in a wax-filled pore (L_{pore} = $6.75 \cdot 10^{-4}$ m → $d_{particle,cyl}$ = $L_{pore} \cdot 4$ = $2.7 \cdot 10^{-3}$ m) at different CO conversion degrees X = 5% (left), 40% (middle), and 80% (right) modeled with Presto Kinetics.

When having a look at variable conversion degrees, the different gas phase concentrations affect the liquid phase concentrations via the solubility equilibrium, as shown in Table 12.3.

For a better understanding of the obtained effectiveness factors in Table 12.2, the results of the simulation at different conversion degrees (concentrations of CO, H_2, and H_2O in a cylindrical catalyst pore) are depicted in Figure 12.3.

According to Figure 12.3, the increase of the effectiveness factor with rising CO conversion (see Table 12.2) is a result of the flattening concentration profiles along the catalyst bed (increasing conversion), meaning the CO concentration drops to zero further in the inner part of the pore for a low conversion (front part of a fixed bed) and increases for a higher conversion (rear part), although the concentration in the inner part of the pore is still lower than at the pore mouth. This effect can be explained by the high H_2O concentration at the gas-liquid boundary layer arising at high CO conversions (see Table 12.3). Since H_2O inhibits the Fischer-Tropsch

TABLE 12.3
Effectiveness Factors Calculated with the Simulation Program Presto Kinetics ($L_{pore} = 6.75 \cdot 10^{-4}$ m → $d_{particle,cyl} = L_{pore} \cdot 4 = 2.7 \cdot 10^{-3}$ m)

CO Conversion %	Effectiveness Factor Simulated with Presto Kinetics
5	0.13
40	0.16
80	0.29

reaction, the reaction rate is already lowered at the pore entrance. By reason of the limited reaction rate, less H_2 and CO are converted, and therefore less H_2O is formed, which yields a smaller concentration gradient inside the pore.

For comparison reasons, the results derived from the simulation were additionally calculated by means of the Thiele modulus (Equation 12.12), i.e., for a simple first-order reaction. The reaction rate used in the model is more complex (see Equation 12.14); thus, the surface-related rate constant k_A in Equation 12.12 is replaced by

$$k_A = \frac{k_{A,H_2,FT}}{1 + 1.6 \dfrac{c_{H_2O,g}}{c_{CO,g}}} \approx const \qquad (12.21)$$

where, to simplify matters, the concentration-dependent denominator is assumed to be constant. The effectiveness factor was calculated as follows:

$$\eta = \frac{\tanh \Phi}{\Phi} \qquad (12.22)$$

The Thiele modulus and the effectiveness factor, respectively, were calculated for the three CO conversions $X = 5$, 40, and 80%. The H_2, CO, and H_2O gas phase concentrations as well as the respective H_2 concentration at the gas-wax phase boundary were taken from Table 12.3. The value of the diffusion coefficient $D_{H2,l}$ is listed in Table 12.1.

The effectiveness factors calculated by the Thiele modulus as well as the findings obtained from the simulation are shown in Table 12.4.

By comparing the values we find that the results are roughly the same, but the data obtained from the Thiele modulus exceed the values of the modeling at

TABLE 12.4
Comparison of the Effectiveness Factors Modeled with Presto Kinetics and Calculated by Means of the Thiele Modulus $L_{pore} = 6.75 \cdot 10^{-4}$ m → $d_{particle,cyl} = L_{pore} \cdot 4 = 2.7 \cdot 10^{-3}$ m)

CO Conversion %	Effectiveness Factor Calculated (Thiele)	Effectiveness Factor Simulated (Presto Kinetics)	Absolute Deviation %	Relative Error %
5	0.15	0.13	2	15
40	0.21	0.16	5	29
80	0.39	0.29	10	33

all conversion degrees. This effect can be explained as follows. In contrast to the Thiele modulus, the simulation software takes into account that the reaction rate reaches zero when the CO concentration approaches a lower limit in the pore (see Figure 12.3). In contrast, by calculating the Thiele modulus, the reaction of H_2 is not limited. Figure 12.4 shows the concentration profiles of H_2 derived from both the modeling (Presto Kinetics) and the Thiele modulus (see Equation 12.9).

In Figure 12.4, we clearly see that the effective reaction rate is smaller in the simulation than that calculated by the Thiele modulus, which causes a higher effectiveness factor (see also Equations 12.11 and 12.20).

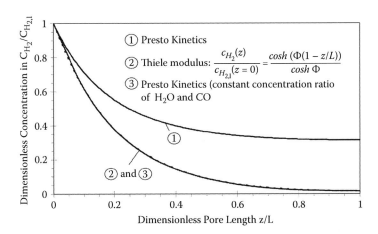

FIGURE 12.4   Concentration profiles of H_2 in a wax-filled pore ($L_{pore} = 6.75 \cdot 10^{-4}$ m → $d_{particle,cyl} = L_{pore} \cdot 4 = 2.7 \cdot 10^{-3}$ m) at a CO conversion of 40% modeled with Presto Kinetics and calculated by the Thiele modulus.

$$\left.\frac{c_{H_2O}}{c_{CO}}\right|_\Phi < \left.\frac{\bar{c}_{H_2O}}{\bar{c}_{CO}}\right|_{sim} \Rightarrow \left.\frac{1}{1+1.6\dfrac{c_{H_2O}}{c_{CO}}}\right|_\Phi > \left.\frac{1}{1+1.6\dfrac{\bar{c}_{H_2O}}{\bar{c}_{CO}}}\right|_{sim}$$

(12.23)

$$\Rightarrow r_{eff,\Phi} > r_{eff,sim} \Rightarrow \eta_{pore,\Phi} > \eta_{pore,sim}$$

Keeping the concentration ratio of H_2O and CO in the simulation model constant (according to the Thiele modulus; see Equation 12.21) leads to equal concentration profiles of H_2, as shown in Figure 12.4, and consequently to equal effectiveness factors for both methods (Thiele modulus and simulation). In fact, the concentrations of H_2, CO, and H_2O change inside the pore, as considered in the simulation. Therefore, the results obtained by the software used represent reality best.

12.4 CONCLUSION

The concentration profiles of CO, H_2, and H_2O in a single cylindrical catalyst pore were modeled by two different simulation programs, Presto Kinetics and Comsol Multiphysics. Both software showed the same result. For a technical particle size in the range of several millimeters ($L_{pore} = 6.75 \cdot 10^{-4}$ m $\rightarrow d_{particle,cyl} = L_{pore} \cdot 4 = 2.7 \cdot 10^{-3}$ m) and a typical reaction temperature of 523 K (conversion X = 5% at p = 20 bar), only 13% of the average pore is actually used for synthesis. This result must be considered in order to minimize the pressure drop and, hence, energy loss in fixed bed reactors. With rising conversion along the catalyst bed, the effectiveness factor increases due to the inhibition of the reaction rate by the reaction product H_2O.

The data derived from modeling at different conversion degrees (X = 5, 40, and 80%) were also compared to the results obtained from the calculation of the classical Thiele modulus. The calculated (by the Thiele modulus) and modeled (by Presto Kinetics) effectiveness factors showed comparable values. Hence, the usage of simulation software is not required to get a first impression of the diffusion limitations in a Fischer-Tropsch catalyst pore. Nevertheless, modeling represents a valuable tool to better understand conditions within a catalyst pore.

REFERENCES

1. Post, M. F. M., Van't Hoog, J. K., Minderhoud, J. K., Sie, S. T. 1989. Diffusion limitations in Fischer-Tropsch catalysts. *AIChE Journal* 35:1107–14.
2. Raak, H. 1995. Reaktionskinetische Untersuchungen in der Anfangsphase der Fischer-Tropsch-Synthese an einem technischen Eisenfällungskatalysator. PhD dissertation, University Karlsruhe, Germany.

3. Kuntze, T. 1991. Kinetik der Fischer-Tropsch-Synthese unter Druck an einem Eisenfällungskatalysator bei Einsatz eines stickstoffreichen Synthesegases. PhD dissertation, University Karlsruhe, Germany.

4. Zhan, X., Davis, B. H. 2002. Assessment of internal diffusion limitation on Fischer-Tropsch product distribution. *Applied Catalysis A: General* 236:149–61.

5. Xu, B. L., Fan, Y. N., Zhang, Y., Tsubaki, N. 2005. Pore diffusion simulation model of bimodal catalyst for Fischer-Tropsch synthesis. *AIChE Journal* 51:2068–76.

6. Guttel, R., Kunz, U., Turek, T. 2007. Reactors for the Fischer-Tropsch synthesis. *Chemie Ingenieur Technik* 79:531–43.

7. Wang, Y. N., Xu, Y. Y., Xiang, H. W., Li, Y. W., Zhang, B. J. 2001. Modeling of catalyst pellets for Fischer-Tropsch synthesis. *Industrial & Engineering Chemistry Research* 40:4324–35.

8. Popp, R. 1996. Ergebnisse halbtechnischer Untersuchungen zur Fischer-Tropsch-Synthese mit stickstoffreichem Synthesegas. PhD dissertation, University Karlsruhe, Germany.

9. Jess, A., Popp, R., Hedden, K. 1999. Fischer-Tropsch-synthesis with nitrogen-rich syngas—Fundamentals and reactor design aspects. *Applied Catalysis A: General* 186:321–42.

10. Wilke, C. R., Chang, P. 1955. Correlation of diffusion coefficients in dilute solutions. *AIChE Journal* 1:264–70.

13 Fe-LTFT Selectivity
A Sasol Perspective

*Matthys Janse van Vuuren, Johan Huyser,
Thelma Grobler, and Godfrey Kupi*

CONTENTS

The use of a Fischer-Tropsch (FT) process to produce long-chain hydrocarbons is well known in industry, and achieving the desired selectivity from the FT reaction is crucial for the process to make economic sense. It is, however, well known that a one-alpha model does not describe the product spectrum well. From either a chemicals or fuels perspective, hydrocarbon selectivity in the FT process needs to be thoroughly understood in order to manipulate process conditions and allow the optimization of the required product yield to maximize the plant profitability. There are many unanswered questions regarding the selectivity of the iron-based low-temperature Fischer-Tropsch (Fe-LTFT) synthesis.

In order to manipulate the selectivity of the FT synthesis, there needs to be an understanding of the parameters that control the selectivity. Within Sasol there are many activities to ensure more accurate measurements, e.g., developments in the two-dimensional gas chromatography (GC×GC) technique. This discussion will give a flavor of what Sasol is doing regarding the understanding of Fe-LTFT selectivity.

13.1 INTRODUCTION

The use of an FT process to produce long-chain hydrocarbons is well known in industry, and achieving the desired selectivity from the FT reaction is crucial for the process to make economic sense. FT selectivity is discussed in detail elsewhere.[1] Assuming an ideal polymerization type mechanism, a single parameter (namely, the propagation factor referred to as the α-value) can be used to describe the carbon number distribution of the product spectrum. A low α-value (<0.7) corresponds to a very light product spectrum (typically obtained in the iron-based high-temperature Fischer Tropsch (Fe-HTFT) synthesis) and a high α-value (>0.9) corresponds to a heavy, waxy product (typically obtained in the Fe- and Co-LTFT processes). Figure 13.1 shows the change in the ideal product distribution with the change in the α-value. There are, however, numerous examples where the product distribution from an Fe- or Co-based FT process cannot be adequately described by using a single α-value, and many attempts have been made to modify this single alpha model to correctly predict the measured selectivity.[2-4]

The FT reaction produces 1 mole of water for every mole of CO converted to aliphatic products. The water produced by the FT reaction can be consumed in the water-gas-shift (WGS) reaction, yielding CO_2 and H_2. The usage ratio of hydrogen to CO (i.e., the rate of hydrogen consumption relative to the rate of CO consumption) is about 2 if only aliphatic FT products are considered and the WGS reaction does not occur. If all the water formed in the FT reaction is consumed by the WGS, then the overall H_2/CO usage ratio would be 0.5. Compared to cobalt catalysts, the Fe-LTFT catalyst is much more active for the WGS reaction, and the resulting usage ratio can be as low as 0.7. Knowledge of the actual usage ratio is important, since the hydrogen content in the reactor tail gas depends on the hydrogen content in the feed gas and the usage ratio. Manipulation of the fresh-feed H_2/CO ratio, conversion, recycle ratio, and usage ratio allows some degree of control over the syngas ratio to which the catalyst is exposed.

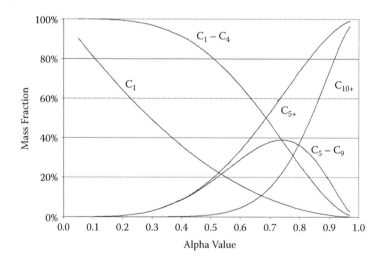

FIGURE 13.1   Influence of α-value on product distribution.

The exit H_2/CO ratio in the FT synthesis can be lower or higher than the H_2/CO ratio at the inlet of the reactor, depending on whether the initial H_2/CO ratio is lower or higher than the usage ratio. Figure 13.2 illustrates this change in the exit H_2/CO ratio of an FT reactor with an iron catalyst for different conversions when the initial H_2/CO ratio is below, at, or above the usage ratio of 1.55. Operating off the usage ratio will cause a shift in the H_2/CO ratio from feed to tail. In a continuously stirred tank reactor (CSTR), the catalyst is only exposed to the tail H_2/CO ratio, as no gradients exist in such a reactor. However, this is not the case for a fixed bed reactor where concentration profiles will exist along the bed length. Hydrocarbon selectivity is dependent on the H_2/CO ratio (with a higher H_2/CO ratio generally producing lighter products). Hence, there is a gradient in FT product with a gradient in H_2/CO ratio. It is thus preferable to either use a CSTR reactor or operate a fixed bed reactor in a gradientless regime when selectivity measurements are made. For reasons like temperature and conversion gradients in fixed bed reactors, caution should be applied when comparing selectivity data measured in a slurry bed reactor with data measured in a fixed bed reactor.

Furthermore, it is sometimes questionable to use literature data for modeling purposes, as small variations in process parameters, reactor hydrodynamics, and analytical equipment limitations could skew selectivity results. To obtain a full product spectrum from an FT process, a few analyses need to be added together to form a complete picture. This normally involves analysis of the tail gas, water, oil, and wax fractions, which need to be combined in the correct ratio (calculated from the drainings of the respective phases) to construct a true product spectrum. Reducing the number of analyses to completely describe the product spectrum is one obvious way to minimize small errors compounding into large variations in

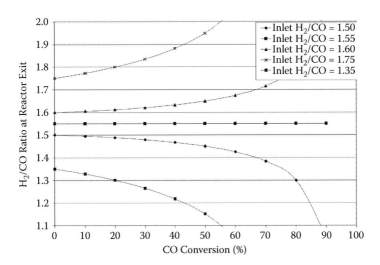

FIGURE 13.2 Change in the exit H_2/CO ratio of an Fe-LTFT reactor for different conversions when the inlet H_2/CO ratio is different from the usage ratio of 1.55.

the final product spectrum. These are only some of the challenges that need to be addressed for the accurate determination of a true product spectrum. It is therefore reasonable to state that the use of in-house selectivity data should be preferred for the development and testing of selectivity models. All selectivity data reported in the following discussions were obtained in a well-mixed laboratory-scale slurry reactor with a fresh-feed syngas ratio close to the usage ratio.[5]

The total product spectrum for a typical precipitated iron catalyst in an LTFT process is shown in Figure 13.3. Constructing an Anderson-Schulz-Flory (ASF) plot from the total product spectrum does not give a straight line and can conveniently be separated in two distinct regions, one from C_1 to C_8 and another from C_{20} onward (as shown in Figure 13.4). The light olefins and

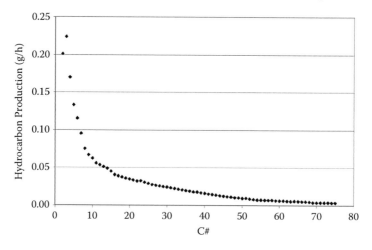

FIGURE 13.3   Typical hydrocarbon product distribution of a Fe-LTFT catalyst.

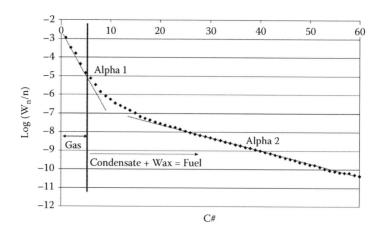

FIGURE 13.4   ASF plot of selectivity data presented in Figure 13.3.

oxygenates require extensive workup to be sold as valuable chemicals, and the light paraffins have a low fuel value that make this section of the product spectrum undesirable. Therefore, for fuels production, the C_{10+} selectivity should be optimized, and from the ASF graph it is clear that the products from the α_1 region should be minimized.

From either a chemicals or fuels perspective, hydrocarbon selectivity in the FT process is something that needs to be thoroughly understood in order to manipulate process conditions and allow the optimization of the required product yield to maximize the plant profitability. There are many unanswered questions regarding the selectivity of the Fe-LTFT synthesis. These include the selectivity of the oxygenates and what can be done to manipulate the oxygenate selectivity. Recent studies also showed some good correlation between CO_2 selectivity and acid selectivity, and highlighted a need for an analytical method for the accurate determination of low-level compounds like branched- and straight-chain olefins, alcohols, aldehydes, ketones, and acids. The developments in GCxGC as an analytical tool for the analysis of very complex product streams have paved the way to start an investigation into the oxygenate selectivity. Subsequently, Sasol's effort in GCxGC will be discussed to illustrate the power of this analysis method. As a conclusion, some interesting relationships between oxygenate selectivity and other parameters will be shown.

13.2 AN INCREASE IN THE ACCURACY OF THE FT PRODUCT ANALYSIS WITH GCxGC

Figure 13.5 shows the ASF plot for the linear acids, n-alcohols, and total product as analyzed with a multidimensional GC. It is clear that the total product shows a two-alpha distribution with a break between C_8 and C_{10}. Furthermore, α_1 for the

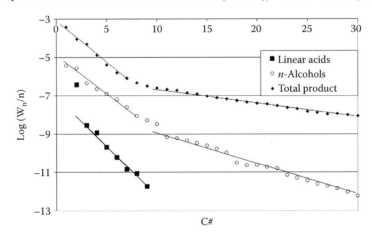

FIGURE 13.5 ASF plot for hydrocarbon classes as determined by full analysis of the FT product.

TABLE 13.1
α-values for Compound Classes

	α₁	α₂
Total product	0.59	0.93
n-alcohols	0.62	0.82
Linear acids	0.59	—

acids and alcohols is similar to that for the total product (see Table 13.1), which may suggest that the same mechanism is at play for the production of oxygenates and light hydrocarbons.[6] If the alcohols are measured up to sufficiently high carbon numbers, they also show a bend in the distribution.

Even more detail can be extracted from the GC×GC analysis, as illustrated in Figure 13.6. If the branching is compared for different product classes (Figure 13.7), it is clear that the acids have a much higher percentage of branching. This may give us a hint as to what is happening on the catalyst surface where the product is made. The detail regarding oxygenate selectivity as presented in Figures 13.5 to 13.7 will lead to a more fundamental understanding of the formation of FT products and, in particular, the mechanism of oxygenate formation.

13.3   IMPROVEMENTS IN THE SAMPLING METHOD FOR ACCURATE ALPHA DETERMINATION

A full product spectrum for the LTFT synthesis as measured in a laboratory CSTR is based upon the products collected in the knockout vessels after leaving

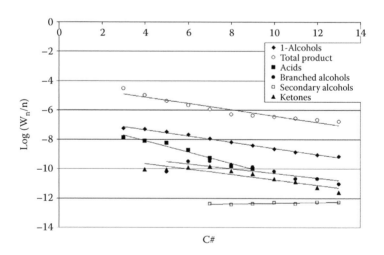

FIGURE 13.6   ASF plot of oxygenates as determined by GC×GC.

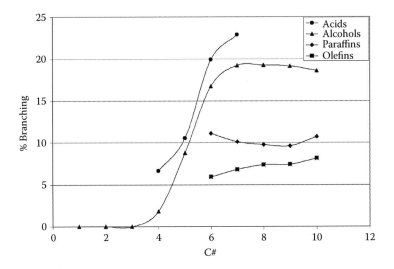

FIGURE 13.7   Branching of different hydrocarbon classes.

the reactor. Heavy FT products have a low vapor pressure, and therefore a long residence time in the reactor. If the products collected are not representative of those formed during the sampling period, the product spectrum will not be an accurate representation of the catalyst selectivity during the period under investigation.[7] Deviations brought about by residence time effects can have a big impact on the value of α, so it is crucial to have a realistic and accurate product distribution in order to calculate true α-values.

One way to get a representative product distribution for a specific period is to remove all FT products in the reactor system and replace them with a substance that will not influence selectivity determination. The FT reaction is then run for a specific period, after which a full analysis can be done that will represent only the products produced during that specific period. In Figure 13.8, data are presented for a run started with the catalyst suspended in a highly paraffinic wax (FT H1 wax, C_{30}–C_{90}). After a certain time of synthesis, the FT run was stopped and the catalyst placed under inert conditions (argon). The reactor content was then displaced with degassed and dried polyalphaolefin oil (Durasyn). After restarting the FT synthesis, the total product spectrum was determined (H1 run after displacement). It was found that the value of α_2 was much lower than before the displacement of the H1 wax. In fact, the α_2 values were quite comparable to those measured when the FT synthesis was started up with Durasyn (compare with Durasyn runs 1, 2, and 3). This clearly illustrates the impact that the reactor medium used to start the FT reaction can have on the determination of the α-value. The results further show that there was no change in the value of α_2 of the iron catalyst up to 500 h on-line.

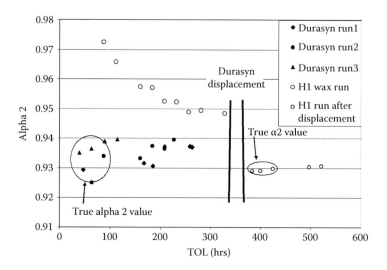

FIGURE 13.8 True α_2 determination. Durasyn runs 1, 2, and 3 were started with Durasyn as solvent.

13.4 OBSERVED CORRELATIONS BETWEEN PRODUCT SELECTIVITIES IN THE FE-LTFT SYNTHESIS

Although the manipulation of process conditions to optimize certain fractions of the product spectrum can significantly increase the economic feasibility of LTFT as a fuel-producing technology, another aspect of the selectivity data is that a better understanding of the mechanism or catalyst surface can be obtained. This will allow certain predictions and further optimizations to be made from a more fundamental viewpoint. One such example is the previously published correlations between CO_2 and acid selectivity for Fe-LTFT catalysts, where the potassium levels on the catalysts were varied,[8] as illustrated in Figure 13.9. A very interesting correlation was also found between the double-bond isomerization and the acid selectivity (Figure 13.10). Accurate selectivity data on low-level minor products like acids, internal olefins, esters, etc., and correlations like those shown in Figures 13.9 and 13.10 opened up a whole new field to investigate interdependent parameters. Questions like the extent to which the CO_2 selectivity is dependent on the acid selectivity (or vice versa) prompted investigations into co-feeding of organic acids and led to a novel way to manipulate the CO_2 selectivity.[9] From Figure 13.11, it is clear that the CO_2 selectivity can be lowered by co-feeding carboxylic acids.

To further understand the possible changes occurring on the catalyst, it is important to investigate the FT product in more detail. Previous efforts to simulate the Fe-LTFT product spectrum mainly focused on the paraffins and olefins in the lighter fraction (C_1-C_{10}).[10] This needs to be expanded to include the heavier fraction, as well as other product classes (such as oxygenates). The possible influence of

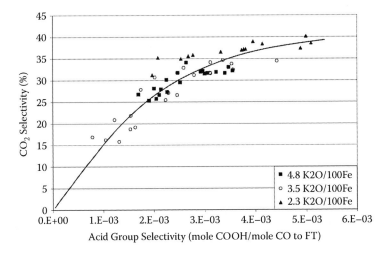

FIGURE 13.9 Correlation between CO_2 selectivity and acid group selectivity.

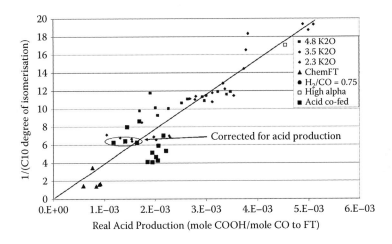

FIGURE 13.10 Inverse of degree of isomerization against acid group selectivity.

secondary reactions, e.g., hydrogenation and double-bond isomerization, must also be accounted for. At Sasol it has proved useful to focus on the oxygenate fraction in order to get a more fundamental understanding of the changes in the catalyst.

13.5 SELECTIVITY CHANGES OVER TIME

There are also changes that occur while the catalyst ages. For instance, it was observed that the methane and CO_2 selectivities of the Fe-LTFT synthesis increase concomitantly with increasing catalyst age, as illustrated in Figure 13.12. From

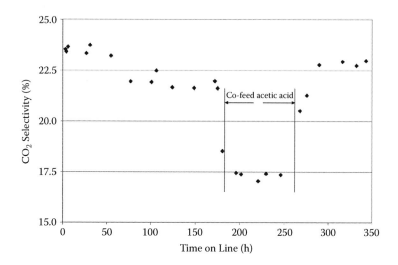

FIGURE 13.11   Effect of acid co-feeding on CO_2 selectivity.

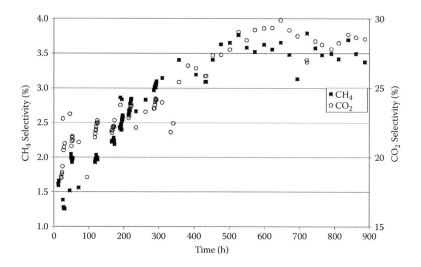

FIGURE 13.12   Change in CH_4 and CO_2 selectivities as the Fe-LTFT catalyst ages.

the same run, it was observed that the ratio between oil and wax changes over synthesis time (Figure 13.13). It is also clear that the product selectivity did not significantly change with a catalyst age beyond 500 h on-line. These variations in the product selectivity indicate that there are changes occurring on the catalyst surface. These changes are possibly related to the conversion of carbidic to oxidic phases, since it is widely accepted that the Fe-LTFT catalyst oxidizes as it ages.[11]

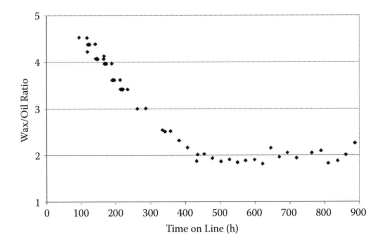

FIGURE 13.13 Change of wax to oil ratio as catalyst ages (wax condensation temperature = 200°C, oil condensation = 25°C).

13.6 CONCLUSION

Several decades of FT selectivity research focused mainly on the paraffins and olefins, as the available GC techniques were able to give good resolution on these compounds. Fortunately, better analysis is now possible with GCxGC, and progress has been made to accurately analyze low-level oxygenates. In the foregoing discussions it has been shown that there is more to FT than just paraffins and olefins. Oxygenates, probably formed via the same primary reaction, contain information about the selectivity behavior of the catalyst. Accurate quantification of oxygenates gives the advantage that different product classes can be compared in terms of branching, α-value, etc. Comparison of these different classes will aid in developing better mechanistic insight into the FT reaction. There are also very interesting correlations between apparently unrelated parameters, especially with respect to oxygenates. These relationships will test our fundamental understanding of the FT process, as they must be explained by the mechanisms proposed for the reaction.

In terms of industrial application, Figure 13.14 gives a good indication of what is required for Fe-LTFT to be really competitive as a diesel process, namely, a low gas and high liquid plus wax selectivity. When we understand what factors really control the FT product spectrum, we will be in a position to manipulate the FT product spectrum as indicated in Figure 13.14.

From the presented data there may be a correlation between the catalyst surface and the observed selectivity changes over time. These changes may be explained by a two-alpha model, where different products are produced on two different sites. Figure 13.15 shows how a product spectrum can be simulated with a change in ratio between the amounts of products produced on each of the two assumed types of sites. For this simulation the values of α_1 and α_2 were kept constant and only

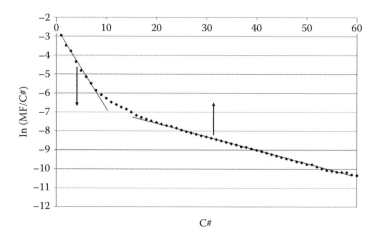

FIGURE 13.14   ASF plot for Fe-LTFT with required manipulations indicated (decrease in gas fraction and increase in liquid and wax fractions).

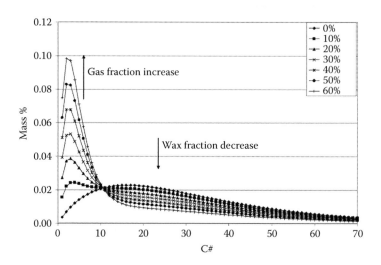

FIGURE 13.15   Effect of weight fraction of double-alpha model on product spectrum.

the relative contributions were changed. With an increase in the α_1 contribution, the gas fraction increases with a corresponding decrease in the wax fraction. This change could explain the observed changes, as shown in Figures 13.12 and 13.13.

   In order to manipulate the selectivity of the FT synthesis, there needs to be an understanding of the parameters that control the selectivity. The preceding discussion gives a flavor of what Sasol is doing regarding the understanding of Fe-LTFT selectivities. There are many activities to ensure more accurate measurements, e.g., developments in the GC×GC technique. There are also attempts to find correlations between apparently unrelated selectivities to get a firmer understanding of the observed product spectrum.

REFERENCES

1. Van Der Laan, G.P. 1999. Kinetics, selectivity and scale up of the Fischer Tropsch process. PhD dissertation, Rijksunivertiteit Groningen.
2. Donnelly, T.J., Yates, I.C., Satterfield, C.N. 1988. Analysis and prediction of product distributions of the Fischer-Tropsch synthesis. *Energy & Fuels* 2:734.
3. Schulz, H., Claeys, M. 1999. Kinetic modelling of Fischer-Tropsch product distributions. *Appl. Catal. A* 186:91.
4. Van der Laan, G.P., Beenackers, A.A.C.M. 1999. Hydrocarbon selectivity model for the gas-solid Fischer-Tropsch synthesis on precipitated iron catalysts. *Ind. Eng. Chem. Res.* 38:1277.
5. Govender, N.S., Janse van Vuuren, M., Claeys, M., Van Steen, E. 2006. Importance of the usage ratio in iron-based Fischer-Tropsch synthesis with recycle. *Ind. Eng. Chem. Res.* 45:8629.
6. Janse van Vuuren, M.J., Grobler, T., Bekker, R. 2007. Comprehensive analysis of FT product with GCxGC. Paper presented at the NACS 20th North American Meeting, Houston, TX, June 17–22.
7. Shi, B., Davis, B. 2004. Fischer-Tropsch synthesis: Accounting for chain length related phenomena. *Appl. Catal. A* 277:61.
8. Janse van Vuuren, M.J., Govender, G.N.S., Kotze, R., Masters, G.J., Pete, T.P. 2005. The correlation between double bond isomerisation, water-gas-shift and acid production during Fischer-Tropsch synthesis. *Prepr. Am. Chem. Soc. Div. Pet. Chem.* 50:200.
9. Janse van Vuuren, M.J. 2007. Introduction of an acid in a Fischer-Tropsch process. US 20070100004.
10. Botes, F.G. 2007. Proposal of a new product characterisation model for the iron-based low-temperature Fischer Tropsch synthesis. *Energy & Fuels* 21:1379.
11. Sirimanothan, N., Hamdeh, H.H., Zhang, Y., Davis, B.H. 2002. Fischer-Tropsch synthesis: Changes in phase and activity during use. *Catal. Lett.* 82:3–4.

14 Fischer-Tropsch Synthesis
Comparison of the Effect of Co-Fed Water on the Catalytic Performance of Co Catalysts Supported on Wide-Pore and Narrow-Pore Alumina

Amitava Sarkar, Gary Jacobs, Peter Mukoma, David Glasser, Diane Hildebrandt, Neil J. Coville, and Burtron H. Davis

CONTENTS

Catalysts with two different cobalt loadings (12 and 25%) supported on wide-pore and narrow-pore alumina (Degussa C-aluminoxide and Condea Vista Catalox SBA150 alumina, respectively) were prepared to study the effect of co-fed water during Fischer-Tropsch synthesis in a continuously stirred tank reactor (CSTR) operated at 220°C, 2.03 MPa, and H_2/CO = 2:1. The amount of co-fed water was varied between 5 and 30 vol%, while total space velocity of the feed and the partial pressure of H_2 and CO in the feed were kept constant during all experiments. The average pore diameters of the wide-pore and narrow-pore supports were 22.9 and 9.5 nm, respectively. In the 12% Co/Degussa C-alumina catalyst, the major fraction of cobalt clusters (size 12.2 nm) are believed to be located within the pore, while in the 25% Co/Degussa C-alumina (cluster diameter, 19.5 nm), 12% Co/SBA150 alumina (cluster diameter, 8.5 nm), and 25% Co/SBA150 alumina (cluster diameter, 9.2 nm), the cobalt clusters are thought to be outside the pore structure. A negative effect of co-fed water on catalytic activity was observed for all four catalysts. However, in the case of 12% Co/Degussa C-alumina catalyst, this negative effect was severe and, in particular, irreversible when 25 vol% water was added, while for the other three catalysts this effect was either marginal (25% Co/Degussa C-alumina) or significant (12 and 25% Co/SBA150 alumina), but always reversible even when up to 30% water was added. The CO_2 selectivity of all four catalysts was found to increase with the addition of water, suggesting the participation of the water-gas shift reaction in parallel with Fischer-Tropsch synthesis. The calculated C_{5+} selectivity of catalysts with 25% cobalt loading (for both wide-pore and narrow-pore supports) increased with co-fed water, while the CH_4 selectivity decreased. The differences in catalytic performance can be explained in terms of relative sizes of cobalt clusters, differences in pore volumes and pore networks of supports, location of cobalt clusters within pore networks (i.e., within the pore or outside the pore), and relatively higher residence time of water vapor within the pore, compared to the outside of the pores.

14.1 INTRODUCTION

Supported cobalt catalysts may be preferable for production of premium middle distillate, a wide variety of hydrocarbon and oxygenates produced by slurry phase Fischer-Tropsch synthesis (FTS) utilizing a syngas with high H_2/CO ratio, such as derived from natural gas. This is mainly due to their low selectivity for the water-gas shift reaction, good catalytic activity, high selectivity toward high molecular weight hydrocarbons, and high attrition resistance, particularly for alumina-supported cobalt catalysts in slurry phase operation.

Various materials, such as SiO_2, TiO_2, and Al_2O_3, have been used as support for the preparation of these catalysts. It has been found that the support has a significant influence on the reducibility of the cobalt species, catalytic activity, and product selectivities.[1] The pore characteristics of the support have a significant effect on the cobalt crystallite size in the catalyst after impregnation and calcination. In some cases, the support interacts strongly with the active phase, leaving a fraction of cobalt chemically inactive after reduction. In the case of supports with a weak interaction with cobalt (e.g., silica), cobalt oxide produces large cobalt clusters after standard reduction (at 623 K with hydrogen), while in the case of supports like TiO_2 and Al_2O_3, much smaller clusters with a smaller fraction of cobalt being reduced are observed.[2] The higher reducibility of wide-pore silica-supported catalysts than of narrow-pore silica-supported cobalt catalysts was explained by the fact that larger particles (in wide pores) are expected to behave more like bulk Co_3O_4 with respect to their reducibility than smaller particles.[3–5] It was reported that the ease of reduction decreases from larger (20–70 nm) to smaller (6 nm) silica-supported Co_3O_4 particles.[6] On the contrary, a negative correlation between the support pore size, Co_3O_4 particle size, and degree of reduction was observed for $Co/ \gamma - Al_2O_3$ catalysts.[7] It has been reported that in Co/Al_2O_3 catalysts, there is a strong interaction between the support and the cobalt oxide phase.[8] Therefore, the catalytic activity and product selectivity for FTS are influenced by the surface properties of the Al_2O_3 support and the nature of the species formed in the calcined and reduced catalyst.

The use of cobalt-based catalysts in Fischer-Tropsch synthesis on a commercial basis requires that the catalysts withstand long-term use at high CO conversion levels. In FTS with cobalt catalysts, water is produced at a significant rate, which may in turn influence the syngas conversion rate, product selectivity, and rate of catalyst deactivation. The rate of catalyst deactivation also depends on the type of reactor employed. With fixed bed reactors, high partial pressure of water at the reactor exit is generally observed. Ideally, in continuously stirred tank reactors (CSTRs), a uniform water concentration and temperature profile exists throughout the entire reactor as a result of extensive back-mixing. High CO conversion is feasible in a CSTR slurry system due to efficient control of the reaction exotherm.[9] High CO conversion also results in a high concentration of water in the reaction environment. In spite of many studies that have been conducted regarding the effect of water on the activity of cobalt-based FTS catalysts, there has been little consensus in the findings. The impact of water on the performance of cobalt catalysts can be traced to the composition, support, preparation procedure, and pretreatment. The effect of water on cobalt catalysts has been reported to be negligible,[10–14] negative,[15–17] or even positive,[18,19] depending on the nature of the support. This is probably due to different support materials, promoters, cobalt precursors, and preparation methods that are used by various researchers.[20–22] The oxidation of cobalt metal to cobalt oxide or cobalt aluminate (in alumina-supported catalysts) by the product water has long been postulated to be a significant cause of deactivation in supported Co catalysts.[23] From the study on the formation of metal-alumina spinel at high temperatures using a model system consisting of thin metal films of Fe, Ni, Co, and Cu evaporated onto flat polycrystalline $\alpha - $ alumina, Bolt[24] reported that water increased the rate of metal-aluminate formation, and the reaction rate for aluminate formation increased in the order: Fe < Ni < Co < Cu. Generally, the extent of deactivation depends on the water partial pressure in the reactor. Low water partial pressures are associated with temporary catalytic activity loss, and high water partial pressures are associated with more permanent loss of the catalyst activity. It was proposed that conditions that favor the formation of low valence ions also favor spinel formation.

γ–Alumina was found to have a much higher reactivity than α–alumina, which may be due to the higher mobility of Co and Al ions along the grain boundaries and surfaces of the internal pores of the γ–Al_2O_3 support in the presence of steam, and the similarity in structure with the cobalt aluminate product.

From an analysis of the results of water co-feeding experiments in a fixed bed reactor and utilizing thermogravimetric analysis (TGA) and x-ray photoelectron spectroscopy (XPS) techniques, Schanke et al.[25–27] suggested the oxidation of cobalt species to cobalt oxide was the main reason for catalyst deactivation, and that the extent of reoxidation depends on the partial pressure of water and the H_2O/H_2 ratio. Similar studies by Hilmen et al.[16,28,29] concluded that surface oxidation of small cobalt crystallites was responsible for catalyst deactivation during FTS. The effect of water during FTS on the number of active sites on a Co/Al_2O_3 and $Co/Re/Al_2O_3$ catalyst using steady-state isotopic transient kinetic analysis (SSITKA) was studied by Rothaemel et al.[30] Co-feeding of water was found to decrease the number of active sites during FTS, which was assigned to the surface oxidation of the metallic cobalt during water treatment. It was concluded that the catalyst surface available for methane formation and reversible CO adsorption decreased after water addition.

A number of studies at the University of Kentucky Center for Applied Energy Research (CAER) focused on testing the stability of low loaded (~15% Co) cobalt catalysts, whereby a reduction promoter was used to facilitate reduction of smaller cobalt oxide species interacting strongly with the support. Jacobs et al.[31,32] analyzed wax-coated used Ru–promoted 15% Co/Al_2O_3 catalysts collected during FTS by x-ray absorption near-edge spectroscopy (XANES) and found oxidation of a fraction of the cobalt in spent catalysts. It was concluded that small cobalt crystallites oxidized during FTS to Co_3O_4 or cobalt aluminate. It was also reported that a clear differentiation between the two cobalt phases was not possible using this technique. It was postulated that the fraction that did oxidize was most likely the small cobalt clusters. Slight oxidation of a fraction of Co was also observed by Das et al.[33] for a Re-promoted 15% Co/Al_2O_3 catalyst tested at 220°C, H_2/CO = 2, P = 20 bar. Deactivation was also accompanied by growth of the peak for Co–Co metal coordination in EXAFS as a function of time on stream, suggesting sintering. The influence of co-fed water on the deactivation of Pt-promoted 15% Co/Al_2O_3 catalysts during FTS (210 °C, 29 bar, H_2/CO = 2) was reported by Li et al.[20] A reversible effect of water was observed at low amounts in the reactor (about $P_{H_2O}/P_{H_2} < 0.5$). However, an irreversible deactivation of the Pt-promoted 15% Co/Al_2O_3 catalyst along with an increase in the CO_2 selectivity was observed with a P_{H_2O}/P_{H_2} ratio of 0.6 in the reactor. It was suggested that the catalyst underwent oxidation to cobalt oxide or cobalt aluminate.

In a similar fashion, from analysis of results of water co-feeding experiments with the Pt-promoted 15% Co/Al_2O_3 catalyst (~5.6 nm cluster size), Jacobs et al.[34] concluded that, at $P_{H_2O}/P_{H_2} < 0.6$ in the reactor, a negative but reversible kinetic effect takes place with no changes in the cobalt structure as observed with XANES. However, when a higher P_{H_2O}/P_{H_2} ratio in the reactor was used ($P_{H_2O}/P_{H_2} > 0.6$), irreversible deactivation of the catalyst took place. XANES analyses revealed that cobalt oxidation took place, including cobalt aluminate formation. In another study, Jacobs et al.[21] reported the oxidation behavior of unpromoted alumina-supported cobalt catalysts, containing 15 and 25 wt% cobalt, using water co-feeding during FTS. It was found that the 25 wt% Co/Al_2O_3 (~11.8 nm cobalt cluster diameter) catalyst did not significantly deactivate irreversibly under the following conditions: 210°C, P_{tot} = 20 bar, P_{H_2O} = 7.7 bar, P_{H_2} = 6.2 bar. At 30% volume % added H_2O, XANES analysis confirmed that a fraction of cobalt oxidized to CoO and a drop in CO conversion took place; however, when the H_2O was switched off, the catalyst recovered. The 15 wt% Co/Al_2O_3 catalyst showed irreversible deactivation during FTS (conditions: 220°C, P_{tot} = 19.7 bar, P_{H_2O} = 7.4 bar, P_{H_2} = 8.1 bar), which was explained by cobalt oxidation and the subsequent formation of cobalt-aluminate, as observed by XANES analyses. The results are consistent with Sasol's view that high loadings of cobalt are important for stabilizing against support-influenced reoxidation phenomena. Catalysts with a larger average cluster size are more robust.

It is evident that the pore structure of the support, the cobalt loading of the catalyst, and the P_{H_2O}/P_{H_2} ratio during FTS significantly affect the catalyst performance and deactivation rate. The objective of the present research is to ascertain the effect of water on the performance of catalysts with two different cobalt loadings (i.e., 12 and 25%) supported on wide-pore and narrow-pore alumina (i.e., Degussa C-aluminoxide and Condea Vista Catalox SBA150, respectively). It is expected that the cobalt clusters will be situated within the pore in the case of the wide-pore support, while in the case of the narrow-pore support a higher fraction of the cobalt clusters will likely be outside the pore. The amount of co-fed water was varied to obtain a wide variation in P_{H_2O}/P_{H_2} ratio. The activity and product selectivity of these catalysts can be compared, and the effect of water will be studied in an attempt to explain it.

14.2 EXPERIMENTAL

14.2.1 CATALYST PREPARATION

A narrow-pore γ-Al_2O_3 (Condea Vista Catalox SBA150, high purity) and a wide-pore γ-Al_2O_3 (Degussa C-aluminoxide, high purity) were used as the supports for catalyst preparation. The Degussa C-aluminoxide support is a narrow particle size material with a high surface area. When wetted with water to the incipient wetness point, dried in air at 375 K for 24 h, and calcined at 623 K for 5 h, the material is transformed. The particles agglomerate into clusters, and this results in a sponge-like structure characterized by cavities with an average diameter of 25.5 nm. The catalysts were prepared by a slurry impregnation method, and cobalt nitrate was used as the cobalt precursor. Two catalysts were prepared from each support with two different cobalt loadings: 12 and 25%.

In the slurry impregnation method, which adheres in part to a patented procedure,[35] the ratio of the volume of loading solution used to the weight of alumina was 1:1, such that approximately 2.5 times the pore volume of solution was used to prepare the catalyst. Multiple impregnations were required to achieve the final loading of the catalyst. Between two impregnation steps, the catalyst was dried under vacuum in a rotary evaporator at 333 K and the temperature was slowly increased to 373 K. After the final impregnation/drying step, the catalyst was calcined at 623 K.

14.2.2 Characterization

Structural characterization of the prepared Co/alumina catalysts was studied by using the following techniques: Brunauer-Emmett-Teller (BET), temperature-programmed reduction (TPR), H_2 chemisorption by temperature-programmed desorption (TPD) with O_2 pulse reoxidation, and powder x-ray diffraction (XRD).

14.2.2.1 Surface Area and Pore Size Distribution

The surface area was calculated using the BET equation,[36] while the total pore volume and the average pore size were calculated from the nitrogen desorption branch applying the Barrett-Joyner-Halenda (BJH) method.[37] BET and BJH adsorption measurements were carried out with a Micromeritics Tri-Star system on both the supports and the calcined catalysts. Prior to measurements, the samples were evacuated at 433 K to approximately 50 mTorr for 4 h.

14.2.2.2 Temperature-Programmed Reduction

TPR profiles of calcined catalysts were recorded using a Zeton-Altamira AMI-200 unit, which employs a thermal conductivity detector (TCD). Calcined samples were first purged at 623 K in flowing Ar to remove traces of water. A liquid nitrogen trap was used to prevent water generated by reduction from interfering with the TCD signal. TPR was performed using a 10% H_2/Ar mixture at a flow rate of 30 cm³/min, using Ar as the reference to maximize the signal-to-noise ratio. The sample was heated from 323 to 1,300 K using a heating ramp of 10 K/min.

14.2.2.3 H_2 Chemisorption by TPD and Percentage Reduction by Pulse Reoxidation

Hydrogen chemisorption was conducted using temperature-programmed desorption, also measured with the Zeton-Altamira AMI-200 instrument. The sample weight was 0.22 g. Catalysts were activated in a flow of 10 cm³/min of H_2 mixed with 20 cm³/min of Ar at 623 K for 10 h, and then cooled under flowing H_2 to 373 K. The sample was held at 373 K under flowing argon to remove weakly bound species prior to increasing the temperature slowly to 623 K, the reduction temperature of the catalyst. The catalyst was held under flowing Ar to desorb any remaining chemisorbed hydrogen until the TCD signal returned to baseline. The TPD spectrum was integrated and the number of moles of desorbed hydrogen determined by comparing its area to the areas of calibrated hydrogen pulses. The loop volume was first determined by establishing a calibration curve with syringe injections of nitrogen in helium flow. Dispersion calculations were based on the assumption of a 1:1 H:Co stoichiometric ratio and a spherical cobalt cluster morphology.

In estimating the cluster size from hydrogen chemisorption measurements, as reported in the literature,[38] it is necessary to estimate the fraction of cobalt existing in the metallic phase. Therefore, after TPD of hydrogen, the sample was reoxidized at 623 K using pulses of oxygen in a helium carrier. The percentage of reduction was calculated by assuming that the metal reoxidized to Co_3O_4. Further details of the procedure are provided elsewhere.[2,38] In summary, the volume component of the calculated dispersion is corrected by the percentage reduction, as follows:

$$D_{corr} (\%) = \frac{\text{number of } Co^0 \text{ atoms on surface}}{\text{total number of } Co^0 \text{ atoms}} \times 100$$

$$D_{corr} (\%) = \frac{\text{number of } Co^0 \text{ atoms on surface} \times 100}{(\text{total number of Co atoms in sample}) (\text{fraction reduced})}$$

Geometrical arguments are used only for the reduced metal to obtain an estimate of cluster size. This is equivalent to the assumption that the unreduced cobalt is interacting with the support or is present as a cobalt aluminate species. Despite the inherent assumptions of the method, the resulting cluster size much more closely approximates the true cluster size in comparison with reported data that erroneously assume that the cobalt is 100% reduced (i.e., excludes the reoxidation step).

14.2.2.4 X-ray Diffraction

Powder x-ray diffraction patterns for calcined catalysts were recorded at room temperature using a Philips X'Pert Diffractometer and CuKα radiation (λ = 1.54 Å). The scans were recorded in the 2θ range between 10° and 90° using a step size of 0.04°. Long scans were made over the intense peak at 2θ of 36.8° so that estimates of Co_3O_4 cluster size could be assessed from line-broadening analysis. The Co_3O_4 particle size was converted to the corresponding cobalt metal particle size according to the relative molar volumes of metallic cobalt and Co_3O_4 with a resulting contraction factor of 0.75 for conversion from the oxide to metallic phase.[32]

14.2.2.5 Reaction System and Procedure

The FTS experiments were conducted in a 1 L CSTR equipped with a magnetically driven stirrer with turbine impeller, a gas inlet line, and a vapor outlet line with an stainless steel (SS) fritted filter (7.0 microns) placed external to the reactor. A tube fitted with an SS fritted filter (2.0 micron opening) extends below the liquid level of the reactor for withdrawing reactor wax to maintain a nearly constant liquid level in the reactor. Another SS dip tube (1/8 inch OD) extends to the bottom of the reactor, and this could be used to withdraw catalyst/wax slurry samples from the reactor at different synthesis times. Separate mass flow controllers were used to control the flow of hydrogen and carbon monoxide at the desired rate. The gases were premixed in a vessel before entering the reactor. Carbon monoxide was passed through a vessel containing lead oxide–alumina to remove traces of iron carbonyl. The mixed gases entered the CSTR below the stirrer operated at 750 rpm. The reactor slurry temperature was maintained constant ($\pm$ 1°C) by a temperature controller.

Prior to loading into the CSTR, the calcined catalyst (ca. 15 g, powdered form) was reduced *ex situ* in a fixed bed reactor at 623 K for 10 h in hydrogen at a flow rate of 1 L/min. The reactor temperature was increased from room temperature to 373 K at the rate of 2 K/min and held at 373 K for 1 h; then the temperature was increased to 623 K at a rate of 1 K/min and kept at 623 K for 10 h. The catalyst was then transferred pneumatically under the protection of helium to the CSTR, which contained 310 g of melted Polywax-3000 (polyethylene fraction with an average molecular weight of 3,000). To facilitate the transfer, the fixed bed reactor was connected to the CSTR using a transfer tube fitted with a ball valve. The fixed bed reactor was pressurized with argon, forcing the catalyst powder out of the reactor through the valve. The reactor was weighed before and after the transfer of the catalyst to ensure that all the catalyst powder was transferred to the CSTR. The catalyst was then reduced *in situ* with hydrogen at a flow rate of 60 SL/h at atmospheric pressure. With the temperature controller programmed in a ramp/ soak mode, the reactor temperature was ramped up to 553 K at a rate of 2 K/min and held at 553 K for 24 h.

After the activation period, the reactor temperature was decreased to 453 K, synthesis gas (H_2:CO = 2:1) was introduced to the reactor, and the pressure was increased to 2.03 MPa (20.7 atm). The reactor temperature was increased to 493 K at a rate of 1 K/min, and the space velocity was maintained at 5 SL/h/g_{cat}. The reaction products were continuously removed from the vapor space of the reactor and passed through two traps, a warm trap maintained at 373 K and a cold trap held at 273 K. The uncondensed vapor stream was reduced to atmospheric pressure through a letdown valve. The gas flow was measured using a wet test meter and analyzed by an online GC. The accumulated reactor liquid products were removed every 24 h by passing through a 2 μm sintered metal filter located below the liquid level in the CSTR. The conversions of CO and H_2 were obtained by gas chromatography (GC) analysis (micro-GC equipped with thermal conductivity detectors) of the reactor exit gas mixture. The reaction products were collected in three traps maintained at different temperatures: a hot trap (200°C), a warm trap (100°C), and a cold trap (0°C). The products were separated into different fractions (rewax, wax, oil, and aqueous) for quantification. However, the oil and wax fractions were mixed prior to GC analysis.

In all experiments, the pressures of CO plus H_2 in the feed were kept constant at 70% of the total pressure, while argon was used to make up the balance before, during, and after H_2O addition experiments. When water was added, a fraction of the argon was replaced by water. The amount of water added was calculated based on the amount of argon to be replaced, thus ensuring that the sum of the water and argon partial pressures remained at 30% of the total pressure. Hence, the sum of partial pressure of water vapor added plus argon partial pressure was always 0.878 MPa (8.67 atm). Water addition was done using a high-precision, high-pressure ISCO syringe pump. The added water, together with the water produced by the reaction, was collected in the warm and cold traps and the fractions were combined to ensure an accurate water balance.

14.3 RESULTS AND DISCUSSIONS

14.3.1 CHARACTERIZATION

To determine any structure modification to the Degussa C-aluminoxide support during wet impregnation and calcinations, adsorption/desorption isotherms obtained before and after modifications were compared. The results as presented in Figure 14.1 are in agreement with our previous results.[39] It is found that the

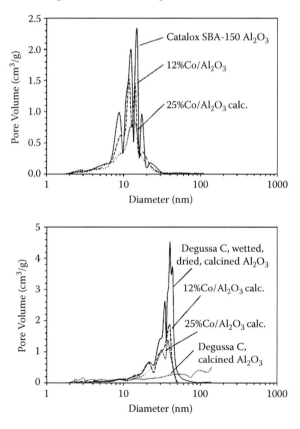

FIGURE 14.1   Pore size distributions by BJH adsorption of nitrogen for (top) Catalox SBA150 and (bottom) Degussa C (wetted, calcined) γ-Al$_2$O$_3$ supports and supported cobalt catalysts.

as-received Degussa C-aluminoxide yields a type II BET adsorption isotherm. The wetted/dried/calcined Degussa C-alumina, on the other hand, yields a type IV adsorption isotherm, indicating the filling of a limited porous structure. The average pore volume of the modified Degussa C-aluminoxide is almost twice that of the as-received material. The BET surface area, BJH pore volume, and average pore radius for both supports and all four catalysts are presented in Table 14.1.

Cobalt catalysts supported on alumina may have limited reducibility due to a strong interaction between the support and cobalt oxides. Calcined Co/Al_2O_3 catalysts have a complicated composition on the surface of the support, and this has been suggested to be a mixture of Co_3O_4 and cobalt support complexes (e.g., cobalt aluminates).[40] Unlike bulk Co_3O_4, which gives a sharp TPR profile centered below 673 K, broad profiles typical of Co/Al_2O_3 Fischer-Tropsch catalysts, consisting of a relatively sharp peak centered about 600 K and a broad peak centered

TABLE 14.1
BET Surface Area, Pore Volume, and Average Pore Diameter of the Supports and Catalysts after Calcination at 623 K

Catalyst Description	Measured BET SA (m²/g)	BJH Des. Pore Vol. (cm³/g)	BJH Des. Ave. Pore Diameter (nm)
Condea Vista γ-Al_2O_3 Catalox SBA150 Narrow pore	149	0.50	9.5
Degussa C γ-Al_2O_3 Dried, calcined	103	0.34	14.0
Degussa C γ-Al_2O_3 Wetted, dried, calcined Wide pore	117	0.80	22.9
12% Co/γ-Al_2O_3 Catalox SBA150 Slurry impregnation Narrow pore	127	0.37	9.3
12% Co/γ-Al_2O_3 Degussa C γ-Al_2O_3 Wetted, dried, calcined Wide pore	89.1	0.48	18.6
25% Co/γ-Al_2O_3 Catalox SBA150 Slurry impregnatio Narrow pore	103	0.26	8.7
25% Co/γ-Al_2O_3 Degussa C γ-Al_2O_3 Wetted, dried, calcined Wide pore	76.8	0.36	16.8

about 850–950 K, are observed for all four catalysts, as presented in Figure 14.2. A number of explanations have been given to explain the broad profile of Co/Al$_2$O$_3$ catalysts.[21] Some favor the idea that atoms of the support are present in the cobalt oxide clusters themselves, yielding a broad reduction profile for a surface overlayer containing, in part, Co^{2+} and Co^{3+} ions; other explanations favor the idea that the small cobalt oxide clusters exhibit varying degrees of surface interaction with the support, and this hinders their reduction to different degrees. One may consider the two peaks of the profile to reflect a two-step reduction process passing through CoO as the intermediate: Co$_3$O$_4$ → 3CoO → 3Co$^\circ$. In such a scenario, it is the CoO that is most sensitive to the surface interaction, and the strong interactions with the support can stabilize the cobalt oxide phase. This view has been confirmed by recent TPR EXAFS/XANES investigations.[41,42] In the case

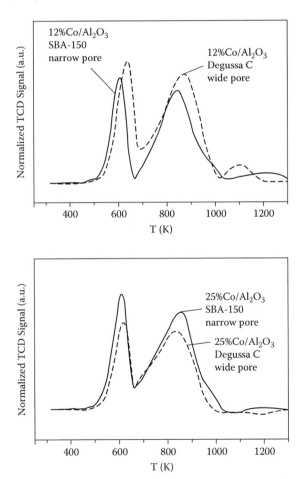

FIGURE 14.2 TPR profiles of the Co/Al$_2$O$_3$ catalysts prepared by the slurry phase impregnation method.

of Degussa C-alumina-supported catalyst, the lower loaded 12% Co catalyst displays a higher fraction of species interacting strongly with the support than the 25% Co catalyst, as can be observed by the presence of a higher shoulder at about 1,100 K. The reduction temperature maxima for the 12% Co catalyst supported on SBA150 alumina is found to be lower than the 12% Co catalyst supported on Degussa C-alumina, suggesting a lower temperature reduction of the cobalt species present in the narrow-pore network for lower Co loading. However, for the 25% Co catalyst, the reduction temperature maxima are similar for both the Degussa C-alumina- and the SBA150 alumina-supported catalyst. The high-temperature shoulder at about 1,100 K found in the case of the 12% Co loading was not observed at 25% Co loading.

To calculate the cluster size and the degree of reduction by chemisorption, a method in which the amount of oxygen consumed by the metallic component during reoxidization following the reduction procedure has been employed to estimate the metallic fraction of cobalt that is present.[2] The extent of reduction term, calculated from the O_2 uptake measurement, is included in the denominator of the dispersion equation so that an accurate estimate of the true Co^o dispersion, and therefore, the average cobalt diameter, can be obtained. Results hydrogen chemisorption, pulse reoxidation for the Co/Al_2O_3 catalysts are presented in Table 14.2 along with the corrected average diameter for the cobalt clusters. Although the TPR profile of 12% Co/SBA150 alumina catalyst indicated an easy reduction of cobalt species compared to the 12% Co/Degussa C-alumina catalyst, results of H_2 chemisorption by TPD of H_2 and pulse reoxidation study clearly suggest that percentages of reduction of cobalt species are much higher for wide-pore supported catalysts than the narrow-pore supported catalysts for a similar cobalt loading (Table 14.2). However, it is found that the percent dispersion of the cobalt crystallites is higher for the narrow-pore supported catalysts than for the wide-pore supported catalysts at corresponding cobalt loading (Table 14.2). The higher reducibility of the wide-pore supported catalysts than of the narrow-pore supported catalyst may be explained by the fact that larger particles (in wide pores) are expected to behave more like bulk Co_3O_4, and the reduction decreases in going from the larger to smaller Co_3O_4 particles.[6]

Powder XRD patterns of the calcined catalyst were utilized to estimate the cluster size of the cobalt crystallites based on a contraction factor. For ease of comparison of cobalt cluster diameter calculated from XRD measurement and H_2 TPD/pulse reoxidation, the results from the above two methods are presented in Table 14.3. It can be seen that results calculated from the data for the above two measurements are in close agreement after considering the percentage reduction. The average diameter of the cobalt cluster for the 25% Co/Degussa C-Al_2O_3 catalyst is almost double that of the 12% metal-loaded catalyst. It can be assumed, therefore, that for the 12% Co/Degussa C-Al_2O_3 catalyst, nearly all of the cobalt metal particles can be present within the pores of the support since the average pore diameter of the support, 22.9 nm, is almost twice as large as the average cobalt cluster size, 12.2 nm. However, for the 25% Co/Degussa C-alumina

TABLE 14.2
Results of H$_2$ Chemisorption by TPD of H$_2$ and Pulse Reoxidation for the Co/Al$_2$O$_3$ Catalysts

Catalyst	Al$_2$O$_3$ BJH Ave. Pore Diameter (nm)	Calc. T (K)	Red. T (K)	H$_2$ Desorb. per g cat (mmol)	Uncorr. % Disp.	Uncorr. Diameter (nm)	O$_2$ Uptake per g cat (mmol)	% Red.	Corr. % Disp.	Corr. Diameter (nm)
12% Co/Al$_2$O$_3$ (Catalox SBA150) (Narrow pore)	9.5	623	623	60.9	6.0%	17.3	665	49.0	12.2%	8.5
12% Co/Al$_2$O$_3$ (Degussa C) (Wide pore)	25.5	623	623	58.3	5.7%	18.0	921	67.8	8.4%	12.2
25% Co/Al$_2$O$_3$ (Catalox SBA150) (Narrow pore)	9.5	623	623	89.2	4.2%	24.5	1058	37.4	11.2%	9.2
25% Co/Al$_2$O$_3$ (Degussa C) (Wide pore)	25.5	623	623	67.3	3.2%	32.5	1698	60.1	5.3%	19.5

TABLE 14.3

Results of XRD in Comparison with H_2 TPD/Reoxidation Data and Average Pore Size and Co° Cluster Size

Catalyst	Co$_3$O$_4$ Cluster Size (nm) from XRD	Calculated Co° Cluster Size (nm)	Co° Cluster Size from H$_2$ TPD/Reoxid.	Support Ave. Pore Diameter (nm)
12% Co/γ-Al$_2$O$_3$ (Catalox SBA150) (Narrow pore)	11.2	8.4	8.5	9.5
12% Co/γ-Al$_2$O$_3$ (Degussa C) (Wide pore)	13.3	10.0	12.2	22.9
25% Co/γ-Al$_2$O$_3$ (Catalox SBA150) (Narrow pore)	14.4	10.8	9.2	9.5
25% Co/γ-Al$_2$O$_3$ (Degussa C) (Wide pore)	23.2	17.4	19.5	22.9

catalyst, a major fraction of cobalt clusters can be assumed to be located near the pore entrance or external to the pore.

14.3.2 EFFECT OF WATER ADDITION ON CATALYST ACTIVITY

Experiments were performed with the four catalysts (mentioned in Tables 14.1 and 14.2) with co-fed water to determine the effect of water on the activity and selectivity for FTS. The amount of co-fed water was varied between 5 and 30 vol% in different experiments by replacing corresponding amounts of argon to maintain constant syngas partial pressure and space velocity. The partial pressures of the inlet H_2 and CO were kept constant during the experiments. After each period of water addition, the reaction conditions were adjusted by stopping the water flow and adding the required amount of argon so as to obtain flow conditions similar to those obtained before water addition (i.e., constant space velocity of feed stream).

The variation in CO conversion during FTS with co-fed water for the 12% Co/Degussa C-alumina catalyst is shown in Figure 14.3, while for 25% Co/Degussa C-alumina catalyst, the CO conversion profile is presented in Figure 14.4. It is evident from Figure 14.3 that with the addition of water along with syngas, the catalytic activity decreases, but the effect is reversible in the 5 to 20 vol% water addition range. The catalyst activity was found to recover once the water addition was terminated. The magnitude of conversion decline with time on stream

(0.543% per day) suggests that the catalyst was not impacted irreversibly by the brief operating periods with additional water present. The temporary decline in CO conversion with the addition of co-fed water may be due to the kinetic effect of water by adsorption inhibition, which may be due in part to reversible oxidation. Most kinetic rate equations for cobalt-catalyzed Fischer-Tropsch synthesis do not include the influence of water.[43] However, when 25 vol% water was added, a severe and irreversible loss in the catalytic activity was observed. The CO conversion decreased from 11.77% to 0.68% with the addition of 25 vol% water. More importantly, the catalyst activity was not recovered to the expected value after the water addition was terminated, indicating that the catalyst had deactivated permanently. In the case of 25% Co/Degussa C-alumina catalyst (Figure 14.4), the rate of decline of CO conversion was higher (i.e., 2.54% per day) than that for the 12% Co/Degussa C-alumina. However, the addition of up to 30 vol% co-fed water did not show any significant additional effect.

It is proposed that in FTS with cobalt catalysts, water does not participate in the FT reaction, and minor reactions, such as water-gas shift, can be ignored. Hence, the water effect is normally not considered in the FT reaction kinetics with cobalt catalysts. In addition, kinetic studies that are performed at low CO conversions do not allow the true determination of the water effect since the water partial pressures in the reactor are low. It is commonly believed that the water effect is related to reoxidation of the cobalt catalysts, which leads to an activity loss for FT reaction, while the presence of hydrogen or CO (i.e., reducing environment) prevents the catalyst from oxidation when the water partial pressure is not too high to cause permanent deactivation. Therefore, it is conceivable that the water effect is such that the amount of catalyst active sites available for FT reaction changes with the partial pressures of water and hydrogen. Three possibilities for the deactivation of cobalt Fischer-Tropsch catalyst have been advanced: surface condensation,[44] sulfur poisoning,[45] and oxidation.[16,17] Thermodynamic calculations indicate that the oxidation of bulk phase metallic cobalt to either CoO or Co_3O_4 is unlikely,[17] particularly under the FTS conditions used in the present study. However, previous investigators[16,21,25,39] have found that irreversible deactivation of cobalt catalysts in Fischer-Tropsch synthesis may be a result of an oxidative process caused by the presence of a high partial pressure of water, and is often attributed to cobalt-support complex formation. However, oxidation of Co to $CoAl_2O_4$ is often considered to be kinetically restricted during typical Fischer-Tropsch synthesis conditions.[46] The higher resistance of the 25% Co/Degussa C-alumina catalyst toward deactivation upon addition of co-fed water may be due to the higher cobalt loading of the catalyst. Another possibility for higher resistance to water may be related to the difference in the relative size of cobalt clusters and their location within the pore network of the wide-pore support. The average pore diameter of Degussa C-alumina support is 22.9 nm, while the diameters of cobalt clusters for catalysts with 12 and 25% metal loading are 12.2 and 19.5 nm, respectively (Table 14.3). Hence, most of the cobalt clusters in 12% Co/Degussa C-alumina catalyst are probably located within the pore, while for 25% Co/Degussa C-alumina catalyst a major fraction of cobalt clusters is expected to be situated outside the pore. The residence time of water vapor within the pore is higher than the corresponding residence time at the catalyst surface. Thus, the exposure time of cobalt clusters to water vapor is higher for the 12% Co/Degussa C-alumina catalyst than for the 25% Co/Degussa C-alumina catalyst, which in turn can result in significant deactivation of the catalyst with co-fed water or even cause permanent deactivation.

The change in the catalytic activity for CO conversion during FTS with co-fed water for the 12% Co/SBA150 alumina catalyst is shown in Figure 14.5, while for 25% Co/SBA150 alumina catalyst, the variation in CO conversion level is presented in Figure 14.6. It can be seen from Figure 14.5 that the addition of co-fed water during FTS results in a significant and reversible change in catalytic activity for the 12% Co/SBA150 alumina catalyst. The decline in CO conversion with time on stream for this catalyst was 1.4% per day. However, the catalyst activity was found to recover once the water addition was terminated. The magnitude of reversible loss in catalytic activity with different amounts of co-fed water was found to be in the order: 25% > 20% > 10%. In the case of the 25% Co/SBA150 alumina catalyst (Figure 14.6), the CO conversion did not change significantly with the addition of up to 20 vol% water. This catalyst exhibited a small decline of 0.11% per day in CO conversion with time on stream. However, the conversion decreased significantly and reversibly during the addition of 25 and 30 vol% water.

The above behavior of narrow-pore supported cobalt catalysts toward co-fed water can also be explained in terms of relative size of cobalt clusters, pore network of support, expected location of cobalt clusters within the pore network, and relative differences in the residence time of water vapor within and outside the

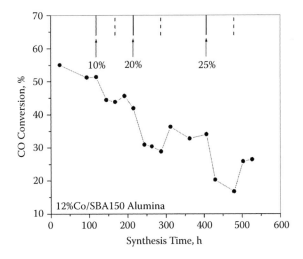

FIGURE 14.5   The variation in CO conversion with co-fed water during FTS with 12% Co/SBA150 alumina catalyst. FTS was effected at 220°C, 2.03 MPa, and $H_2/CO = 2$. The solid line represents the start of water addition, and the broken line represents when the corresponding water addition was terminated.

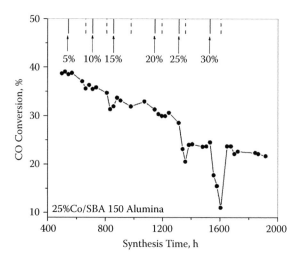

FIGURE 14.6   The variation in CO conversion with co-fed water during FTS with 25% Co/SBA150 alumina catalyst. FTS was effected at 220°C, 2.03 MPa, and $H_2/CO = 2$. The solid line represents the start of water addition, and the broken line represents when the corresponding water addition was terminated.

pore. The pore diameter of the narrow-pore support is 9.5 nm, and the diameters of cobalt clusters for 12% Co/SBA150 alumina and 25% Co/SBA150 alumina catalysts are 8.5 and 9.2 nm, respectively (Table 14.3). Hence, it can be expected that the major fraction of cobalt clusters in both of these catalysts will be located outside the pore (i.e., at the surface of the pore). Since the residence time of water vapor at the pore/catalyst surface is lower than that within the pore, the adverse effect of water vapor to cobalt clusters located at the pore surface is probably less severe and reversible. In addition, if a cobalt cluster is situated at the pore/catalyst surface, it can experience a more reducing environment (presence of H_2 and CO) immediately once the addition of co-fed water is terminated, compared to cobalt clusters situated within the pore where higher partial pressure of water vapor may continue to exist even when the addition of co-fed water is stopped (depending on the diffusion time of water vapor from pore to surface). This in turn may facilitate re-reduction of the cobalt cluster, and the resulting water effect can be reversible (as can be observed for 25% Co/Degussa C-alumina, 12% Co/SBA150 alumina, and 25% Co/SBA150 alumina in Figures 14.4 to 14.6, respectively). However, when the cobalt cluster is located within the pore, the residence time of water vapor within the pore is relatively higher, and the cluster may not experience the reducing environment even after a long time once the addition of co-fed water is terminated. Such a scenario may result in a more severe and irreversible water effect and, in turn, catalyst deactivation (which can be seen for 12% Co/Degussa C-alumina catalyst presented in Figure 14.3).

The cobalt metal cluster size of the catalyst is believed to be related to the catalyst deactivation,[21,47] as the number of surface active sites varies with cobalt cluster size for a fixed cobalt loading of the catalyst. The pore structure of the support and the size/appearance of the cobalt crystallites are believed to be important factors in determining the effect of co-fed water. It has been reported that the appearance of cobalt particles depends on the support.[48] In the narrow-pore support, cobalt seems to exist as clusters of small particles, whereas for wide-pore supports, cobalt exists as larger particles when prepared by the incipient wetness impregnation technique. Jacobs et al.[21] suggested that, in the case of noble metal–promoted catalysts with lower Co loading, a significant fraction of the cobalt clusters that reduce arises from those cobalt species interacting strongly with the support (presumably with a smaller cluster size); these clusters are expected to display a higher sensitivity to the water effect. Instead of an unfavorable thermodynamic calculation for bulk oxidation of cobalt metal to CoO or Co_3O_4 under normal FTS conditions, it is conceivable that the interaction of the clusters with the support may allow for surface oxidation to CoO to occur. Krishnamoorthy et al.[15] also pointed out that the metal-oxygen bonds are stronger at metal surfaces than in the bulk. It was also indicated[21] that unlike the smaller cobalt clusters on unpromoted Co/Al_2O_3 catalyst, the larger crystallites (>10 nm diameter, measured by chemisorption and XRD measurement) undergo oxidation by water to CoO and are most likely confined to the surface. However, these clusters are re-reduced when addition of co-fed water is stopped and the catalytic activity displays a reversible recovery. This fact, along with the pore diameter of narrow-pore alumina support used in the present study (i.e., 9.5 nm), may explain why the 12% Co/SBA150 alumina (cluster diameter, 9.3 nm) and 25% Co/SBA150 alumina (cluster diameter, 8.7 nm) catalysts are less sensitive to the water effect and exhibit a relatively reversible deactivation with co-fed water.

14.3.3 Effect of Water Addition on Product Selectivity in FTS

The effects of co-fed water on the CO_2 and CH_4 selectivities during FTS for 12% and 25% Co/Degussa C-alumina are presented in Figures 14.7 and 14.8, respectively, while for 12% and 25% Co/SBA150 alumina the corresponding variations in CO_2 and CH_4 selectivity are shown in Figures 14.9 and 14.10, respectively. In the case of 12% Co/Degussa C-alumina (Figure 14.7), both CO_2 and CH_4 selectivity increased steadily with an increase in the vol% of water added to the feed. As mentioned earlier the CO conversion dropped to 0.68% with the addition of 25 vol% co-fed water. The highest magnitude of CO_2 and methane selectivity was also observed with the addition of 25 vol% water. Generally, a higher CO_2 selectivity in FTS suggests a higher rate of the water-gas shift reaction. Hence, the above observation also suggests that some surface cobalt atoms would have been oxidized to form cobalt oxide or some other oxidized form of cobalt (e.g., cobalt-support complex), and that this form is active for the water-gas shift reaction, resulting in a higher selectivity to CO_2. A similar trend in CO_2 selectivity was observed for the 25% Co/Degussa C-alumina catalyst with the addition of co-fed water (Figure 14.8), suggesting an increase in water-gas shift activity similar to that discussed for the 12% Co/Degussa C-alumina catalyst. However, the

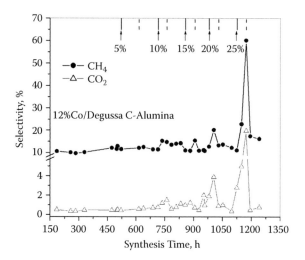

FIGURE 14.7   The variation of CH_4 and CO_2 selectivity with co-fed water during FTS with 12% Co/Degussa C-alumina catalyst. FTS was effected at 220°C, 2.03 MPa, and H_2/$CO = 2$. The solid line represents the start of water addition, and the broken line represents when the corresponding water addition was terminated.

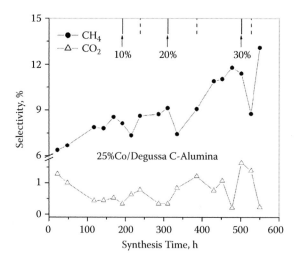

FIGURE 14.8   The variation of CH_4 and CO_2 selectivity with co-fed water during FTS with 25% Co/Degussa C-alumina catalyst. FTS was effected at 220°C, 2.03 MPa, and H_2/$CO = 2$. The solid line represents the start of water addition, and the broken line represents when the corresponding water addition was terminated.

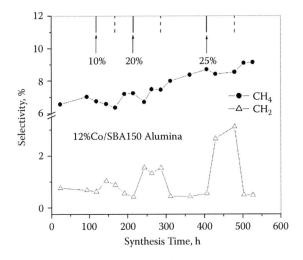

FIGURE 14.9   The variation of CH_4 and CO_2 selectivity with co-fed water during FTS with 12% Co/SBA150 alumina catalyst. FTS was effected at 220°C, 2.03 MPa, and H_2/$CO = 2$. The solid line represents the start of water addition, and the broken line represents when the corresponding water addition was terminated.

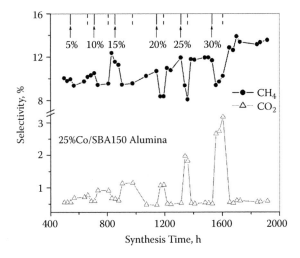

FIGURE 14.10   The variation of CH_4 and CO_2 selectivity with co-fed water during FTS with 25% Co/SBA150 alumina catalyst. FTS was effected at 220°C, 2.03 MPa, and H_2/$CO = 2$. The solid line represents the start of water addition, and the broken line represents when the corresponding water addition was terminated.

methane selectivity for the 25% Co/Degussa C-alumina was found to decrease with the addition of co-fed water (with the highest decrease in the magnitude of CH_4 selectivity when 25 vol% water was added). Nevertheless, the variations in the CH_4 and CO_2 selectivities were found to be reversible.

A small decrease in CH_4 selectivity was also observed with the addition of co-fed water for 12% Co/SBA150 alumina catalyst, while a significant and reversible increase in CO_2 selectivity was observed (Figure 14.9). A similar trend for CO_2 selectivity was observed with the 25% Co/SBA150 alumina catalyst (Figure 14.10). The CH_4 selectivity during FTS with the 25% Co/SBA150 alumina catalyst showed a significant decrease with addition of co-fed water, and then the conversion returned to the expected value when water addition was terminated (Figure 14.10). The variations in CH_4 selectivity for the 12% Co/SBA150 alumina and 25% Co/SBA150 alumina catalysts follow the same trend as the catalytic activity (i.e., CO conversion) in FTS. Among the four catalysts used in FTS in the present study, three (i.e., 12% Co/SBA150 alumina, 25% Co/SBA150 alumina, and 25% Co/Degussa C-alumina) displayed a reversible decrease in CH_4 selectivity upon the addition of co-fed water. In FTS, methane is believed to be produced via hydrogenation of a hydrocarbon precursor.[49] Hence, methane selectivity in FTS can be considered to be an indicator of hydrogenation selectivity for the catalyst. The shift in the above-mentioned product selectivity during water addition from methane (i.e., hydrogenation product) to CO_2 (water-gas shift product) suggests that a fraction of cobalt was oxidized, thereby enhancing the water-gas shift activity. The large increase in CH_4 selectivity for 12% Co/Degussa C-alumina with the addition of 25 vol% water (Figure 14.7) can be explained by the very low FTS activity (with a CO conversion of only 0.68%) of the catalyst.

14.4 CONCLUSION

Cobalt FTS catalysts with comparable metal loadings were prepared with wide-pore (average pore diameter, 22.9 nm) and narrow-pore (average pore diameter, 9.5 nm) alumina supports. Wet impregnation followed by drying and calcination transforms the nonporous Degussa C-alminoxide into a porous alumina support having a wide-pore network. The cobalt cluster size in the case of wide-pore supported catalyst (19.5 nm diameter) was found to be almost double that of the narrow-pore supported catalyst (9.2 nm) of comparable cobalt loading (25%). A reversible negative effect of co-fed water on catalytic activity was observed for 12% Co/Degussa C-alumina (cobalt cluster diameter, 12.2 nm) catalyst up to 20 vol% of co-fed water. However, when 25 vol% of water was added, a severe and irreversible catalyst deactivation resulted. In contrast, the 25% Co/Degussa C-alumina (cobalt cluster diameter, 19.5 nm) catalyst did not exhibit any significant effect during water addition. This difference can be related to the difference in cobalt loading of the catalyst and relative size and location (i.e., inside or outside the pore) of the cobalt clusters. In the case of narrow-pore support, both the 12% Co/SBA150 alumina (cobalt cluster diameter, 8.5 nm) and 25% Co/SBA150 alumina (cobalt cluster diameter, 9.2 nm) displayed a reversible and significant negative effect with the addition of co-fed water. The CO_2 selectivity of all four catalysts was found to increase with addition of co-fed water, indicating an enhancement of water-gas shift activity of the catalysts. The C_{5+} selectivity of catalysts with 25% cobalt loading (for both wide-pore and narrow-pore support) increased with co-fed water, while the CH_4 selectivity decreased simultaneously. The variations in above selectivities are believed to due to a shift of catalyst behavior from hydrogenation activity to water-gas shift activity. However, 12% Co/Degussa C-alumina catalyst showed a reversible increase in CH_4 selectivity with the addition of water, whereas a slight negative effect on CH_4 selectivity was observed for the 12% Co/SBA150 catalyst with water addition. The difference in water effect toward catalytic performance of cobalt catalysts supported on wide-pore and narrow-pore alumina supports is explained in terms of the relative size and location (within or external to the pore) of cobalt clusters which in turn leads to marked differences in the residence time of water vapor in contact with the Co particles.

REFERENCES

1. Reuel, R.C., and Bartholomew, C.H. 1984. Effects of support and dispersion on the carbon monoxide hydrogenation activity/selectivity properties of cobalt. *J. Catal.* 85:78–88.
2. Jacobs, G., Das, T.K., Zhang, Y., Li, J., Racoillet, G., and Davis, B.H. 2002. Fischer-Tropsch synthesis: Support, loading, and promoter effects on the reducibility of cobalt catalysts. *Appl. Catal. A Gen.* 233:263–81.
3. Saib, A.M., Claeys, M., and van Steen, E. 2002. Silica supported cobalt Fischer–Tropsch catalysts: Effect of pore diameter of support. *Catal. Today* 71:395–402.
4. Khodakov, A.Y., Griboval-Constant, A., Bechara, R., and Villain, F. 2001. Pore-size control of cobalt dispersion and reducibility in mesoporous silicas. *J. Phys. Chem. B* 105:9805–11.
5. Khodakov, A.Y., Griboval-Constant, A., Bechara, R., and Zholobenko, V.L. 2002. Pore size effects in Fischer Tropsch synthesis over cobalt-supported mesoporous silicas. *J. Catal.* 206:230–41.
6. Khodakov, A.Y., Lynch, J., Bazin, D., Rebours, B., Zanier, N., Moisson, B., and Chaumette, P. 1997. Reducibility of cobalt species in silica-supported Fischer–Tropsch catalysts. *J. Catal.* 168:16–25.
7. Xiong, H., Zhang, Y., Wang, S., and Li, J. 2005. Fischer–Tropsch synthesis: The effect of Al_2O_3 porosity on the performance of Co/Al_2O_3 catalyst. *Catal. Commun.* 6:512–16.
8. Chin, R.L., and Hercules, D.M. 1982. Surface spectroscopic characterization of cobalt-alumina catalysts. *J. Phys. Chem.* 86:360–67.
9. Fox, J.M. 1993. The different catalytic routes for methane valorization: An assessment of processes for liquid fuels. *Catal. Rev. Sci. Eng.* 35:169–212.
10. Sarup, B., and Wojciechowski, B.W. 1989. Studies of the Fischer-Tropsch synthesis on a cobalt catalyst. II. Kinetics of carbon monoxide conversion to methane and to higher hydrocarbons. *Can. J. Chem. Eng.* 67:62–74.
11. Yates, I., and Satterfield, C.N. 1991. Intrinsic kinetics of the Fischer-Tropsch synthesis on a cobalt catalyst. *Energy & Fuels* 5:168–73.

12. Iglesia, E., Reyes, S., Madon, R., and Soled, S. 1993. Selectivity control and catalyst design in the Fischer-Tropsch synthesis: Sites, pellets, and reactors. *Adv. Catal.* 39:221–302.

13. Jager, B., and Espinoza, R. 1993. Advances in low temperature Fischer-Tropsch synthesis. *Catal. Today* 23:17–28.

14. Schulz, H., Claeys, M., and Harms, S. 1997. Effect of water partial pressure on steady state Fischer-Tropsch activity and selectivity of a promoted cobalt catalyst. *Stud. Surf. Sci. Catal.* 107:193–200.

15. Krishnamoorthy, S., Tu, M., Ojeda, M.P., Pinna, D., and Iglesia, E. 2002. An investigation of the effects of water on rate and selectivity for the Fischer–Tropsch synthesis on cobalt-based catalysts. *J. Catal.* 211:422–33.

16. Hilmen, A.M., Schanke, D., Hanssen, K.F., and Holmen, A. 1999. Study of the effect of water on alumina supported cobalt Fischer–Tropsch catalysts. *Appl. Catal. A Gen.* 186:169–88.

17. van Berge, P.J., van deLoosdrecht, J., Barradas, S., and van derKraan, A.M. 2000. Oxidation of cobalt based Fischer–Tropsch catalysts as a deactivation mechanism. *Catal. Today* 58:321–34.

18. Kim, C.J. 1993. U.S. Patent 5227407, July 13.

19. Iglesia, E. 1997. Design, synthesis, and use of cobalt-based Fischer-Tropsch synthesis catalysts. *Appl. Catal. A Gen.* 16:59–78.

20. Li, J., Zhan, X., Zhang, Y., Jacobs, G., Das, T., and Davis, B.H. 2002. Fischer–Tropsch synthesis: Effect of water on the deactivation of Pt promoted Co/Al_2O_3 catalysts. *Appl. Catal. A Gen.* 228:203–12.

21. Jacobs, G., Patterson, P.M., Das, T.K., Luo, M., and Davis, B.H. 2004. Fischer–Tropsch synthesis: Effect of water on Co/Al2O3 catalysts and XAFS characterization of reoxidation phenomena. *Appl. Catal. A Gen.* 270:65–76.

22. Jacobs, G., Das, T.K., Li, J., Luo, M., Patterson, P.M., and B.H. Davis. 2007. Fischer–Tropsch synthesis: Influence of support on the impact of co-fed water for cobalt-based catalysts. In *Fischer-Tropsch synthesis: Catalysts and catalysis*, ed. B.H. Davis and M.L. Occelli, 217–53 Amsterdam, The Netherlands: Elsevier.

23. van deLoosdrecht, J., Balzhinimaev, B., Dalmon, J.-A., Niemantsverdriet, J.W., Tsybulya, S.V., Saib, A.M., van Berge, P.J., and Visagie, J.L. 2007. Cobalt Fischer-Tropsch synthesis: Deactivation by oxidation? *Catal. Today* 123:293–302.

24. Bolt, H. 1994. Transition metal-aluminate formation in alumina-supported model catalysts. PhD thesis, University of Utrecht.

25. Schanke, D., Hilmen, A.M., Bergene, E., Kinnari, K., Rytter, E., Adnanes, E., and Holmen, A. 1995. Study of the deactivation mechanism of Al_2O_3-supported cobalt Fischer-Trospch catalysts. *Catal. Lett.* 34:269–84.

26. Schanke, D., Hilmen, A.M., Bergene, E., Kinnari, K., Rytter, E., Adnanes, E., and Holmen, A. 1996. Reoxidation and deactivation of supported cobalt Fischer-Tropsch catalysts. *Energy & Fuels* 10:867–72.

27. Schanke, D., Hilmen, A.M., Bergene, E., Kinnari, K., Rytter, E., Adnanes, E., and Holmen, A. 1995. Reoxidation and deactivation of supported cobalt Fischer-Tropsch catalysts. *Prepr. ACS Div. Fuel Chem.* 40:167–71.

28. Hilmen, A.M., Lindvag, O.A., Bergene, E., Schanke, D., Eri, S., and Holmen, A. 2001. Selectivity and activity changes upon water addition during Fischer-Tropsch synthesis. *Stud. Surf. Sci. Catal.* 135:295–300.

29. Hilmen, A.M., Schanke, D., and Holmen, A. 1997. Reoxidation of supported cobalt Fischer-Tropsch catalysts. *Stud. Surf. Sci. Catal.* 107:237–42.

30. Rothaemel, M., Hanssen, K.F., Blekkan, E.A., Schanke, D., and Holmen, A. 1997. The effect of water on cobalt Fischer-Tropsch catalysts studied by steady-state isotopic transient kinetic analysis (SSITKA). *Catal. Today* 38:79–84.

31. Jacobs, G., Zhang, Y., Das, T.K., Patterson, P.M., and Davis, B.H. 2001. Deactivation of a Ru promoted Co/Al$_2$O$_3$ catalyst for FT synthesis. *Stud. Surf. Sci. Catal.* 139:415–22.

32. Jacobs, G., Patterson, P.M., Zhang, Y., Das, T., Li, J., and Davis, B.H. 2002. Fischer–Tropsch synthesis: Deactivation of noble metal-promoted Co/Al$_2$O$_3$ catalysts. *Appl. Catal. A Gen.* 233:215–26.

33. Das, T.K., Jacobs, G., Patterson, P.M., Conner, W.A., Li, J., and Davis, B.H. 2003. Fischer–Tropsch synthesis: Characterization and catalytic properties of rhenium promoted cobalt alumina catalysts. *Fuel* 82:805–15.

34. Jacobs, G., Das, T.K., Patterson, P.M., Li, J., Sanchez, L., and Davis, B.H. 2003. Fischer–Tropsch synthesis XAFS: XAFS studies of the effect of water on a Pt-promoted Co/Al$_2$O$_3$ catalyst. *Appl. Catal. A Gen.* 247:335–43.

35. Espinoza, R.L., Visagie, J.L., van Berg, P.J., and Bolder, F.H. 1998. U.S. Patent 5733839.

36. Brunauer, S., Emmett, P.H., and Teller, E. 1938. Adsorption of gases in multimolecular layers. *J. Am. Chem. Soc.* 60:309–19.

37. Barrett, E.P., Joyner, L.G., and Halenda, P.P. 1951. The determination of pore volume and area distributions in porous substances. I. Computations from nitrogen isotherms. *J. Am. Chem. Soc.* 73:373–80.

38. Vada, S., Hoff, A., Adnanes, E., Schanke, D., and Holmen, A. 1995. Fischer-Tropsch synthesis on supported cobalt catalysts promoted by platinum and rhenium. *Topics Catal.* 2:155–62.

39. Brenner, A.M., Adkins, B.D., Spooner, S., and Davis, B.H. 1995. Porosity by small-angle x-ray scattering (SAXS): Comparison with results from mercury penetration and nitrogen adsorption. *J. Non-crystal. Solids* 185:73–77.

40. Wang, W.J., and Chen, Y.W. 1991. Influence of metal loading on the reducibility and hydrogenation activity of cobalt/alumina catalysts. *Appl. Catal. A Gen.* 77:223–33.

41. Borg, Q., Ronning, M., Storsaeter, S., van Beek, W., and Holmen, A. 2007. Identification of cobalt species during temperature programmed reduction of Fischer-Tropsch catalysts. *Stud. Surf. Sci. Catal.* 163:255–72.

42. Jacobs, G., Ji, Y., Davis, B.H., Cronauer, D., Kropf, A.J., and Marshall, C.L. 2007. Fischer–Tropsch synthesis: Temperature programmed EXAFS/XANES investigation of the influence of support type, cobalt loading, and noble metal promoter addition to the reduction behavior of cobalt oxide particles. *Appl. Catal. A Gen.* 333:177–91.

43. Wojciechowsky, B. 1988. The kinetics of the Fischer-Tropsch synthesis. *Catal. Rev. Sci. Eng.* 30:629–702.

44. Iglesia, E., Soled, S.L., Fiato, R.A., and Via, G.H. 1993. Bimetallic synergy in cobalt-ruthenium Fischer-Tropsch synthesis catalysts. *J. Catal.* 141:345–68.

45. van Berge, P.J., and Everson, R.C. 1997. Cobalt as an alternative Fischer-Tropsch catalyst to iron for the production of middle distillates. *Stud. Surf. Sci. Catal.* 107:207–12.

46. Lapidus, A., Krylova, A., Kazanskii, V., Borovkov, V., and Zaitsev, A. 1991. Hydrocarbon synthesis from carbon monoxide and hydrogen on impregnated cobalt catalysts. Part I. Physico-chemical properties of 10% cobalt/alumina and 10% cobalt/silica. *Appl. Catal.* 73:65–81.

47. Iglesia, E. 2001. In *Proceedings of the 17th Meeting of the North American Catalysis Society*, Toronto, Canada, p. 32.

48. Storsaeter, S., Tøtdal, B., Walmsley, J.C., Tanem, B.S., and Holmen, A. 2005. Characterization of alumina-, silica-, and titania-supported cobalt Fischer–Tropsch catalysts. *J. Catal.* 236:139–52.

49. Claeys, M., and van Steen, E. 2004. In *Fischer-Tropsch technology. Studies in surface science and technology*, ed. A.P. Steynberg and M.E. Dry, 601–13. San Diego: Elsevier.

15 Fischer-Tropsch Synthesis

A Continuous Process for Separation of Wax from Iron Nano-Catalyst Particles by Using Cross-Flow Filtration

Amitava Sarkar, James K. Neathery,
Robert L. Spicer, and Burtron H. Davis

CONTENTS

A continuous cross-flow filtration process has been utilized to investigate the effectiveness in the separation of nano sized (3–5 nm) iron-based catalyst particles from simulated Fischer-Tropsch (FT) catalyst/wax slurry in a pilot-scale slurry bubble column reactor (SBCR). A prototype stainless steel cross-flow filtration module (nominal pore opening of 0.1 μm) was used. A series of cross-flow filtration experiments were initiated to study the effect of mono-olefins and aliphatic alcohol on the filtration flux and membrane performance. 1-hexadecene and 1-dodecanol were doped into activated iron catalyst slurry (with Polywax 500 and 655 as simulated FT wax) to evaluate the effect of their presence on filtration performance. The 1-hexadecene concentrations were varied from 5 to 25 wt% and 1-dodecanol concentrations were varied from 6 to 17 wt% to simulate a range of FT reactor slurries reported in literature. The addition of 1-dodecanol was found to decrease the permeation rate, while the addition of 1-hexadecene was found to have an insignificant or no effect on the permeation rate.

The separation efficiency of pilot-scale integrated SBCR cross-flow filtration module was studied with composite FT catalyst/wax slurry. A filtration loop with a moyno-type progressive cavity pump to circulate the catalyst slurry through the primary hydrocyclone separator, with one output (bottom stream) returning to the reactor with the other output (top stream) flowing to the cross-flow filtration module, was developed. The wax permeate flow from the filter module was controlled by a valve actuated by a reactor-level controller. It was found that a passive flux maintenance procedure, in which the transmembrane pressure (and flux) is suspended for a selected period, can increase the flux by over 100%, but the flux decays back to the original steady-state value over time on stream. On the other hand, an active flux maintenance system, where the filter membrane is back-flushed by a cleaned permeate stream via a piston pump triggered by a computer-controlled timer, results in significant flux recovery that decays off during the next 24 h. A better flux stability was achieved by increasing the permeate off-cycle to 1 h per day in addition to 30 s off per half-hour cycle. The permeate flux was found to vary linearly with transmembrane pressure within the range studied. The iron concentration in permeate wax was found to be always less than 35 ppm, with over 85% below 16 ppm level.

15.1 INTRODUCTION

Iron-based Fischer-Tropsch synthesis (FTS) catalysts are preferred for synthesis gas with a low H_2/CO ratio (e.g., 0.7) because of their excellent activity for the water-gas shift reaction, lower cost, lower methane selectivity, high olefin selectivity, lower sensitivity toward poisons, and flexible product slate. Use of SBCR for slurry phase FTS is advantageous to control highly exothermic heat of reaction in large-scale industrial operation. In a commercial SBCR operation, heavy wax products must be continuously separated and removed from the reactor to maintain the required conversion, to avoid catalyst loss, and for downstream wax upgrading.[1]

However, use of iron-based catalysts in SBCR has been limited by their high rate of attrition to ultrafine particles, leading to catalyst loss and high slurry viscosity, and the resulting difficulty in catalyst/wax separation. Phase transformation of iron catalyst during activation/FTS plays an important role in determining the structural integrity of the catalyst particles. A detailed study on the variation of particle size distribution, and morphological and phase transformation of ultrafine iron nano-catalyst particles (with initial diameters of 3 to 5 nm) during carbidization/FTS, has been reported[1] that can be utilized for design and optimization of a continuous flow filtration module for catalyst/wax separation.

A number of processes have been disclosed in the literature for catalyst/wax separation during FTS.[2-7] Most of these techniques can be classified as either internal (filter elements located inside the reactor vessel) or external. Internal methods have the disadvantage of being inaccessible during normal reactor operation. Therefore, in anticipation of plugging problems, duplicate filtration systems must be installed to achieve high reliability in commercial settings. However, internal separation systems are difficult or sometimes impractical to service without taking the reactor off-line. Most of the proposed filtration schemes also suffer from high transmembrane pressure drop and low flux limitations. Many of the separation techniques described in the literature employ only simple hydrocyclones irrespective of their low separation efficiency. Brennan et al. disclosed a process for removing catalyst fines from Fischer-Tropsch (FT) catalyst/wax slurry by using a magnetic separator[4] for which the operational cost was relatively low; however, the capital costs of the magnet assemblies were reported to be expensive. Roberts et al.[7] developed a method of extracting catalysts from the catalyst/wax slurry of an SBCR that involves supercritical hydrocarbon solvent. The disadvantage with this system is that it requires process equipment in addition to the reactor, thus exploiting a considerable residence time of catalysts outside the reactor system. Achieving an efficient wax separation method for iron-based catalysts is one of the most challenging technical problems associated with slurry phase (i.e., SBCR) iron-based FTS and is a key factor for optimizing operating costs of such a process.

The cross-flow filtration method is applied mainly to hyper- and ultrafiltration as well as to some microfiltration.[8] In cross-flow filtration the slurry solution or suspension fed to the filter flows parallel to the filter medium or membrane. The filtration product (permeate or filtrate) leaves the filtration module at right angles to the filter medium (the membrane). The traditional perpendicular flow filtration (where the flow of the suspension is directed at right angles to the filter medium and the permeate leaves the filter medium in the same direction) entails filter cake buildup, whereas cross-flow filtration is intended to prevent such filter cake buildup. A perpendicular filtration mode is favored when the filter cake is to be collected for the purpose of solids recovery, while the main rationale for cross-flow filtration is to maximize the recovery of the liquids while retaining the solid content of the system with minimum or no deposit of solids on the filter medium.[8] Various advantages of cross-flow filtration can be exploited to develop an efficient method to continuously separate FT wax from ultrafile iron particles while retaining the catalyst loading of the slurry in the reactor to maintain the steady-state conversion. To minimize the degree of membrane surface fouling in continuous operation, a constant permeate flux maintenance procedure can be developed. The constant permeate flux maintenance procedure would ensure that the cross-flow filtration module operates at the optimum transmembrane pressure while maintaining the desired permeate flux.

Wax products from FTS utilizing an iron-based catalyst composition contain a significant amount of olefins and oxygenates, depending on the catalyst formulation and operating conditions. The presence of olefins and alcohols in catalyst/wax slurry may influence the rheology of wax to some degree, which in turn may affect the hydrodynamics of slurry flow and degree of catalyst/wax separation. The effect of olefins and alcohols present in FT wax products on the performance of catalyst/wax separation by filtration process has not been documented in open literature.

The objective of the present study is to develop a cross-flow filtration module operated under low transmembrane pressure drop that can result in high permeate flux, and also to demonstrate the efficient use of such a module to continuously separate wax from ultrafine iron catalyst particles from simulated FTS catalyst/wax slurry products from an SBCR pilot plant unit. An important goal of this research was to monitor and record cross-flow flux measurements over a long-term time-on-stream (TOS) period (500+ h). Two types (active and passive) of permeate flux maintenance procedures were developed and tested during this study. Depending on the efficiency of different flux maintenance or filter media cleaning procedures employed over the long-term test to stabilize the flux over time, the most efficient procedure can be selected for further development and cost optimization. The effect of mono-olefins and aliphatic alcohols on permeate flux and on the efficiency of the filter membrane for catalyst/wax separation was also studied.

15.2 CONCEPTUAL BASIS OF CROSS-FLOW FILTRATION FOR FISCHER-TROPSCH CATALYST/WAX SEPARATION

The design of a cross-flow filter system employs an inertial filter principle that allows the permeate or filtrate to flow radially through the porous media at a relatively low face velocity compared to that of the mainstream slurry flow in the axial direction, as shown schematically in Figure 15.1.[9] Particles entrained in the high-velocity axial flow field are prevented from entering the porous media by the ballistic effect of particle inertia. It has been suggested that submicron particles penetrate the filter medium and form a dynamic membrane or submicron layer, as shown in

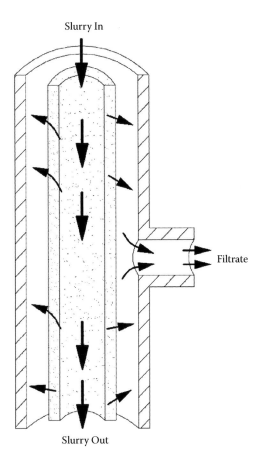

Slurry In

Filtrate

Slurry Out

FIGURE 15.1 Cross-sectional view of a cross-flow filter element.

Figure 15.2(a). The membrane impedes further penetration of even smaller particles through the porous filter media. In many filtration applications, this filtration mechanism is valid for an axial velocity greater than about 4 to 6 m/s.

To continuously separate FT wax products from ultrafine iron catalyst particles in an SBCR employed for FTS, a modified cross-flow filtration technique can be developed using the cross-flow filter element placed in a down-comer slurry recirculation line of the SBCR. Counter to the traditional cross-flow filtration technique described earlier, this system would use a bulk slurry flow rate below the critical velocity, thereby forcing a filter cake of solids to form between the filter media and the bulk slurry flow, as depicted in Figure 15.2b. In this mode, multiple layers of catalyst particles that deposit upon the filter medium would act as a prefilter layer.[10] Both the inertial and filter cake mechanisms can be effective; however, the latter can be unstable if the filter cake depth is allowed to grow indefinitely. In the context of the SBCR operation, the filter cake could potentially occlude the slurry recirculation flow path if allowed to grow uncontrollably.

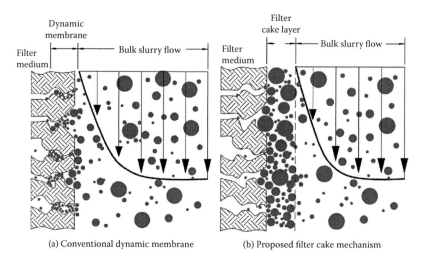

(a) Conventional dynamic membrane   (b) Proposed filter cake mechanism

FIGURE 15.2   Schematic models for filter cake mechanism in cross-flow filtration.

Therefore, when operating in the filter cake mode, the axial velocity should be maintained at a level such that an adequate shear force exists along the filter media to prevent excessive caking of the catalyst that could cause a blockage in the down-comer circuit. For the separation of ultrafine catalyst particles from FT catalyst/wax slurry, the filter medium can easily become plugged using the dynamic membrane mode filtration. Also, small iron carbide particles (less than 3 nm) near the filter wall are easily taken into the pores of the medium due to their low mass and high surface area. Therefore, pure inertial filtration near the filter media surface is practically ineffective.

Results from constant differential pressure filtration tests have been analyzed according to traditional filtration science techniques with some modifications to account for the cross-flow filter arrangement.[11] Resistivity of the filter medium may vary over time due to the infiltration of the ultrafine catalyst particles within the media matrix. Flow resistance through the filter cake can be measured and correlated to changes in the activation procedure and to the chemical and physical properties of the catalyst particles. The clean medium permeability must be determined before the slurries are filtered. The general filtration equation or the Darcy equation for the clean medium is defined as

$$Q_c = \frac{\Delta P \cdot A}{\mu \cdot R_m} \tag{15.1}$$

where Q_c is the clean medium filtrate flow rate (m³/s), μ is the liquid viscosity (Pa-s), ΔP is the driving pressure force or differential pressure across the filter medium (Pa), R_m is the clean medium resistance or the permeability per unit filter cake depth (m⁻¹), and A is the filter medium surface area (m²). Once the solids build up on the filter medium, the filter cake resistance in series with the medium must be considered. Hence, the Darcy equation becomes

$$Q = \frac{\Delta P \cdot A}{\mu \cdot \left(R_m + R_c \right)} \qquad (15.2)$$

where R_c is the filter cake resistance (m⁻¹). If it is assumed that the catalyst filter cake is incompressible, then the filter cake resistance would depend on the mass of the catalyst deposited on the filter medium as follows:

$$R_c = \alpha \cdot w \qquad (15.3)$$

where w is the mass of catalyst per unit area (kg/m²) and α is the specific filter cake resistance (m/kg). In the context of the proposed cross-flow filtration technique, it is assumed that w is constant after some measured steady-state period t_{ss}, while the filter medium is exposed to a constant differential pressure and a constant axial velocity through the filter tube (Figure 15.3). Initially (at $t = 0$), the permeate stream exiting the filter module only encounters flow resistance from the medium. As the catalyst particles begin to accumulate on the filter, the cake resistance term increases until the rate of catalyst deposited (on the filter media) and the rate of catalyst entrained into the bulk flow field (i.e., leaves filter media) are equal. Thus, a steady-state permeate flow, Q_{ss} , is attained. A mass balance of

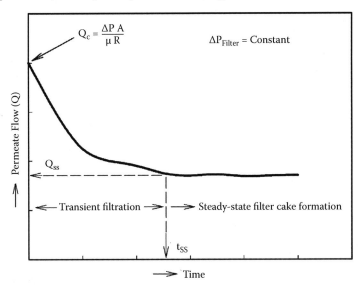

FIGURE 15.3 Variation of permeate flow against time in a typical cross-flow filtration test.

the filter cake during the un-steady-state period yields the following equation for the steady-state filter loading:

$$w_{ss} = \frac{C_s}{A} \cdot \int_0^{t_{ss}} Q(t) \cdot dt - r_{ent} \int_0^{t_{ss}} dt \qquad (15.4)$$

where C_s is the catalyst concentration in the bulk slurry (kg/m³), $Q(t)$ is the transient permeate flow rate (m³/s), t_{ss} is the approximate time to reach equilibrium in the filter cake (s), and r_{ent} is the reentrainment rate of catalyst particles from the filter cake into the bulk slurry (kg/m²/s). In Equation 15.4 it is assumed that the reentrainment rate of catalyst from filter into the bulk slurry is constant throughout the filtration process. After the steady-state filtration is achieved (i.e., $t > t_{ss}$), the re-entrainment rate of slurry will be equal to the rate of catalyst captured on the filter cake:

$$r_{ent} = \frac{Q_{ss}}{A} \cdot C_s \qquad (15.5)$$

where Q_{ss} is the steady-state permeate flow rate. Once the equilibrium filter cake mass is known, the specific resistance, α, can be calculated using Equations 15.2 and 15.3. After each constant differential pressure filtration test, the filter medium can be backwashed to remove the filter cake. Filtration of the slurry will then be restarted in order to check clean filter permeate flow rate, Q_c, and thus the filter medium resistance factor, R_m.

It is anticipated that the equilibrium filter cake mass would depend strongly on the axial velocity through the cross-flow filter assembly. The shear rate at the filter surface will increase the entrainment of the catalyst solids for a given permeate flow rate. Therefore, for each differential pressure condition, the axial velocity will be varied in order to quantify the effect of the wall shear on the filter cake resistance term.

15.3 EXPERIMENTAL DETAILS

15.3.1 Materials

15.3.1.1 Reagents

Polywax 500 and 655 (polyethylene fraction with average molecular weights of 500 and 655, respectively), purchased from Baker Petrolite, Inc., was used to prepare simulated FT wax (i.e., solvent) for the evaluation of filtration performance with and without the presence of aliphatic alcohol and mono-olefin in the FT wax. 1-dodecanol (ACS reagent, $\geq 98\%$, Sigma-Aldrich, Inc.) and 1-hexadecene (GC standard grade reagent, $\geq 99.8\%$, Sigma-Aldrich, Inc.) were used as model compounds to represent aliphatic alcohol and mono-olefin in the simulated FT wax. A test slurry containing C_{30} oil (hydrogenated polyalphaoleffin, CAS 68037-01-4), FT wax obtained from previous pilot-scale SBCR and bench-scale stirred reactor tests, and a carbidized commercial ultrafine iron catalyst (NANOCAT® superfine iron oxide, Mach I, Inc.) was prepared. The properties reported for the commercial catalysts are: average particle size, 3–5 nm; surface area, 250 m²/g; and bulk density, 0.05 g/cm³. The molecular weight distribution of this composite wax has been reported in the literature.[12]

15.3.1.2 Prototype Cross-Flow Filtration Module

The prototype shell-and-tube type cross-flow filtration modules (Pall Corp.) used for filtration tests are welded into a stainless steel shell enclosure. The modules have an inlet (filtrate) and outlet (retentate) port (both at tube sides) with ½-inch tubing ends, and a permeate port, located near the midpoint of the shell side of the unit. The stainless steel filter membranes have a nominal pore size of 0.1 μm. The surface of the filter media is coated with a proprietary submicron layer of zirconia.

15.3.2 METHOD

15.3.2.1 Cross-Flow Filtration Test

The cross-flow filtration test unit is depicted schematically in Figure 15.4; it allows several types of cross-flow filter media to be studied under simulated FTS conditions. The filtration piping and instrumentation are heated via several circuits

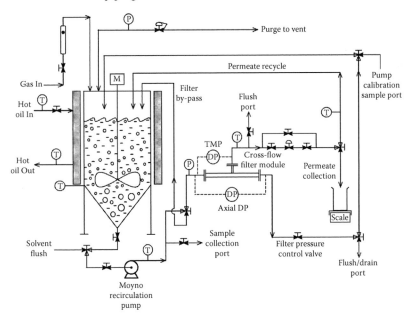

FIGURE 15.4 Schematic of the cross-flow filtration test platform.

of copper heat-trace tubing. A Therminol 66 (Solutio, Inc.) high-temperature heat transfer fluid is circulated through the heat-trace tubing from an electrically heated hot oil system. The temperature controller was calibrated to operate over a temperature range of 453 to 523 K. Data acquisition and process control functions are accomplished by a National Instruments real-time computer system. A 98 L (26-gallon) slurry mixing tank is heated by a hot-oil circulation jacket. Slurry mixtures of catalyst, Polywax 500, Polywax 655, and additives (aliphatic alcohols and mono-olefins) are loaded batch-wise into the system. A moyno (Liberty progressive cavity) pump is used to circulate the slurry mixture through the cross-flow filter element. A manually actuated valve, located downstream of the filter element, maintains a slurry flow rate set point of 2 to 40 lpm.

Unfiltered slurry (or retentate) passing through the filter tube is recycled to the mixing tank. The differential pressure across the filter medium or transmembrane pressure (TMP) is automatically controlled by a letdown valve. The permeate can be recycled to the slurry tank for continuous filtration simulation (in order to maintain a constant solids concentration in the system) or can be collected and removed from the system to test semibatch filtration schemes. The permeation rate is periodically measured by diverting the stream into a collection flask over a measured time interval. Slurry samples can be collected before and after the filtration for characterization purposes. In tests with catalyst, a gas stream of CO or syngas can be applied to the system; otherwise, the system vapor space is purged with inert gas such as argon or nitrogen. Slurry temperature, simulating the FTS activation conditions, can be controlled up to 543 K. Modular filtration media can be tested under various filtration rates, differential pressures, and operating modes. The system is designed in such a fashion that the filter unit can be bypassed in order to change filters while the slurry continues its recirculation path. A variety of precipitated catalyst slurries mixed with various molecular weight waxes (C_{30} to C_{100}) allows a range of effective slurry viscosities to be studied.

15.3.2.2   Laboratory Pilot-Scale SBCR Unit

The pilot-scale SBCR unit with cross-flow filtration module is schematically represented in Figure 15.5. The SBCR has a 5.08 cm diameter and 2 m height with an effective reactor volume of 3.7 L. The synthesis gas passes continuously through the reactor and is distributed by a sparger near the bottom of the reactor vessel. The product gas and slurry exit at the top of the reactor and pass through an overhead gas/liquid separator, where the slurry is disengaged from the gas phase. Vapor products and unreacted syngas exit the gas/liquid separator and enter a warm trap (373 K) followed by a cold trap (273 K). A dry flow meter downstream of the cold trap measures the exit gas flow rate.

The down-comer from the gas/liquid separator that collects the liquid slurry product is connected to the suction side of a moyno-type (Liberty progressive cavity) pump. The pump discharge is connected to a primary separation device (an inertial separator similar to a hydrocyclone). Since the slurry pump is a positive displacement device (i.e., no slurry slippage inside the pump), the total flow

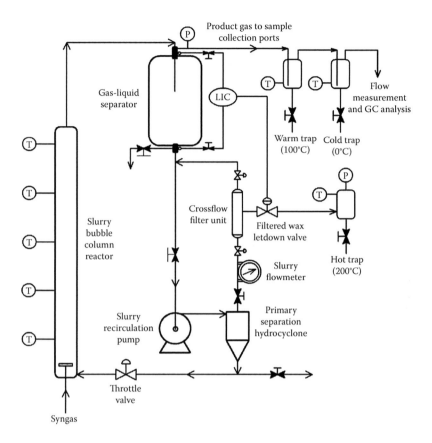

FIGURE 15.5 Schematic of the pilot-scale integrated SBCR unit with cross-flow filtration module.

should remain the same regardless of pressure changes incurred by the temporary switching of the reactor circuit. A catalyst-rich stream (i.e., bottom stream from the primary separation device) is recycled to the reactor vessel while the lean catalyst/slurry stream (i.e., top stream from the primary separation device) is diverted to a secondary filtration loop. The fraction of clarified slurry entering the secondary loop is controlled by a throttle valve. Quantifying the secondary flow is important, as the slurry velocity is crucial in cross-flow filtration. Hence, the secondary slurry flow rate is measured by a coriolis flow meter. Fine polishing of the clarified slurry is achieved by the cross-flow filter element.

Ideally, the axial velocity through the cross-flow unit should be greater than about 4–6 m/s to minimize the boundary layer of particles near the membrane surface. The wax permeate flow from the filter is limited by a control valve actuated by a reactor-level controller. Hence, a constant inventory of slurry is maintained within the SBCR system as long as the superficial gas velocity remains constant. Changes in the gas holdup due to a variable gas velocity are calculated so that the space velocity can be accurately quantified. The flux of the clean permeate through the cross-flow filtration module is controlled by the pressure in the wax letdown vessel or hot trap (473 K). Therefore, TMP is fixed for a given filtration event. The TMP can be changed manually by varying the set point of the pressure regulator connected to the letdown vessel. The flux rate is measured by weighing the mass of permeate collected in the collection vessel hot trap. A filtration event is initiated by the overhead vessel-level controller. The wax permeate flow from the filter module is switched on by a control valve between the permeate discharge of the cross-flow unit and the collection vessel. Hence, a relatively constant inventory of slurry is maintained within the SBCR system as long as the superficial gas velocity remains constant. The level or volume of the slurry within the overhead gas/liquid separator is continuously monitored by measuring the differential pressure across the height of the vessel. Argon is purged through each of the pressure legs to keep the lines free of slurry. Slurry volume within the receiver is controlled to be no more than 1.3 L by removing wax from the reactor system via the level control valve.

15.3.2.3 Constant Filter Flux Maintenance Procedure

A flux maintenance system or filter membrane cleaning (back-flush) procedure is developed and integrated to the permeate side of the cross-flow filtration module (Figure 15.6). The flux maintenance system is capable of back-flushing the filter membrane with a piston pump that is triggered by a computer-controlled timer. The back-flush fluid consists of cleaned permeate stored in a 40 ml vessel located near the suction side of the piston pump. However, back-flushing with clean permeate is only used as a last resort. Preferably, the maintenance procedure is to turn off the permeate flow on a short but regular period. This system was used throughout the study to develop an optimum cleaning program that can sustain a permeate flux rate over many days.

15.3.2.4 Chemical Analysis

The concentration of iron in permeates and retentates is determined by atomic absorption spectroscopy. The concentration of added 1-dodecanol and 1-hexadecene in permeate and retentate samples is determined by gas chromatography (GC) analysis. Approximately 0.25 g of wax and 7 ml of o-xylene (HPLC grade, Sigma-Aldrich, Inc.) are placed in a 16 × 125 mm culture tube and then put in a Thermoclyne dry heating bath. The sample is heated until the wax is dissolved, and then transferred to a sample vial for analysis by GC. The sample is analyzed on an HP 5890 Series II Plus GC equipped with an Flame Ionization Detector (FID). A 25 m SGE aluminum-clad HT5 column of 25 m. HT5 column (aluminum clad, SGE, Inc.) of 0.53 mm inner diameter and 0.15 μm film thickness was used. Identification of the compounds of interest is accomplished by running standards and by spiking the sample with the standard compounds of interest and comparing the corresponding retention times.

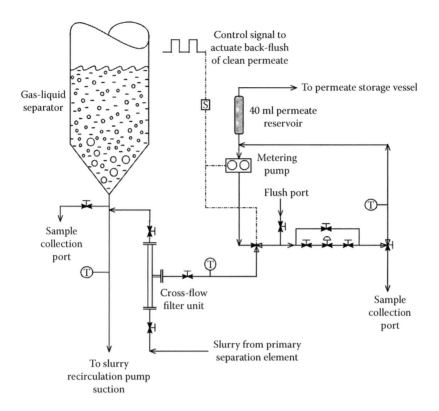

FIGURE 15.6 Schematic of constant flux maintenance manifold, which can be used with SBCR: utilization of active back-flushing of membrane surface with permeate solution.

15.4 RESULTS AND DISCUSSIONS

15.4.1 Effect of Added 1-Dodecanol on the Performance of Cross-Flow Filtration with Polywax 500/655-Activated Catalyst Slurry

Traditionally, iron-based catalysts have been used for FT synthesis when the syngas is coal derived, because of their activity in both FTS and WGS reactions. Complex mixtures of straight-chain paraffins, olefins, and oxygenate (in substantial proportions) compounds are known to be formed during iron-based FTS. Olefin selectivity of iron catalysts is typically greater than 50% of the hydrocarbon products at low carbon numbers, and more than 60% of the produced olefins are α-olefins.[13] For iron-based catalysts, the olefin selectivity decreases asymptotically with increasing carbon number.

A blend of Polywax 500 and 655 (81.3 and 18.7 wt%, respectively) with a CO-activated ultrafine iron catalyst was used for the evaluation of catalyst/wax slurry filtration performance of the filter module with and without an alcohol compound. All of the filtration tests were conducted with a TMP of

1.4 bar, 473 K, and a slurry containing 5 wt% iron. The axial velocity of the slurry within the filter was maintained at 13 m/s. 1-dodecanol was chosen as the alcohol additive because its low vapor pressure would ensure a sufficient residence time in the slurry phase. Since alcohols are known to be reactive in the presence of an iron catalyst,[14,15] an excessive quantity of 1-dodecanol was added once the filtration flux reached the equilibrium value. The initial dosages of 1-dodecanol were 6, 11, and 17 wt%. In order to maintain a constant iron concentration of 5 wt% in the slurry, a mass of filtered permeate was extracted from the slurry that was comparable to the mass of the 1-dodecanol to be added.

The variation of permeation flux against time on stream for filtration tests with added 1-dodecanol addition is shown in Figure 15.7. Before adding the 6 wt% 1-dodecanol dosage, the permeate flux was allowed to reach an equilibrium during the filter membrane/slurry induction period. During this period the flux was found to decline from 0.65 to 0.40 lpm/m^2. Once the first dose of 1-decanol was added, the flux increased to 0.46 lpm/m^2 and subsequently dropped to less than 0.2 lpm/m^2 over a 2-day period. It was found that the permeate color changed from a bright white to brownish once the 1-dodecanol was added. This change in the permeate color can be attributed to the increase of iron concentration (from 50 to about 250 ppm or more) and possibly the existence of reaction products of 1-dodecanol being converted to other oxygenates, such as ketones, aldehydes, or acids. However, it is not clear if these reaction products caused the initial decrease of the permeate flux. Subsequent additions of the 1-dodecanol (11 and 17 wt%) were found to have only a marginal effect on the permeate flux. Figure 15.8

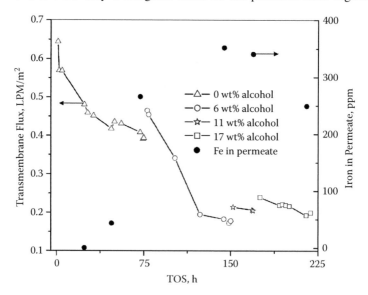

FIGURE 15.7 Variation of permeate flux against time on stream when 1-dodecanol was added to catalyst/wax slurry.

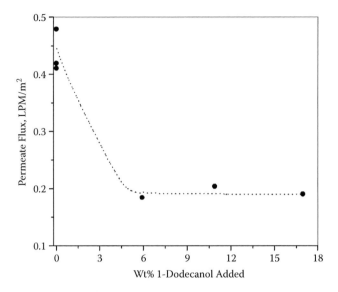

FIGURE 15.8 Variation of permeate flux against concentration of added 1-dodecanol.

presents the variation of equilibrium permeation flux against the concentration of 1-dodecanol. Once the initial 6 wt% dose of 1-dodecanol was added, the steady-state flux was maintained at 0.2 lpm/m^2 even with the further addition of a higher concentration of 1-dodecanol. The permeate quality, in terms of steady-state iron concentration, did not vary substantially with increasing alcohol concentration, as presented in Figure 15.9.

15.4.2 EFFECT OF ADDED 1-HEXADECENE ON THE PERFORMANCE OF CROSS-FLOW FILTRATION WITH POLYWAX 500/655-ACTIVATED CATALYST SLURRY

The specifications of catalyst/wax slurry and operating conditions for filtration test with added 1-hexadecene were identical to those of the previously mentioned 1-dodecanol doping tests. The method for doping the mono-olefins in the slurry was similar in that a constant 5 wt% concentration of iron was maintained throughout the test run. 1-hexadecene was selected as the doping agent since the vapor pressure of lower molecular weight olefins can cause material balance uncertainties. The target concentration levels of the olefin were 0, 14, 26, and 29 wt%. This range of concentrations is typical for various iron-based FT processes.[13]

As shown in Figure 15.10, the baseline flux, without 1-hexadecene addition, stabilized to 0.30 lpm/m^2 at 473 K with a TMP of 1.4 bar. Similar to previous filtration runs using an activated ultrafine iron catalyst slurry, the duration of the induction period for the catalyst particles and membrane was approximately 48 h TOS. Initially, the appearance of permeate was bright white. With the first dosage

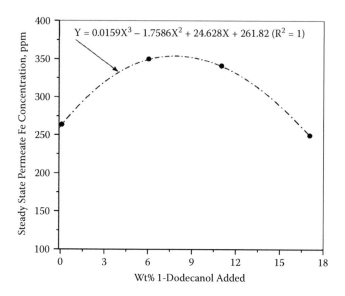

FIGURE 15.9 Variation of iron concentration in the permeate (at steady state) against concentration of added 1-dodecanol.

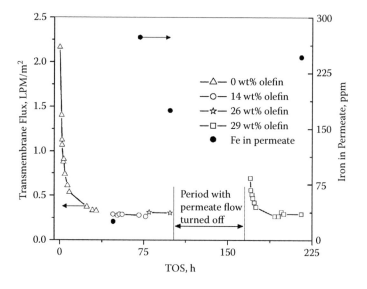

FIGURE 15.10 Effect of co-fed olefin (1-hexadecene) concentration on the permeate flux at different time on stream.

of 1-hexadecene (14 wt%), the permeate flux remained essentially unchanged and was stable after 48 h TOS at 0.31 lpm/m^2. A slight permeate color change (from white to brownish) was observed with increasing TOS. The iron content of the permeates was found to increase from 25 to 272 ppm with the addition of 1-hexadecene, as presented in Figure 15.11. However, this continuous increase can be attributed to the formation of finer catalyst particulate caused by attrition, as was observed in the alcohol doping study. Likewise, as presented in Figure 15.12, no appreciable change in the permeate flux was observed at the 1-hexadecene concentration level of either 26 or 29 wt%.

15.4.3 DEVELOPMENT OF FILTER FLUX MAINTENANCE (PASSIVE) PROCEDURE

Cross-flow filtration systems utilize high liquid axial velocities to generate shear at the liquid-membrane interface. Shear is necessary to maintain acceptable permeate fluxes, especially with concentrated catalyst slurries. The degree of catalyst deposition on the filter membrane or membrane fouling is a function of the shear stress at the surface and particle convection with the permeate flow.[16] Membrane surface fouling also depends on many application-specific variables, such as particle size in the retentate, viscosity of the permeate, axial velocity, and the transmembrane pressure. All of these variables can influence the degree of deposition of particles within the filter membrane, and thus decrease the effective pore size of the membrane.

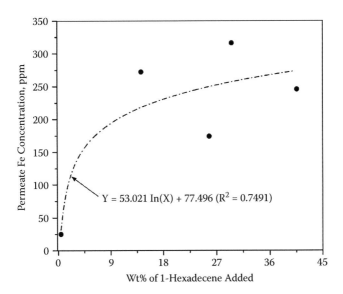

FIGURE 15.11 Effect of added olefin (1-hexadecene) concentration on the iron concentration in the permeate.

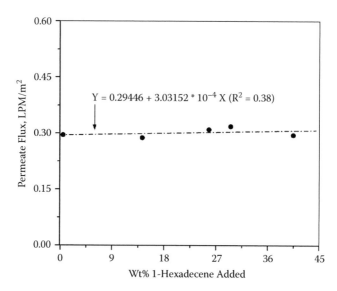

$$Y = 0.29446 + 3.03152 * 10^{-4} X \; (R^2 = 0.38)$$

FIGURE 15.12   Effect of co-fed olefin (1-hexadecene) concentration on the permeate flux.

Several operating techniques have been used in industrial ultrafiltration applications to improve permeation flux performance:[16–19] back-flushing or back-pulsing with clear permeate, back-flushing an inert gas through the membrane, and the use of mechanical devices to increase turbulence near the membrane. Each of these techniques essentially disrupts the boundary layer of concentrated particles at the membrane surface. Generally, the permeate flux in cross-flow filtration can be maintained by occasional back-flushing of clear permeate, which allows removal of a fraction of particles deposited at the boundary layer of the filtering medium.

A more passive technique to maintain the permeate flux is to disrupt the permeation flow while allowing the retentate to circulate axially through the cross-flow module. This would in theory allow inertial lift of the particles on the membrane surface in the absence of the permeate convective force component. Unlike active back-pulsing, this passive approach would not expend cleaned permeate product or require high-pressure inert gas handling. The approach of switching the filter tube off momentarily would be simple and, thus, likely more economical to implement. To quantify the degree of filter media fouling, static permeation tests were performed to determine the permeability of the membranes before and after slurry filtration tests. The variation of the membrane flux against TMP provides a quick snapshot of the surface fouling condition of the filter membrane.[12]

As presented in Figure 15.10, after 100 h TOS the permeate valve was closed, and thus the transmembrane pressure fell to zero. The pilot plant remained in a standby mode and unmanned for approximately 75 h. During this period, the catalyst slurry was circulated through the cross-flow filter, without permeate flow radially through the filter membrane (i.e., test conditions were constant with the exception of the absence of a transmembrane pressure). After 75 h TOS, when the 1.4 bar transmembrane pressure was reinstated, the flux momentarily increased by 116% (from 0.3 to 0.65 lpm/m^2); however, over the next 24 h the flux returned to the baseline value before interruption of the permeate flow. Even though the flux returned to the steady-state value before the permeate valve was closed, there was a 24 h period of increased flux. Several options to determine the optimum period of filter downtime were analyzed. It is found that simply interrupting the permeate flow for 30 s per half hour was effective in recovering the initial membrane fouling; however, the long-term steady-state flux was not achieved with this method. Flux stability was attained only after increasing the permeate off-cycle to 1 h per day in addition to 30 s off per half-hour cycle. Results from these studies were utilized for laboratory pilot-scale SBCR tests.

15.4.4 LABORATORY PILOT-SCALE TESTS IN SBCR WITH CROSS-FLOW FILTRATION MODULE USING FT CATALYST/WAX SLURRY

15.4.4.1 Technical Difficulties Encountered for FTS in Pilot-Scale Tests

During the start-up of the pilot-scale SBCR system, several problems were encountered with the moyno-type pump used to circulate the catalyst/wax slurry though the cross-flow filter element and reactor. The progressive cavity pump is designed to be capable of delivering 20 lpm of FT slurry at 503 K. The pump has a stainless steel rotor with a Viton® (fluoroelastomer) stator designed for pumping slurries at Fischer-Tropsch synthesis and activation conditions.

Initially, the as-received pump could not pass a 13.8-bar (200 psig) static pressure test at ambient temperature. Consequently, the pump was returned to the manufacturer for extensive seal modifications. Upon receiving the modified pump, the unit was placed into service for a "hot" shakedown test with C$_{30}$ oil at 175 psig pressure under the present experimental setup. The modified seal continued to leak at a small but acceptable rate on the order of 1 g/min. However, leak problems were encountered in the threaded seal of the stator and the pump housing. The stator was resealed with a thick Teflon tape and a high-temperature pipe sealant that initially slowed the total system leaks to less than 50 g/h.

At this point, the system was tested with catalyst for activation and FTS, in the hopes that the seal leak rates would be impeded by the presence of small catalyst particles. The FTFE 20-B catalyst (L-3950) (Fe, 50.2%; Cu, 4.2%; K, 1.5%; and Si, 2.4%) was utilized. This is part of the batch used for LaPorte FTS run II.[20] The catalyst was activated at 543 K with CO at a space velocity (SV) of 9 sl/h/g catalyst for 48 h. A total of 1,100 g of catalyst was taken and 7.9 L of C$_{30}$ oil was used as the start-up solvent. At the end of the activation period, an attempt was made for Fischer-Tropsch synthesis at 503 K, 175 psig, syngas SV = 9 sl/h/g catalyst, and H$_2$/CO = 0.7 . However, the catalyst was found to be completely inactive for Fischer-Tropsch synthesis. Potential reasons for catalyst poisoning under present experimental conditions were investigated. Sulfur and fluorine are known to poison iron-based Fischer-Tropsch catalysts.[21,22] Since the stator of the pump is made of Viton, which is a fluoroelastomer, it was suspected that activation of the catalyst with elevated temperature results in release of fluorine, which acted as severe catalyst poison. The sulfur content of the catalyst/wax slurry was found to be 0.3 wt% (by atomic absorption spectroscopy analysis). The source of the sulfur poisoning was uncertain, as the elastomers' inside pumps are commercial products and detailed compositions are not available. Catalyst particles from slurry samples were recovered after Soxhlet extraction with HPLC-grade o-xylene. The fluorine content of the extracted catalyst was found to be 7 wt%. It is possible that during the activation of the catalyst at elevated temperature (greater than 543 K), the elastomer inside the pump released sulfur and fluorine compounds, thereby totally deactivating the iron-based catalyst. Additionally, difficulties in maintaining an acceptable leak rate from the pump seal and stator housing continued after an exhaustive effort of implementing counter-measures. Consequently, the system leak rate always exceeded the expected production rate of wax; therefore, no on-line filtration could be accomplished.

In order to test the cross-flow filtration scheme during a long-term test, the cross-flow filtration module was operated independent of the SBCR pilot system at low pressure (1.7 bar at the pump discharge). In lieu of FT wax produced directly from the SBCR, we prepared a test slurry batch containing FT composite wax obtained from previous pilot-scale SBCR and bench-scale (1 L) continuous stirred tank reactor tests and activated ulrafine Mach I iron catalyst (0.26 wt% as raw catalyst). The molecular weight distribution of the composite wax is reported in the literature.[12] The objective of this test was to evaluate the performance of the cross-flow filter module over long TOS periods (500+ h). To stabilize the permeate flux over time, active as well as passive flux maintenance procedures, as evaluated earlier, were employed over this long-term test.

15.4.4.2 Long-Term Cross-flow Filtration Test with Fischer-Tropsch Composite Catalyst/Wax Slurry

In order to develop a continuous flux maintenance procedure, the present study examined the transmembrane flux values from the cross-flow filtration module with a filtration media area of 0.0198 m² (0.213 ft²), a slurry density of approximately 0.69 g/cm³ at 200°C, 17 kg of simulated FT wax with a catalyst loading of 0.26 wt%, and a TMP between 0.68 and 1.72 bar (10–25 psig). The filtration process was run in a recycle mode, whereas clean permeate was added back to the slurry mixture, thus allowing the catalyst concentration to remain approximately constant over the course of the run (given minor adjustments for about 5 ml permeate and slurry samples collected throughout the test).

After initial start-up, it was found that the magnitude of transmembrane flux decreased dramatically, indicating that a mass transfer boundary layer may form on the filter media. This boundary layer appears to remain somewhat constant after about 6 h on-line, whereas the slope of the flux versus time plot (Figure 15.13) becomes fairly linear. This linear decrease in flux was found to continue over the next 43 h. This could be attributed to fouling of the membrane by the small iron/

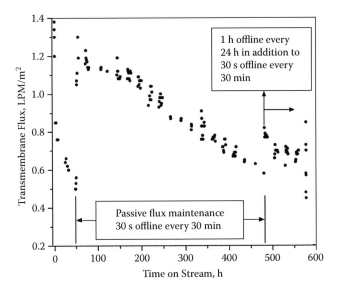

FIGURE 15.13   Variation of permeate flux against time on stream at a TMP of 1.5 bar.

iron carbide particles. At this point, a passive flux maintenance procedure was employed (49.33 h TOS), introducing a 30 s flux shutoff period every 30 min. This allowed the catalyst slurry to recirculate through the cross-flow filter media with no radial permeate flow through the filter membrane. This axial flow was designed to relax the filter membrane, possibly by releasing any embedded particles and aiding in increasing the flux magnitude. The flux did not increase significantly initially after the TMP was reinstated to 1.4 bar; however, the flux was found to slowly increase after a few of the 30 s flux shut off cycles. After twelve of the 30 s flux shut off cycles (about 6 h), the overall flux increased to 161% (from 0.5 to 1.3 lpm/m², or 17.6 to 46.0 Gallons per day (GPD)/ft²). Thereafter, the flux slowly decreased over the next 280 h TOS, down to 0.76 lpm/m², or 27.0 GPD/ft² (a 41.3% loss in flux over this span).

An active flux maintenance procedure was initiated at this point (about 330 h TOS), beginning with a 2 s back-flush of clean permeate through the filter membrane. This active flux maintenance cycle was continued every 30 min for just over 24 h. The flux initially recovered to 0.90 lpm/m² (32.0 GPD/ft²), but declined again within 24 h to a baseline value of 0.76 lpm/m² (26.7 GPD/ft²) without clean permeate back-flush. The flux maintenance method was then returned to passive (no back-flush with clean permeate) mode, only increasing the flux off-time to 60 s every 30 min. Thereafter, the flux steadily declined over the next 120 h TOS from 0.77 to 0.58 lpm/m² (27.3 to 20.4 GPD/ft²). At 480 h TOS, a 1 h flux off-cycle was attempted, resulting in an increase of the flux back to 0.82 lpm/m² (29.1 GPD/ft²), a 42.6% increase. When the flux off-cycle was returned to the 60 s off-cycle for the next 48 h, it was found that the permeate flux decreased to 0.62 lpm/m² (21.9 GPD/ft²). Applying another 1 h flux off-cycle returned the flux to 0.72 lpm/

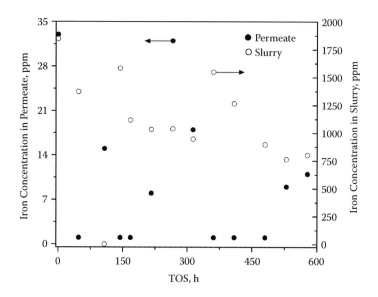

FIGURE 15.14   Concentration of iron in permeate and slurry samples collected at different times.

m² (25.3 GPD/ft² only, a 15.5% increase). It was found that applying a 1 min flux off-cycle every 30 min for the next 23 h resulted in a much lower rate of decline in permeate flux (down to 0.65 lpm/m², or 23.1 GPD/ft²). A final 1 h flux off-cycle exhibited minimal flux increase (0.04 lpm/m², or 5.6%).

The variation of iron content in both the slurry and permeate samples against time on stream is represented in Figure 15.14. The permeate purity (in terms of iron concentration) was consistently below 35 ppm (as Fe) for the entire experiment, with over 85% below the 16 ppm level. The variation over iron content could be due to sampling during or after flux maintenance cycles, which can disturb the boundary layer of submicron particles on the membrane surface.

At a TOS of about 560 h the flux was shut off for the overnight period to clean the filter media to get an estimation of variation of flux magnitude against TMP. The plot of permeate flux against TMP, as represented in Figure 15.15, indicated a linear relationship between the two variables, with the highest flux (0.85 lpm/m², or 29.9 GPD/ft²) obtained at the upper end of the TMP range, 1.72 bar (25 psig).

15.5 CONCLUSIONS

Application of cross-flow filtration for the removal of FT wax products can be a useful technique to maintain a constant catalyst loading in an FTS reactor in continuous operation. Addition of 1-dodecanol (at a concentration of 6 wt%) was found to decrease the permeation rate of the cross-flow filter used for the separation of simulated FT wax and activated iron catalyst slurry. However, additional

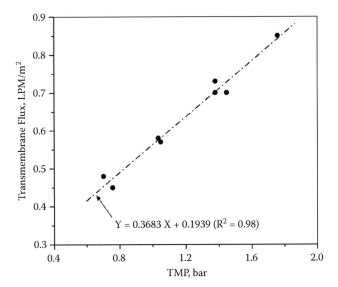

FIGURE 15.15 Variation of permeate flux against TMP (data collected after about 560 h of continuous operation).

increases in the 1-dodecanol concentration (at 11 and 17 wt% levels) did not affect the permeation rates. No significant variation in the concentration of iron in the permeate was found when a higher concentration of alcohol was added. Addition of 1-hexadecene was found to have an insignificant to no effect on the permeate flux and iron concentration in the permeate.

The concentration of iron present in the permeate wax was found to be consistently less than 35 ppm, with over 85% below the 16 ppm level. Following an active flux maintenance procedure results in short-term recovery of flux, which declines to base value within 24 h. The passive flux maintenance procedure of interrupting the permeate flow for 30 or 60 s per 30 min was effective in recovering the initial membrane fouling temporarily. Better flux stability was attained only after increasing the permeate off-cycle to 1 h per day in addition to 30 s off per half-hour cycle. Variation of flux magnitude with TMP was found to follow a linear relationship within the range studied.

ACKNOWLEDGMENTS

This work was supported by U.S. DOE Contract DE-FC26-98FT40308 and the Commonwealth of Kentucky.

REFERENCES

1. Sarkar, A., Seth, D., Dozier, A. K., Neathery, J. K., Hamdeh, H. H., and Davis, B. H. 2007. Fischer–Tropsch synthesis: Morphology, phase transformation and particle size growth of nano-scale particles. *Catalysis Letters* 117:1–17.

2. Jager, B., Steynberg, A. P., Inga, J. R., Kelfkens, R. C., Smith, M. A., and Malherbe, F. E. J. 1997. Process for producing liquid and optionally gaseous products from gaseous reactants. U.S. Patent 5599849.
3. Lorentzen, G. B., Westvik, A., and Myrstad, T. 1996. Solid/liquid slurry treatment apparatus and catalytic multi-phase reactor. U.S. Patent 5520890.
4. Brennan, J. A., Chester, A. W., and Chu, Y.-F. 1986. Separation of catalyst from slurry bubble column wax and catalyst recycle. U.S. Patent 4605678.
5. Engel, D. C., and Van Der Honing, G. 1999. Method for separating liquid from a slurry. U.S. Patent 5900159.
6. Benham, C. B., Yakobson, D. L., and Bohn, M. S. 2000. Catalyst/wax separation device for slurry Fischer-Tropsch reactor. U.S. Patent 6068760.
7. Roberts, G. W., and Kilpatrick, P. K. 2001. Methods and apparatus for separating Fischer-Tropsch catalysts from liquid hydrocarbon product. U.S. Patent 6217830.
8. Murkes, J. 1990. Fundamentals of cross-flow filtration. *Separation and Purification Methods* 19:1–29.
9. Svarovsky, L. 1993. Filtration. In *Kirk-Othmer encyclopedia of chemical technology*, ed. J. I. Kroschmitz and M. Howe-Grant, 788–853. Vol. 10. New York: John Wiley & Sons.
10. Shoemaker, W. 1977. The spectrum of filter media. *AIChE Symposium Series 171* 73:26–32.
11. Svarovsky L. (Ed.). 1981. *Solid-liquid separation*, 242–64. London: Butterworths.
12. Sarkar, A., Neathery, J. K., and Davis, B. H. 2006. *Separation of Fischer-Tropsch wax products from ultrafine iron catalyst particles*. U.S. DOE Final Technical Report, Contract DE-FC26-03NT41965.
13. Van Der Laan, G. P., and Beenackers, A. A. C. M. 1999. Kinetics and selectivity of the Fischer-Tropsch synthesis: A literature review. *Catalysis Reviews—Science and Engineering* 41:255–318.
14. Wang, Y., and Davis, B. H. 1999. Fischer-Tropsch synthesis: Conversion of alcohols over iron oxide and iron carbide catalysts. *Applied Catalysis A: General* 180:277–85.
15. Davis, B. H. 2003. Fischer-Tropsch synthesis: Relationship between iron catalyst composition and process variables. *Catalysis Today* 84:83–98.
16. Mietton-Peuchot, M., Condat, C., and Courtois T. 1997. Use of gas-liquid porometry measurements for selection of microfiltration membranes. *Journal of Membrane Science* 133:73–82.
17. Poslethwaite, J., Lamping, S., Leach, G., Hurwitz, M., and Lye, G. 2004. Flux and transmission characteristics of a vibrating microfiltration system operated at high biomass loading. *Journal of Membrane Science* 228:89–101.
18. Bhave, R. 1991. *Inorganic membranes: Synthesis, characteristics and applications*, 95–107, 129–54. New York: Van Nostrand Reinhold.
19. Cakl, J., Bauer, I., Doleček, P., and Mikulášek, P. 2000. Effects of backflushing conditions on permeate flux in membrane cross-flow microfiltration of oil emulsion. *Desalination* 127:189–98.
20. Bhatt, Bharat L. 1995. *Liquid phase Fischer-Tropsch. II. Demonstration in the LaPorte Alternative Fuels Development Unit—Topical report*. Final (Vol. I/II: Main Report), U.S. DOE Final Technical Report, DOE Contract DE-AC22-91PC90018.
21. Dry, M. E. 1981. The Fischer-Tropsch synthesis. In *Catalysis—Science and technology*, ed. J. R. Anderson and M. Boudart, 201–2. Vol. 1. New York: Springer-Verlag.
22. Madon, R. J., and Shaw, H. 1977. Effects of sulfur on the Fischer-Tropsch synthesis. *Catalysis Reviews—Science and Engineering* 15:69–105.

16 Detailed Kinetic Study and Modeling of the Fischer-Tropsch Synthesis over a State-of-the-Art Cobalt-Based Catalyst

Carlo Giorgio Visconti, Zuzana Ballova, Luca Lietti, Enrico Tronconi, Roberto Zennaro, and Pio Forzatti

CONTENTS

In this work, a detailed kinetic model for the Fischer-Tropsch synthesis (FTS) has been developed. Based on the analysis of the literature data concerning the FT reaction mechanism and on the results we obtained from chemical enrichment experiments, we have first defined a detailed FT mechanism for a cobalt-based catalyst, explaining the synthesis of each product through the evolution of adsorbed reaction intermediates. Moreover, appropriate rate laws have been attributed to each reaction step and the resulting kinetic scheme fitted to a comprehensive set of FT data describing the effect of process conditions on catalyst activity and selectivity in the range of process conditions typical of industrial operations.

The developed model allows the simultaneous prediction of both the reactants conversion and n-paraffins and α-olefins selectivity from C_1 to C_{49} as a function of process conditions.

16.1   INTRODUCTION

The detailed kinetic description of a chemical process is a primary feature for both the industrial practice and the comprehension of the reaction mechanism. The development of a kinetic model able to predict at the same time the reactants conversion and the products distribution (i.e., a detailed kinetic model) is a prerequisite for the design, optimization, and simulation of the industrial process. Also, the detailed description of process kinetics allows the *ex post* evaluation of the goodness of the mechanistic scheme on the basis of which the model itself is developed, making possible the collection of further insight in the chemistry of the process.

The detailed kinetics of the FTS have been studied extensively over several catalysts since the 1950s, and many attempts have been reported in the literature to derive rate equations describing the FT reacting system. A major problem associated with the development of such kinetics, however, is the complexity of the related catalytic mechanism, which results in a very large number of species (more than two hundred) with different chemical natures involved in a highly interconnected reaction network as reaction intermediates or products.

The details of the FT mechanism are still debated. On the basis of the nature of the CO adsorption, popular mechanistic proposals include the carbide mechanism,[1,2] wherein CO adsorbs dissociatively, and the enolic mechanism,[3] involving the molecular adsorption of CO. Also, hydrogen-assisted CO dissociation theories, in which CO forms the chain initiator intermediate CH* passing through the intermediates HCO* and HCOH*, have been recently reproposed[4,5] as alternatives to the more classical hydrogen-unassisted CO dissociative adsorption, where the carbide species C* is supposed to be the common intermediate for the formation of all the reaction products. Depending on the nature of surface chain growth species, in addition, it is possible to distinguish between the alkyl mechanism,[6] based on the insertion of a methylene species into the metal-alkyl bond, and the alkenyl mechanism,[2] wherein a surface vinyl species reacts with a surface methylene to form an allyl species.

On account of this, the difficulties associated with developing a detailed kinetic model for the FTS, able to describe at the same time the rate of formation of all the reaction products, are obvious. It is therefore not surprising that several efforts have been devoted through the years to simplify the kinetic mechanism of the FTS.

A typical approach has relied on lumping the individual mechanistic steps together with the related complexities into a limited number of overall molecular reactions. Such reactions are usually selected in such a way as to account for the formation of the most relevant FT products or classes of products (paraffins, olefins, alcohols, acids, waterm and carbon dioxide); see, e.g., Kravtsov et al. [7] The lumped approach to FT kinetics has even been brought to the extreme by reducing the full chemistry of FTS into a single global stoichiometry, where CO and H_2 react forming the imaginary hydrocarbon C_nH_m (where n and m are two numbers that represent the average FT product) and H_2O, as reported by Zennaro et al.[8] and Lox and Froment.[9] Obviously, such a reaction does not provide any information on the Fischer-Tropsch product distribution, but can be used to describe effectively the overall syngas conversion rate, which can be adequate enough for some specific purposes, like the modeling of the thermal effects in the FT reactors.

However, the detailed description of the FT product distribution together with the reactant conversion is a very important task for the industrial practice, being an essential prerequisite for the industrialization of the process. In this work, a detailed kinetic model developed for the FTS over a cobalt-based catalyst is presented that represents an evolution of the model published previously by some of us.[10] Such a model has been obtained on the basis of experimental data collected in a fixed bed microreactor under conditions relevant to industrial operations (temperature, 210–235°C; pressure, 8–25 bar; H_2/CO feed molar ratio, 1.8–2.7; gas hourly space velocity, (GHSV) 2,000–7,000 cm^3 (STP)/h/$g_{catalyst}$), and it is able to predict at the same time both the CO and H_2 conversions and the hydrocarbon distribution up to a carbon number of 49. The model does not presently include the formation of alcohols and CO_2, whose selectivity is very low in the FTS on cobalt-based catalysts.

16.2 EXPERIMENTAL

The Co/Al_2O_3 catalyst used in this study was a bench-scale prepared sample obtained by incipient wetness impregnation of a commercial γ-Al_2O_3 with an aqueous Co nitrate solution, according to the procedure reported by Oukaci et al.[11] The Co loading was 15% w/w. The catalyst had a surface area near 120 m^2/g and a pore volume of 0.31 cm^3/g, with a monomodal pore size distribution centered near 85 Å as determined by nitrogen adsorption-desorption at 77 K using a Micromeritics Tristar 3000 instrument.

The effects of the process conditions on the kinetics of the process were investigated in the fully automated lab-scale plant sketched in Figure 16.1. High-purity H_2 (99.995% H_2, Sapio), N_2 (99.999% N_2, Sapio), and CO (CO/Ar, 98/2 v/v, Ar internal standard, Sapio) were fed separately to the unit by means of three

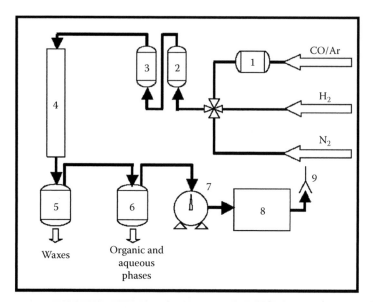

FIGURE 16.1 Sketch of the adopted FT lab-scale plant. (1) Inlet CO purifying trap from iron-carbonyl, (2,3) inlet gas mixture purifying traps, (4) packed bed reactor, (5,6) condensable products vessels, (7) tail gas volumetric totalizer, (8) gas chromatograph, and (9) vent.

different mass flow controllers (Brooks Instrument model 5850S) and mixed in a proper mixing volume. Great attention was dedicated to the reactant purification before feeding to the catalyst. For that reason, the CO stream fed to the unit was first purified from potentially contained iron carbonyls by means of a series of traps containing molecular sieves (5 Å), activated charcoal, and alkalinized alumina (at 200°C). The purified mixture was then fed to the packed bed reactor, consisting of a single-pass stainless steel tube (I.D., 1.1 cm; length, 85 cm; SS-306) with plug-flow hydrodynamics. In order to measure the catalyst axial temperature profile, the reactor was equipped with an axial thermowell (2 mm O.D. tubing, welded on one side) containing a sliding J-type thermocouple.

The unconverted reactants and the reaction products leaving the reactor were sent first to a hot vessel heated at 110°C for the collection of the waxes, followed by a second cold vessel cooled at 0°C for the separation of liquid aqueous and organic products. All the transfer lines between the reactor and the hot trap were kept at 150°C to prevent the solidification of the waxes and the condensation of gasoline and diesel range hydrocarbon products outside of the proper traps.

Noncondensable gases leaving the condensation vessels were depressurized (by means of an electronic back-pressure, Brooks Instrument model 5866), totalized (by means of an on-line flow gas meter, Ritter model TG05-5), and periodically analyzed with an on-line GC (Hewlett-Packard model 6890) equipped with three columns and two detectors for the analysis of C_1–C_{10} hydrocarbons (Al_2O_3 plot capillary column connected to a flame ionization detector), H_2, CH_4,

CO (molecular sieve packed column connected to a thermal conductivity detector), and CO_2 (porapak Q packed column connected to a thermal conductivity detector).

The waxes, the organic phase, and the aqueous products, on the contrary, were unloaded daily from the collection traps and analyzed with an off-line GC (Hewlett-Packard model 6890) equipped with two flame ionization detectors and two identical columns (Hewlett-Packard HP-5), one connected to an on-column injector and dedicated to the analysis of waxes (dissolved in CS_2 before the injection), and the other connected to a split/splitless injector and used for the analysis of the liquid reaction products (aqueous and organic phases). CH_3CN was added to the aqueous sample prior to the injection as internal standard.

This analytical procedure allowed the detection of C_1–C_{49} hydrocarbons and C_1–C_{22} alcohols. Carbon balance, calculated as moles of C contained in the reaction products divided by the moles of CO converted, always closed within ±10%.

In a typical run, 2 g of catalyst, sieved in the range of 75 to 100 μm to minimize intraparticle mass transfer limitations and diluted with α-Al_2O_3 (1:4 v/v) to prevent strong temperature gradients along the reactor bed caused by the exothermic FTS, was loaded in the reactor and reduced overnight (16 h) at 400°C (heating ramp from ambient temperature to 400°C at 2°C/min) and atmospheric pressure under a flow of pure H_2. After H_2 treatment, the process conditions were slowly moved to the so-called standard point (T = 230°C, P = 20 bar, H_2/CO inlet molar ratio = 2.1, GHSV = 5,000 cm³ (STP)/h/g_{cat}), which represents the central point around which the process conditions have been varied to perform the kinetic tests. Accordingly, N_2 was added to the H_2 feed, the reactor was cooled down to 180°C, the syngas flow was fed to the catalyst, and the pressure was slowly raised to 20 bar. The reactor temperature was then increased to 230°C in about 14 h and the N_2 in the feed was gradually removed. A time on stream (T.o.S) of 0 was assigned to the instant at which the unit reached the standard operating conditions.

Experimental conditions were varied in the following ranges: P = 8–25 bar, T = 210–235°C, H_2/CO feed molar ratio = 1.8–2.7, GHSV = 2,000–7,000 cm³(STP)/h/ g_{cat}. In particular, forty-six steady-state runs were performed at twelve different process conditions, following the experimental plan reported in Table 16.1. Each experimental condition was replicated several times in order to verify the experimental data reproducibility. Moreover, the standard point was replicated twelve times at different T.o.S. as an activity check. In order to describe the activity of the catalyst at steady-state conditions for both CO conversion and the product selectivity, all the tests selected for the kinetic modeling were performed at T.o.S. higher than 300 h. As shown in Figure 16.2a, the adopted catalyst is characterized by an initial decrease in activity that reaches a stable level only after about 300 h on stream. On the contrary, only slight changes are evident in the product selectivity as a function of time on stream after 200 h on stream, as shown in Figure 16.2b in terms of an Anderson-Schulz-Flory (ASF) diagram for a typical run. In fact, if one neglects the minor deviations observed for the data collected at 101 h, the data collected between 220 and 461 h on stream are almost superimposed.

TABLE 16.1
Experimental Data Set

No.	T (°C)	P (bar)	H₂/CO Inlet Ratio (mol/mol)	GHSV (cm³(STP)/h/g_cat)
1	230	20	2.1	5,000
2	230	8	2.1	5,000
3	230	25	2.1	5,000
4	230	20	2.1	5,000
5	230	20	1.8	5,000
6	230	20	2.3	5,000
7	230	20	2.7	5,000
8	230	20	2.1	5,000
9	220	20	2.1	5,000
10	235	20	2.1	5,000
11	230	20	2.1	5,000
12	230	20	2.1	4,000
13	230	20	2.1	7,000
14	230	20	2.1	5,000
15	210	20	2.1	2,000
16	220	20	2.1	2,000

Product selectivity to the ith species (S_i) reported in the following has been calculated according to Equation 16.1:

$$S_i = \frac{F_i^{out} \cdot n_i}{\sum\limits_{j=1}^{NP} F_j^{out} \cdot n_j} \tag{16.1}$$

where F_i^{out} is the molar productivity of ith species, n_i is the carbon atom number of the ith species, and NP is the number of the products formed in the process.

16.3 RESULTS AND DISCUSSION

16.3.1 EFFECT OF THE PROCESS CONDITIONS ON THE CATALYST ACTIVITY AND SELECTIVITY

In order to allow the estimation of the kinetic parameters involved in the development of a detailed FT kinetic model, the effects of operating conditions on both

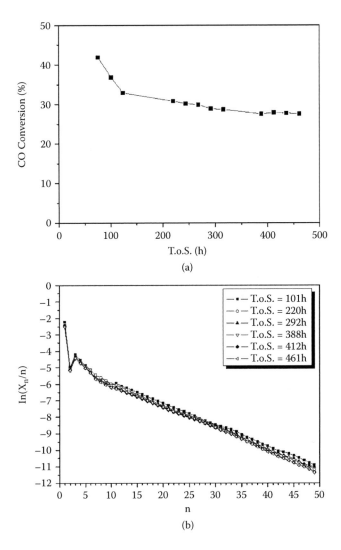

FIGURE 16.2   (a) CO conversion and (b) hydrocarbon distribution (in terms of ASF plot) evolutions during the first hours on stream (T = 230°C, P = 20 bar, H_2/CO feed molar ratio = 2.1, GHSV = 5,000cm³ (STP)/h/g_{cat}).

CO conversion and product distribution were investigated. The experimental data used for the fitting were collected in two different long-term runs of the unit previously described, adopting a new batch of catalyst each time. The reproducibility of the results obtained in the two runs was checked by comparing the catalyst reactivity at standard conditions. A good agreement was found in terms of both CO conversion and product distribution (data not reported).

16.3.1.1 Pressure Effects

The effect of pressure on the FTS was studied, according to Table 13.1, at three different pressure levels (8, 20, and 25 bar), keeping constant the other operating conditions at the level of the standard test (T = 230°C, H_2/CO inlet molar ratio = 2.1, GHSV = 5,000 cm^3 (STP)/h/g_{cat}).

As shown in Figure 16.3a, the increase of the total pressure enhances slightly the CO conversion, from 25% at 8 bar to 29% at 25 bar. In contrast, the selectivity to the heaviest hydrocarbons (C_{25+}) strongly increases upon increasing the pressure, while the selectivity to both methane and olefins decreases. The ASF distributions obtained at the three analyzed pressure levels are compared in Figure 16.3b. In agreement with selectivity data, the increase leads to a significant variation of the slope of the corresponding ASF plot. However, this effect seems to vanish above 20 bar. The ASF product distributions obtained at the highest investigated pressures are very similar, whereas the ASF plot corresponding to the test performed at 8 bar clearly deviates, especially for C_{25+} species.

The effects of pressure on process selectivity can be interpreted by considering the olefins reactivity during low-temperature FTS. At the typical reaction conditions, a thin layer of liquid waxes surrounds the catalyst particles. As a result, reactants have to diffuse inside this layer before reaching the catalyst surface, while reaction products have to do the same in the opposite direction before being desorbed. It is well known that olefins, in contrast to paraffins, can be readsorbed on the catalyst active sites and can be reinserted in the chain growth process (forming heavier hydrocarbons) or can be hydrogenated to the corresponding paraffins.[12–14] The increase of pressure enhances the olefins solubility in the liquid phase surrounding the catalyst pellets, so it induces an increase of olefin readsorption probability (that is proportional to the concentration of olefins in the liquid phase surrounding the catalyst pellets). This clearly explains the increase of C_{25+} selectivity (due to the olefin reinsertion in the chain growth) and the decrease of olefins selectivity (due to their hydrogenation to paraffins). The observed pressure effects are well in agreement with literature data for cobalt-based catalysts.[15–17] For example, when working with a 15 wt% Co/Al_2O_3 catalyst promoted with a small amount (<1% wt%) of rhenium, Das et al.[15] observed that CO conversion increases with pressure, and that this growth is greater at low pressures (CO conversion at 8 atm = 23%, at 21 atm = 39%, at 35 atm = 47.5%), while Jacobs et al.[16] observed that increasing the total pressure increases heavy products selectivity.

16.3.1.2 Effects of Syngas Composition

The effect of syngas composition on FTS is shown in Figure 16.4a (CO conversion and main reaction products selectivity) and 16.4b (ASF product distribution), at the investigated H_2/CO inlet molar ratios (1.8, 2.1, 2.3, and 2.7).

The data plotted in Figure 16.4a clearly show that CO conversion increases linearly upon increasing the H_2/CO inlet ratio. Methane selectivity also increases with the H_2/CO ratio, but the effect is less pronounced. On the contrary, the selectivity to both the heaviest hydrocarbons and olefins decreases upon increasing the

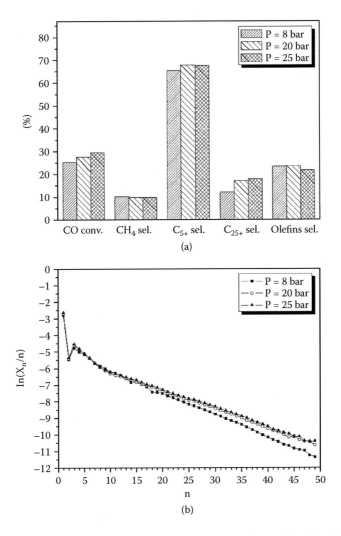

FIGURE 16.3 Effect of pressure (a) on both the CO conversion and the selectivity to the main reaction products and (b) on the hydrocarbon distribution (T = 230°C, H_2/CO feed molar ratio = 2.1, GHSV = 5,000 cm³(STP)/h/g_{cat}).

H_2 content of the feed gas. As a consequence, the corresponding ASF distributions (Figure 16.4b) are strongly affected by the change in the H_2/CO inlet molar ratio, so that an increase of the chain growth probability (corresponding to the exponential of the slope of the ASF diagram) is well evident as the H_2/CO feed ratio decreases.

The observed H_2/CO effect on CO conversion can be explained considering the strong CO adsorption ability on cobalt-based catalysts. At the typical FT process conditions, in fact, it has been reported that the catalyst surface is almost

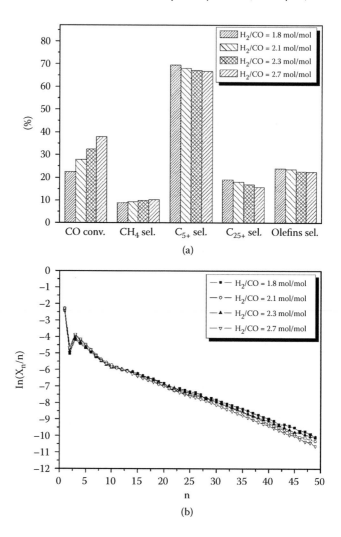

FIGURE 16.4   Effect of H_2/CO inlet molar ratio (a) on both the CO conversion and the selectivity to the main reaction products and (b) on the hydrocarbon distribution ($T = 230°C$, $P = 20$ bar, GHSV $= 5,000$ cm^3(STP)/h/g$_{cat}$).

entirely covered with adsorbed CO* (or the corresponding hydrogenated species CH_2*). This inhibits the H_2 dissociative adsorption, so that the global reaction rates (depending on the hydrogenation rate of the intermediate species, i.e., the carbide in the case of the unassisted CO dissociation theories and CO in the case of the H-assisted theories) are limited. By increasing the H_2/CO inlet ratio, the lower CO partial pressure induces a lower concentration of adsorbed CO, so that more H_2 can be adsorbed and dissociated, and the reaction rates of all hydrogenation reactions increased.

The observed H_2/CO effects on product distribution can be taken into account by considering that high H_2 concentration favors both chain termination and the methanation reaction (both hydrogenation reactions), thus resulting in a decrease of the chain growth probability together with a corresponding decrease of the high molecular weight products selectivity. Also, olefins are readily hydrogenated to the corresponding paraffin at high H_2 feed content, and as a consequence, the olefin selectivity significantly decreases upon increasing the H_2/CO ratio.

The observed effects of syngas composition on both the CO conversion and the product selectivity, again, are in agreement with data already reported.[11,12,14,18,19] For example, Schulz and Claeys[18] observed that when increasing CO partial pressure by decreasing the H_2/CO inlet ratio, both the rate of CO consumption (and thus CO conversion) and the selectivity to methane decrease, while the selectivity to the heavier products increases.

16.3.1.3   Temperature Effects

CO conversion values and hydrocarbon selectivity measured during experiments carried out at different temperatures (220, 230, and 235°C) are plotted in Figure 16.5a. As apparent from this figure, in the analyzed interval, CO conversion strongly increases with temperature, while almost no effect on chain growth probability (i.e., on the selectivity to methane, C_{5+}, and C_{25+}) is observed. As a result, the ASF distributions in term of total hydrocarbons obtained at different temperatures are almost perfectly superimposed (Figure 16.5b). By contrast, a well-defined effect is evident in Figure 16.5a concerning the selectivity to olefins. The fraction of unsaturated products significantly decreases upon increasing the reaction temperature. This observation, along with the indication that the hydrogenation reaction of the olefins to the corresponding paraffins is very far from chemical equilibrium at the actual process conditions, indicates that olefins hydrogenation reaction is kinetically limited during FTS.

These results are in agreement with the literature results, especially with data concerning temperature effects on CO conversion.[12,16,19,20] In case of temperature effect on product distribution, there are many studies that, in apparent disagreement with what is presented here, report an increase of selectivity to the lighter products with increasing temperature. These data, however, are compatible with our results if one considers the narrow temperature interval (220–235°C) investigated in this study.

16.3.1.4   Syngas Space Velocity Effects

Finally, the effect of gas hourly space velocity was investigated in the range of 4,000 to 7,000 $cm^3(STP)/h/g_{cat}$, while keeping constant the other operating conditions at the standard level (T = 230°C, P = 20 bar, H_2/CO inlet molar ratio = 2.1). Since the amount of catalyst loaded in the reactor was kept constant during all experiments, the GHSV was changed by varying the syngas flow rate. Tests with a space velocity of 2,000 $cm^3(STP)/h/g_{cat}$ were also performed (see Table 16.1), but

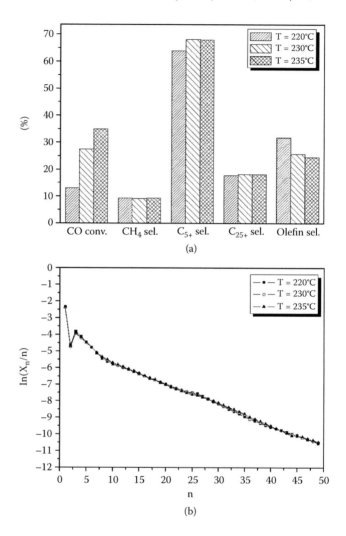

FIGURE 16.5   Effect of temperature (a) on both the CO conversion and the selectivity to the main reaction products and (b) on the hydrocarbon distribution ($P = 20$ bar, H_2/CO feed molar ratio = 2.1, GHSV = 5,000 cm^3(STP)/h/g$_{cat}$).

in order to limit CO conversion, they were carried out at a temperature of 210°C. For this reason, these data cannot be considered here for studying GHSV effects on catalyst performance; instead, these data have been adopted for the estimation of kinetic parameters used in developing the proposed model.

The results of the experiments, in terms of both CO conversion and light hydrocarbons selectivity, are shown in Figure 16.6a. As expected, CO conversion strongly decreases upon increasing space velocity. Moreover, a slight increase of methane selectivity is also observed, while selectivity to C_{5+}, C_{25+}, and olefinic

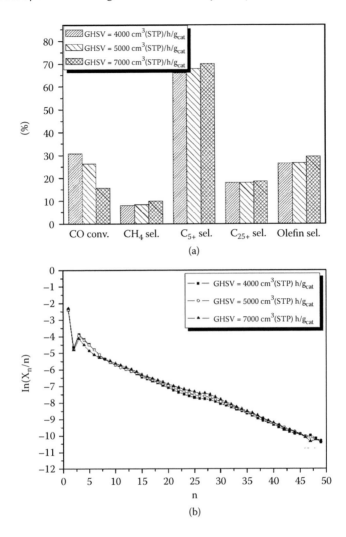

FIGURE 16.6   Effect of GHSV (a) on both the CO conversion and the selectivity to main reaction products and (b) on the hydrocarbon distribution (T = 230°C, P = 20 bar, H_2/CO feed molar ratio = 2.1).

products slightly increases upon increasing the space velocity. The effects of space velocity (in the investigated range) are, however, very small, so that no significant differences are apparent in ASF plots (Figure 16.6b).

Other authors have studied the effect of the gas space velocity, and their results are in agreement with those reported in this work concerning the effect of space velocity on CO conversion.[13,15] In contrast, it is difficult to make a comparison between literature and our GHSV effects on product distribution. As reported by van der Laan and Beenackers,[14] in fact, the effect of the space velocity on product distribution is complex and often controversial, so that contrasting results can be found in literature. For this reason, more detailed studies concerning space velocity effect on FTS are currently under development in our labs.

16.3.2 Development of a Detailed Kinetic Model

16.3.2.1 The Reaction Scheme

Although great efforts have been devoted to elucidate the FT reaction mechanism, there are still many controversies on this point, and many theories have been proposed in the literature in recent years.[21]

On the basis of the nature of CO adsorption and of the nature of chain initiator intermediates, popular mechanistic proposals include the carbide mechanism,[1,2] wherein CO adsorbs dissociatively and the carbide (C*) is the chain initiator intermediate, and the enolic mechanism,[3] involving molecular adsorption of CO and the formation of an oxygen intermediate, the enol (HC*OH).

Depending on the nature of surface chain growth species, on the other hand, one is confronted mainly with the alkyl mechanism,[6] based on the insertion of a methylene species (*CH$_2$) into the metal-alkyl bond, or with the alkenyl mechanism,[2] wherein a surface vinyl species (*CH=CH$_2$) reacts with a surface methylene (*CH$_2$) to form an allyl species (*CH$_2$CH=CH$_2$).

However, many other proposals have been published slightly modifying the original mechanisms described above. For example, Inderwildi et al.[4] and Ojeda et al.[5] recently proposed a modification of the alkyl theory in which the intermediate *CH$_2$ is formed through the so-called H-assisted CO dissociation mechanism as an alternative to the classical unassisted CO dissociation.

In the present work the unassisted CO dissociation version of the carbide theory has been adopted to describe CO conversion, while the alkyl mechanism has been adopted for describing the chain growth process. Accordingly, as proposed in our previous work,[10] the FT reaction pattern has been detailed as follows:

$$H_2 + 2* \leftrightarrows 2H* \qquad Keq_{H2} = \vartheta_{H*}^{2} \cdot P_{H2}^{-1} \cdot \vartheta^{-2} \qquad (16.2)$$

$$CO + * \leftrightarrows CO* \qquad Keq_{CO} = \vartheta_{CO*} \cdot P_{CO}^{-1} \cdot \vartheta^{-1} \qquad (16.3)$$

$$\left.\begin{array}{l} CO* + * \rightarrow C* + O* \\ C* + H* \rightarrow CH* + * \\ CH* + H* \rightarrow CH_2* + * \\ O* + H* \rightarrow OH* + * \\ OH* + H* \rightarrow H_2O + 2* \end{array}\right\} \quad r_M = k_M \cdot \vartheta_{CO*} \cdot \vartheta$$

$$(16.4a\text{--}e)$$

$$CH_2* + H* \rightarrow CH_3* + * \qquad r_{IN} = k_{IN} \cdot \vartheta_{CH2*} \cdot \vartheta_{H*} \qquad (16.5)$$

$$R_n^* + CH_2^* \rightarrow R_{n+1}^* + * \qquad r_{G,n} = k_G \cdot \vartheta_{Rn^*} \cdot \vartheta_{CH2^*} \qquad n: 1 \rightarrow 49 \qquad (16.6)$$

$$CH_3^* + H^* \rightarrow CH_4 + 2^* \qquad r_{CH4} = k_{CH4} \cdot \vartheta_{CH3^*} \cdot \vartheta_{H^*} \qquad (16.7)$$

$$R_n^* + H^* \rightarrow P_n + 2^* \qquad r_{P,n} = k_P \cdot \vartheta_{Rn^*} \cdot \vartheta_{H^*} \qquad n: 2 \rightarrow 48 \qquad (16.8)$$

$$C_2H_5^* \leftrightarrows C_2H_4 + H^* \qquad r_{O,2} = k_{O,dx} \cdot \vartheta_{C2H5^*} - k_{C2H4} \cdot x_{C2H4} \cdot \vartheta_{H^*} \qquad (16.9)$$

$$R_n^* \leftrightarrows O_n + H^* \qquad r_{O,n} = k_{O,dx} \cdot \vartheta_{Rn^*} - k_{O,sx} \cdot x_{On} \cdot \vartheta_{H^*} \qquad n: 3 \rightarrow 49 \qquad (16.10)$$

where P_n and O_n are the generic linear paraffin and α-olefin with n carbon atoms, respectively; R_n^* is the generic linear growing adsorbed hydrocarbon species (with the generic formula $*CH_2(CH_2)_{n-1}H$, with $n = 1 \rightarrow 49$); ϑ is the fraction of free catalytic sites; ϑ_i is the fraction of the catalytic sites occupied by species i; and x_{On} is the molar fraction of the α-olefin with n carbon atoms in the liquid phase surrounding the catalyst pellets. The last is evaluated by means of an explicit correlation, expressing the x_{On} dependence on temperature, number of carbon atoms, and partial pressure of the corresponding olefin in the gas phase, obtained on the basis of *a priori* vapor-liquid equilibrium calculations using the Soave-Redlich-Kwong cubic equation of state. This introduces an implicit dependence of the chain growth probability on the chain length, resulting from higher solubilities of olefins with higher carbon atom numbers.

In the proposed mechanism, H_2 adsorbs reversibly on two different free catalytic sites (*) in the dissociated state (Equation 16.2), while CO is first adsorbed reversibly in the molecular state (Equation 16.3) and then dissociates (Equation 16.4a). Also, the formation of the monomeric species CH_2^* (in accordance with the carbide theory) was assumed to occur via two steps in series, that is, the reaction between the carbide and the surface hydrogen to form the species CH^* (Equation 16.4b), and the reaction between this species and the surface hydrogen to form the monomer CH_2^* (Equation 16.4c). The oxygen formed from CO* dissociation (Equation 16.4a), instead, is removed as water in two steps (Equations 16.4d and 16.4e) involving the intermediate OH^*. Concerning the chain growth mechanism, the alkyl mechanism has been adopted; i.e., we assumed that the reaction is initiated by the formation of a methyl species (Equation 16.5) and that the chain growth takes place by the successive insertion of methylene into the active site–alkyl bond (Equation 16.6). Termination of the chain growth has been described as the result of two different routes: a dual-site reaction between the intermediate R_n^* and an adsorbed hydrogen atom (Equations 16.7 and 16.8) for n-paraffins formation, or a reversible β-hydride elimination reaction (Equations 16.9 and 16.10) for α-olefins synthesis.

16.3.2.2 The Kinetic Model

An elementary rate law or an equilibrium constant was assigned to each step involved in the detailed reaction mechanism of FTS. Both the H_2 dissociative adsorption and CO molecular adsorption were assumed to approach equilibrium at the actual process conditions. For this reason, an equilibrium constant was assigned to Equations 16.2 and 16.3. In contrast, for Equations 16.4a to 16.4e, we assumed that CO* dissociation is rate determining in the sequence of consecutive, nonreversible, and kinetically controlled steps that lead to the formation of methylene species and water. Accordingly, the overall rate of all such steps was described by the rate expression for CO dissociation. It was also assumed that the rate constants describing the elementary steps for the growth of the adsorbed species R_n* (Equation 16.6), the formation of the paraffins C_{2+} (Equation 16.8), and the formation of the olefins C_{3+} (Equation 16.10) are independent of the carbon atom number of the intermediates involved in the elementary reactions. In order to describe the experimental deviation of methane and ethylene from ASF product distribution, specificity was assumed in the kinetic constants involved in the kinetic expressions for the formation of these species (Equations 16.7 and 16.9), extending the approach used by Wang et al.[22] for methane formation on an iron-based FT catalyst.

These assumptions are partially different from those introduced in our previous model.[10] In that work, in fact, in order to simplify the kinetic description, we assumed that all the steps involved in the formation of both the chain growth monomer CH_2* and water (i.e., Equations 16.3 and 16.4a to 16.4e) were a series of irreversible and consecutive steps. Under this assumption, it was possible to describe the rate of the overall CO conversion process by means of a single rate equation. Nevertheless, from a physical point of view, this hypothesis implies that the surface concentration of the molecular adsorbed CO* is nil, with the rate of formation of this species equal to the rate of consumption. However, recent *in situ* Fourier transform infrared (FT-IR) studies carried out on the same catalyst adopted in this work, at the typical reaction temperature and in an atmosphere composed by H_2 and CO, revealed the presence of a significant amount of molecular CO adsorbed on the catalysts surface.[17] For these reasons, in the present work, the hypothesis of the irreversible molecular CO adsorption has been removed.

Hydrogen adsorption was also described as irreversible in our previous mechanism,[10] and an empirical kinetic law was used to describe the rate of this step. However, a deeper analysis of literature data revealed that this step is likely in equilibrium, too. On the basis of this evidence, the previously developed model has been modified in this work in order to improve the physical consistency of the proposed mechanism.

On the basis of the experimental observation that the reaction temperature, in the investigated range (210–235°C), affects CO conversion and the olefin/paraffin ratio (due to the enhancing of the kinetically controlled hydrogenation reactions; see Section 16.3.1), but not the chain growth probability, we introduced in the model activation energies for Equations 16.4, 16.7, and 16.8, while the chain growth step (Equation 16.6) was regarded as nonactivated. In this way, CO and H_2 heats of adsorption were lumped in the activation energy calculated for Equations 16.4, 16.7, and 16.8.

Also, concerning the effect of the temperature on the reaction rates, different assumptions were made here with respect to our previous work.[10] In that case, only the hydrogen and CO adsorption were regarded as activated steps, in order to describe the strong temperature effect on CO conversion. In contrast, due to the insensitivity of the ASF product distribution to temperature variations (see Section 16.3.1), other steps involved in the mechanism were considered as non-activated. In the present work, however, this simplification was removed in order to take into account the temperature effect on the olefin/paraffin ratio. For this reason, Equations 16.7 and 16.8 were considered as activated.

16.3.2.3   The Reactor Model

The reactor model adopted for describing the lab-scale experimental setup is an isothermal homogeneous plug-flow model. It is composed of 2NP + 2 ordinary differential equations of the type of Equation 16.11 with the initial condition of Equation 16.12, NP + 3 algebraic equations of the type of Equation 16.13, and the catalytic sites balance (Equation 16.14):

$$\frac{dF_i}{dW_{cat}} = \sum_{k=1}^{NR} \alpha_{i,k} \cdot r_k \tag{16.11}$$

$$W_{cat} = 0 \quad F_i = F_{i,0} \tag{16.12}$$

$$0 = \sum_{k=1}^{NR} \alpha_{j,k} \cdot r_k \tag{16.13}$$

$$1 = \theta + \theta_{H^*} + \theta_{CO^*} + \theta_{CH_2^*} + \sum_{n=1}^{NP} \theta_{R_n^*} \tag{16.14}$$

where F_i and $F_{i,0}$ are the molar flows of the generic species i (i = CO, H_2, H_2O, CH_4, P_n, C_2H_4, O_n) along the reactor axis and at the reactor inlet, respectively; W_{cat} is the catalyst mass; $\alpha_{i,k}$ and $\alpha_{j,k}$ are the stoichiometric coefficients for the ith and jth (j = H*, CO*, CH_2*, R_n* n:1→NP) species, respectively, in the kth reaction; r_k is the rate of the kth reaction; and NR is the number of the elementary steps (or group of elementary steps if described with the same rate expression) involved in the process.

16.3.2.4   Optimization Method

The adaptive parameters in the model were estimated by nonlinear and multi-response regression, performed using the Fortran subroutine BURENL[23] based on the least-squares method. The algebraic-differential system constituted by Equations 16.11, 16.13, and 16.14 was integrated numerically with the Fortran subroutine LSODI,[24] which allows us to solve stiff problems using Gear's implicit integration method with variable step.

In order to obtain the best fit of the forty-six experimental data sets considered, we adopted thirteen adaptive parameters. The regression was performed using as experimental responses the CO conversion and the CH_4, C_2H_4, C_2H_6, C_3H_6, C_3H_8, C_6H_{12}, C_6H_{14}, $C_{10}H_{20}$, $C_{10}H_{22}$, $C_{12}H_{26}$, $C_{15}H_{32}$, $C_{20}H_{42}$, $C_{25}H_{52}$, $C_{30}H_{62}$, $C_{35}H_{72}$, $C_{45}H_{92}$, C_{5+}, and olefins selectivity. Extradiagonal terms in the correlation matrix were lower than 0.9 in over 80% of the cases.

16.3.2.5 Model Fit

The parity plot in Figure 16.7 shows the ability of the model to estimate the CO conversion at all investigated conditions. The model can satisfactorily predict all the collected experimental data. The average relative error is in fact 9.5%, a satisfactory value that is better than those reported for other FT complete kinetic models by Yang et al.[25] in terms of CO conversion and by Teng et al.[26] in terms of syngas consumption rate, and is significantly lower than that obtained in our previous work.[10] Figure 16.8 shows in addition that the developed model is able to correctly describe the effect of the process conditions on CO conversion, that is, the increase of catalyst activity upon increasing pressure, temperature, and H_2/CO inlet ratio, and upon decreasing space velocity. In the case of the effect of the H_2/CO ratio, the model tends to underestimate the catalyst activity at high hydrogen content and to overestimate it at high CO partial pressures. A deeper analysis of the experimental and calculated data is presently under development in order to improve the fit quality.

In terms of ASF product distribution, Figure 16.9 shows, e.g., one typical model fit at the standard conditions in terms of n-paraffins, α-olefins, and total hydrocarbons. The model satisfactorily describes the products selectivity and also accounts for the typical deviations of the product distribution from the ASF model, namely, the high methane selectivity, the low selectivity to ethylene, and the change of the ASF slope with growing carbon atoms number. This last feature is accounted for by the model due to the adopted dependence on the carbon atoms number of the vapor-liquid equilibrium involving olefins in the liquid surrounding the catalyst pellets. In contrast, the anomalous high selectivity of the reaction to methane and the low selectivity to ethylene are correctly described by the model thanks to the introduction of two specific rate parameters for the reactions involving the formation of these two species (see Section 16.3.2.1).

Notably, with a single set of rate parameter estimates, the present model can also correctly describe the effects of all the investigated process conditions on product distribution. Figure 16.10 compares experimental and calculated ASF product distributions in five of the investigated process conditions. It is worth noticing also that the model predicts the hydrocarbons selectivity up to $n = 49$,

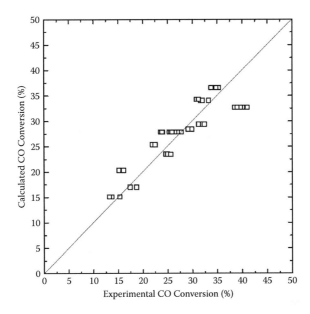

FIGURE 16.7   Parity plot for calculated and experimental CO conversion.

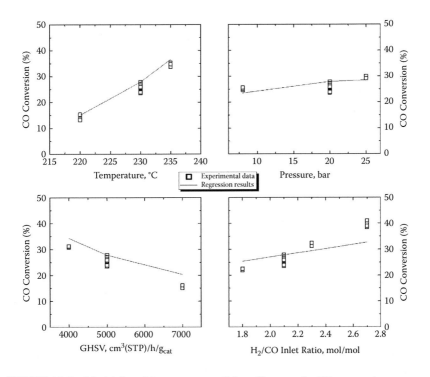

FIGURE 16.8   Model fits of the process condition effects on the CO conversion.

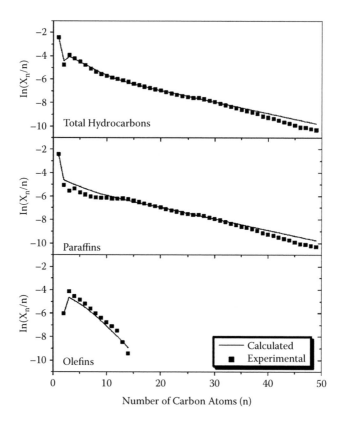

FIGURE 16.9 Model fits at the standard process conditions (T = 230°C, P = 20 bar, H_2/CO feed molar ratio = 2.1, GHSV = 5,000 cm³(STP)/h/g_{cat}) in terms of n-paraffins, α-olefins, and total hydrocarbons.

a range of product distribution that is wide if compared to those found in other literature works.

Inspection of the calculated surface coverage of the intermediate species finally reveals that the surface concentration of the species R_n^* is typically of the same order of magnitude as that of CH_2^*, i.e., the C_1 species associated with CO adsorption/conversion. This implies that the coverage of catalytic sites by the synthesis products has a significant influence on CO conversion rate, which conflicts with the traditional approach of developing separate models for CO conversion and products distribution.

16.4 CONCLUSION

A complete kinetic model of FTS over a state-of-the-art Co catalyst has been derived on the basis of the carbide theory and of CH_2 insertion alkyl mechanism.

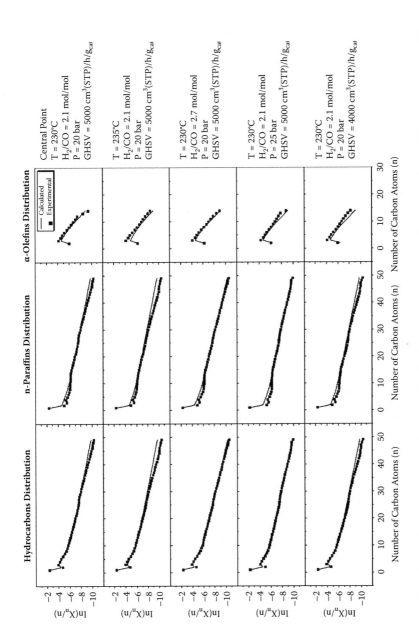

FIGURE 16.10 Experimental and calculated ASF product distributions in terms of total hydrocarbons at five different process conditions.

The adoption of new hypotheses for the reactants adsorption, the removal of all the empiric laws and parameters, and a reevaluation of the temperature effect on product distribution have allowed us to obtain significant improvements with respect to our previous work,[10] in terms of both fitting ability and model consistency.

For the range of industrially relevant conditions, the developed model could accurately predict both the observed CO conversion and the products distribution up to $n = 49$, in terms of total hydrocarbons, n-paraffins, and α-olefins. In particular, using thirteen adaptive parameters, the model is able to describe the typical deviations of the product distribution from the ASF model, i.e., the methane high selectivity, the low selectivity to C_2 species, and the change of the slope of the ASF plot with growing carbon number. Accordingly, the present model can be applied to identify optimized process conditions that are suitable to grant the desired conversion with the requested products distribution.

REFERENCES

1. Fischer, F., and Tropsch, H. 1926. The direct synthesis of petroleum hydrocarbons with standard pressure. (First report). *Berichte der Deutschen Chemischen Gesellschaft* 59:830–31.
2. Maitlis, P.M. 1989. A new view of the Fischer-Tropsch polymerisation reaction. *Pure Appl. Chem.* 61:1747–54.
3. Storch, H.H., Golumbic, N., and Anderson, R.B. 1951. The *Fischer-Tropsch and related synthesis*. New York: Wiley.
4. Inderwildi, O.R., Jenkins, S.J., and King, D.A. 2008. Fischer-Tropsch mechanism revisited: Alternative pathways for the production of higher hydrocarbons from synthesis gas. *J. Phys. Chem. C* 112:1305–7.
5. Ojeda, M., Ishikawa, A., Iglesia, E., Nabar, R., Nilekar, A., and Mavrikakis, M. 2007. Paper presented at the 20th NAM, Houston, TX, June 17–22.
6. Brady, R.C., and Pettit, R. 1980. Reactions of diazomethane on transition-metal surfaces and their relationship to the mechanism of the Fischer-Tropsch reaction. *J. Am. Chem. Soc.* 102:6181–82.
7. Kravtsov, A.V., Moizes, O.E., Usheva, N.V., and Yablonskii, G.S. 1988. Kinetic model for hydrocarbon synthesis from CO and H_2 accounting for its intragroup distribution. *React. Kinet. Catal. Lett.* 36:201–6.
8. Zennaro, R., Tagliabue, M., and Bartholomew, C.H. 2000. Kinetics of Fischer–Tropsch synthesis on titania-supported cobalt. *Catal. Today* 58:309–19.
9. Lox, E.S., and Froment, G.F. 1993. Kinetics of the Fischer-Tropsch reaction on a precipitated promoted iron catalyst. 2. Kinetic modeling. *Ind. Eng. Chem. Res.* 32:71–82.
10. Visconti, C.G., Tronconi, E., Lietti, L., Zennaro, R., and Forzatti, P. 2007. Development of a complete kinetic model for the Fischer–Tropsch synthesis over Co/Al$_2$O$_3$ catalysts. *Chem. Eng. Sci.* 62:5338–43.
11. Oukaci, R., Singleton, A.H., and Goodwin Jr., J.G. 1999. Comparison of patented Co F–T catalysts using fixed-bed and slurry bubble column reactors. *Appl. Catal. A* 186:129–44.
12. Bechara, R., Balloy, D., and Vanhove, D. 2001. Catalytic properties of Co/Al$_2$O$_3$ system for hydrocarbon synthesis. *Appl. Catal. A* 207:343–53.

13. Kuipers, E.W., Scheper, C., Wilson, J.H., Vinkenburg, I.H., and Oosterbeek, H. 1996. Non-ASF product distributions due to secondary reactions during Fischer–Tropsch synthesis. *J. Catal.* 158:288–300.
14. van der Laan, G.P., and Beenackers, A.A.C.M. 1999. Kinetics and selectivity of the Fischer–Tropsch synthesis: A literature review. *Catal. Rev. Sci. Eng.* 41:255–318.
15. Das, T.K., Jacobs, G., Patterson, P.M., Conner, W.A., Li, J., and Davis, B.H. 2003. Fischer–Tropsch synthesis: Characterization and catalytic properties of rhenium promoted cobalt alumina catalysts. *Fuel* 82:805–15.
16. Jacobs, G., Patterson, P.M., Zhang, Y., Das, T., Li, J., and Davis, B.H. 2002. Fischer–Tropsch synthesis: Deactivation of noble metal-promoted Co/Al_2O_3 catalysts. *Appl. Catac. A* 233:215–26.
17. Visconti, C.G., Lietti, L., Tronconi, E., Forzatti, P., Zennaro, R., and Finocchio, E. 2008. Fischer-Tropsch synthesis on a Co/Al_2O_3 catalyst with CO_2 containing syngas. *Appl. Catal. A.* 355:61–8.
18. Schulz, H., and Claeys, M. 1999. Reactions of α-olefins of different chain length added during Fischer–Tropsch synthesis on a cobalt catalyst in a slurry reactor. *Appl. Catal. A* 186:71–90.
19. Jager, B., and Espinoza, R. 1995. Advances in low temperature Fischer-Tropsch synthesis. *Catal. Today* 23:17–28.
20. Dry, M.E. 1983. The Sasol Fischer-Tropsh process. In *Applied industrial catalysis*, 167–213. Vol. 2. New York: Academic Press.
21. Davis, B.H. 2001. Fischer–Tropsch synthesis: Current mechanism and futuristic needs. *Fuel Proc. Tech.* 71:157–66.
22. Wang, Y.N., Ma, W.P., Lu, Y.J., Yang, J., Xu, Y.Y., Xiang, H.W., Li, Y.W., Zhao, Y.L., and Zhang, B.J. 2003. Kinetics modelling of Fischer–Tropsch synthesis over an industrial Fe–Cu–K catalyst. *Fuel* 82:195–213.
23. Buzzi-Ferraris, G., and Donati, G. 1974. A powerful method for Hougen-Watson model parameter estimation with integral conversion data. *Chem. Eng. Sci.* 29:1504–9.
24. Hindmarsh, A.C. 1983. ODEPACK, a systematized collection of ODE solvers. *Scientific Computing*, pp. 55–64.
25. Yang, J., Liu, Y., Chang, J., Wang, Y.N., Bai, L., Xu, Y.Y., Xiang, H.W., Li, Y.W., and Zhong, B. 2003. Detailed kinetics of Fischer-Tropsch synthesis on an industrial Fe-Mn catalyst. *Ind. Eng. Chem. Res.* 42:5066–90.
26. Teng, B.T., Chang, J., Zhang, C.H., Cao, D.B., Yang, J., Liu, Y., Guo, X.H., Xiang, H.W., and Li, Y.W. 2006. A comprehensive kinetics model of Fischer-Tropsch synthesis over an industrial Fe-Mn catalyst. *Appl. Catal. A* 301:39–50.

17 Reduction of CO_2 Emissions from CTL Processes

Use of a Novel FT-Based Chemistry

Diane Hildebrandt, David Glasser, and Bilal Patel

CONTENTS

Coal is the main energy reserve for many countries. In order to ensure energy security, many of these countries are looking at implementing large-scale coal-to-liquid (CTL) plants. Current CTL Fischer-Tropsch (FT)-based processes at best have a carbon efficiency near 45%; that is, less than half the carbon that is fed to the process ends up as hydrocarbon fuels. This has large implications for the environment, in that more than half of the carbon in the feed to the CTL ends up as CO_2. Thus, there is potentially a very serious global environmental problem when these CTL-based processes come on line.

In this paper we look at a novel chemical route that theoretically shows significantly improved carbon utilization in a CTL process. We look at the target CO_2 emissions and thus the potential for reduction of CO_2 emissions from such a CTL process. We also look at the implications for the FT chemistry and the opportunities and potential problems for implementing this chemistry.

17.1   INTRODUCTION

The use of coal as a feedstock and energy source has become more attractive in recent years, particularly in countries where there are little oil reserves or oil or gas is expensive (e.g., China, India, South Africa, and Australia). There is therefore a need for technologies that utilize coal in an efficient and environmentally friendly manner. In this regard, a coal-to-liquid (CTL) process based on Fischer-Tropsch (FT) synthesis is a potential technology to utilize coal reserves in order to achieve energy independence and security. A major challenge facing this technology is the environmental impact, in particular, the carbon dioxide emissions, associated with the process.

The Centre of Material and Process Synthesis (COMPS) has embarked on quantifying the efficiency of processes by setting targets for the process, especially in the very early stages of the design process. Once the target of a process is quantified, the potential for material and energy saving can be identified and inefficiencies in the processes can be identified and eradicated or reduced.

This manuscript will look at the targets for CTL processes, especially in terms of carbon dioxide emissions, with the aim of identifying the major sources of emissions and ways in which these can be reduced.

17.2   FINDING TARGETS FOR PROCESSES: THE APPROACH

A major focus is to develop tools that permit the rapid evaluation, conceptualization, and design of new processes. The purpose of these tools is to gain insight into the process and recognize opportunities for improvement in terms of profitability, mass and energy efficiency, and environmental impact. These tools are most useful in the earliest stage of the design process, where minimum process information is available. The decisions taken in the early stage of the design process or the conceptual phase is of vital importance, as the economics of the process is usually set at this stage. Biegler et al. (1997) estimates that the decisions made during the conceptual design phase fix about 80% of the total cost of the process. Once the process structure has been fixed, only minor cost improvements can be achieved. Thus, the success of the process is largely determined by the conceptual design. Therefore, it is imperative that systematic and comprehensive conceptual design tools be developed.

Instead of focusing on unit operations, we characterize chemical processes by flows of mass, heat, and work (Patel et al., 2005). The flow of mass is usually subjected to the conservation law, i.e., a mass balance. The flow of energy (heat and work) is subjected to both the first and second laws of thermodynamics. The first law describes the conservation of energy, while the second law is used to describe the magnitude of the loss work or irreversibility of the process, thus giving an indication of where designers need to concentrate their efforts to improve the performance of the process. Therefore, we believe, by analyzing the flows of these three variables (mass, heat, and work) in a process from the early stages to the flow-sheet development stage, one can gain insights into the process, understand the implications of particular design decisions, and develop strategies for improving processes with regard to these three variables. Poor decisions at the flow-sheet generation stage and inefficiencies in general manifest themselves as extra carbon dioxide production. This approach thus provides a method to choose between different flow sheets at the earliest stage of plant design or as a method to look at upgrading current flow sheets. These three variables also provide targets for the process, which are based on the overall process, as it considers only the inputs and outputs from the process; i.e., it is independent of the structure of the process or the flow sheet. Ways of calculating the minimum amount of mass inputs or outputs (for example, zero CO_2 emissions), minimum amount of energy (or maximum amount of energy produced), and minimum work required (or maximum work produced) can be determined.

17.3   AN INITIAL LOOK AT THE CTL PROCESS

17.3.1   MASS, ENERGY, AND ENTROPY ANALYSIS

17.3.1.1   Overall Mass Balance

If one considers the overall CTL process, considering the input to and outputs from the process, the overall mass balance for the system can be written as follows:

$$3/2C + H_2O = -CH_2- + 1/2CO_2 \qquad (17.1)$$

It is clear from the mass balance that CO_2 is inevitably produced and that for every mole of product, half a mole of CO_2 must be produced. In terms of a carbon efficiency (defined as the amount of carbon that ends up in product), this means a 67% carbon efficiency. This is simply the result of the wrong ratio of the atoms (C and H) in the feedstock.

For comparison, consider a methane feed. A possible mass balance could be as follows:

$$0.75CH_4 + 0.25CO_2 = -CH_2- + 0.5H_2O \qquad (17.2)$$

If methane is used as a feedstock, it is apparent that, in this case, the possibility exists that carbon dioxide can be used as feedstock and that water can be produced as by-product. Since methane is rich in hydrogen (whereas coal is not), it appears that methane is the cleaner raw material, and also it applies that there should be certain opportunities in terms of combining a coal- and natural gas–based process (Patel et al., 2007a and b).

Thus, by considering the overall mass balance, i.e., looking at the inputs and outputs of the process, one can gain many insights into a process and also identify opportunities for CO_2 emissions reduction and enhancing feedstock utilization.

17.3.1.2 Overall Energy Balance

The energy balance is useful in providing one with information regarding the quantity of energy required or produced by the process. By performing an energy balance over the entire process, one can determine the energy flows into and out of the process.

Consider the process given by Equation 17.1. Assume initially that the inputs enter and outputs leave the process as pure components at temperature T_0 (298.15 K) and pressure P_0 (1 atm).

The energy aspects can be considered in terms of the enthalpy change of the process, ΔH to produce 1 mole of product. The energy balance for the process is as follows, assuming all the energy is initially provided by heat:

$$\Delta H = Q$$

The enthalpy change, ΔH, can be calculated for a steady-state process, using H°_f, which is the enthalpy of formation of the various output and input components. Under the assumption that the inputs and outputs are at ambient conditions, the enthalpy of the components corresponds to the standard enthalpy of formation of each component. The kinetic and potential energy terms are neglected from the energy balance. It is also assumed that water enters the process as a liquid and hydrocarbon products leave the process as a liquid. All other components are in the gas phase.

The enthalpy change for Equations 17.1 (CTL process) and 17.2 (gas-to-liquid (GTL) process) are given in Table 17.1.

The CTL process, as described by Equation 17.1, required approximately 58 kJ/mol of energy, whereas the process based on a methane feed, described by Equation 17.2, produces approximately 20 kJ/mol of energy. The energy required by the CTL process can be supplied by combusting a further quantity of coal (considered as C). This will also mean a further production of carbon dioxide.

TABLE 17.1
Energy Requirements for CTL Process

Process	$\Delta H_{process}$ (kJ/mol)
$3/2C + H_2O = -CH_2- + 1/2CO_2$	57.8
$0.75CH_4 + 0.25CO_2 = -CH_2- + 0.5H_2O$	−19.67

Taking the energy requirements into account results in a reduction of the carbon efficiency for the CTL process from 67% to 60%.

17.3.1.3 Overall Work Balance

It can be shown that the maximum theoretical work produced (or minimum work required) for a process is related to the change in Gibbs energy of the process, assuming again the inputs and outputs of the process are pure components at standard conditions (Denbigh, 1956; De Nevers and Seader, 1980).

The Gibbs energy change for a process can be calculated using $G°_f$ as the Gibbs energy of formation of the outputs and inputs. Under the assumption that the inputs and outputs are at ambient conditions, the Gibbs energy of the components corresponds to the standard Gibbs energy of formation of each component.

The Gibbs energy change for Equations 17.1 (CTL process) and 17.2 (GTL process) are given in Table 17.2.

The CTL process as described by Equation 17.1 requires approximately 41 kJ/mol of work to be input into the process, while the process described by Equation 17.2 requires approximately 19 kJ/mol of work to be input.

Work can be supplied in various ways, for example, via high-temperature heat or compression. This aspect will be discussed later in this manuscript.

17.4 TARGETS FOR CTL PROCESSES

17.4.1 OVERALL TARGET FOR CTL

The overall coal-to-liquid fuels process can be described as follows:

$$a \text{ Coal (C)} + b \text{ Water } (H_2O) + c \text{ Oxygen } (O_2) \rightarrow \text{Hydrocarbons}$$
$$(-CH_2-) + d \text{ Carbon Dioxide } (CO_2) \quad (17.3)$$

The inputs considered are coal, water, and oxygen. Hydrocarbons and carbon dioxide are considered as products. The variables a, b, c, and d represent the amounts of the various components required/produced. Figure 17.1 shows a schematic of the process that is being considered.

Initially it is assumed that coal is pure carbon, although coal usually contains other elements, such as hydrogen and nitrogen. It is also assumed that 1 mol of hydrocarbons is being produced.

TABLE 17.2
Change in the Gibbs Energy for CTL Process

Process	$\Delta G_{process}$ (kJ/mol)
$3/2C + H_2O = -CH_2- + 1/2CO_2$	40.7
$0.75CH_4 + 0.25CO_2 = -CH_2- + 0.5H_2O$	18.8

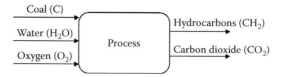

FIGURE 17.1   Schematic representation of an overall CTL process.

Note that we are considering the overall process, beginning from coal and ending with hydrocarbons, thus setting a target for the entire process and not for specific units in the process.

There are three main species in the process: carbon, hydrogen, and oxygen. We thus can write a balance for each species as follows:

$$\text{Carbon constraint: } a = 1 + d \qquad (17.4\text{i})$$

$$\text{Hydrogen constraint: } 2b = 2 \qquad (17.4\text{ii})$$

$$\text{Oxygen constraint: } b + 2c = 2d \qquad (17.4\text{iii})$$

From the above species constraints (Equations 17.4i to 17.4iii), we also notice that we have four unknown variables, and that the constraints provide us with only three equations; we therefore have one degree of freedom in our process. This allows us to evaluate various options for the process. From the above equality constraints (Equations 17.4i to 17.4iii), we also note that the amount of water is fixed simply by the species balance, and that these species (constraints) relationships are linear.

In order to evaluate various options, one can now consider energy requirements (in terms of enthalpy and Gibbs energy). It has been shown that CTL processes are limited by enthalpy change of the process (Patel et al., 2007b). If we consider the process where $\Delta H = 0$, the overall balance is given by (in terms of moles)

$$1.647C + H_2O + 0.147O_2 \Rightarrow CH_2- + 0.647CO_2 \qquad (17.5)$$

Because of the nature of the feedstock to the CTL process, CO_2 production is inevitable. We notice from the above mass balance that the minimum amount of CO_2 that we produce is 0.647 mol per mol $-CH_2-$ produced (or 2.03 tons per ton $-CH_2-$ produced).

The change in Gibbs energy for this process (Equation 17.5) is $\Delta G = -27.78$ kJ/mol. This process has a negative ΔG, meaning that the process is thermodynamically possible and can be a work producer.

This process is the *idealized process mass balance*, representing the process with the highest carbon efficiency and lowest possible CO_2 emissions. The process may consist of various subsystems (reactions, phase change, etc.). Achieving this

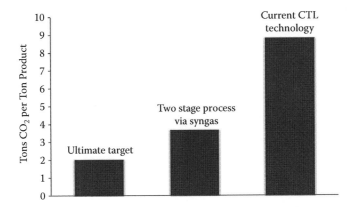

FIGURE 17.2 Comparison of the carbon dioxide emissions from an idealized CTL process and current CTL processes.

target may require mass, heat, and work integration of the various subsystems, which has been dealt with in previous publications (Patel et al., 2007a).

Figure 17.2 compares the carbon dioxide emissions from an idealized process to the carbon dioxide emissions from current processes. Current processes produce approximately 9 tons of CO_2 per ton of product. For example, an 80,000 bbl/d plant will produce approximately 30 million tons of CO_2 per annum. The target is about 2 tons per ton of product.

It is clear that current processes produce 7 tons of carbon dioxide per ton of product more than the idealized target. This suggests that there is room for improvement in terms of reducing the carbon dioxide emissions of the process.

A major reason for the difference between the target and current technology is the chemistry chosen and the way the flow sheet is put together. At present, CTL processes produce syngas (CO and H_2) as an intermediate by gasification of coal, and then convert the syngas into hydrocarbons by the Fischer-Tropsch process. The target for this route can also be calculated.

17.4.2 TARGET FOR TWO-STEP PROCESS

The idealized target for a two-stage process based on current processes can be determined. First, coal is gasified to produce syngas in a two: one ratio (H_2 and CO), and second, the syngas is converted to hydrocarbons (Fischer-Tropsch synthesis):

$$a\ C + b\ H_2O + c\ O_2 \rightarrow CO + 2H_2 + d\ CO_2 \rightarrow -CH_2- + H_2O + d\ CO_2$$

Again, we have mass balance relationships that provide three equations (C, H, O balance) and one free variable.

The first step, the gasification step, can be made thermally neutral by setting a target of $\Delta H = 0$. The second step, the synthesis step, is exothermic and therefore produces heat. We can now solve for a, b, c, and d.

The process is given as

$$2.17C + 2H_2O + 0.67O_2 \rightarrow CO + 2H_2 + 1.17CO_2 \rightarrow -CH_2- + H_2O + 1.17CO_2$$

$$\Delta H = 0$$

Therefore, the minimum amount of CO_2 that can be produced from such a process is 3.7 tons per ton of $-CH_2-$ produced, as shown in Figure 17.2. Thus, by choosing this route, the target in terms of CO_2 emissions is increased from 2 tons per ton product to 3.7 tons per ton product.

17.5 SUPPLYING WORK TO A PROCESS

A major way of supplying work to a process is by setting the temperature of the heat that is added to the process. It is known from the thermodynamic study of Carnot engines that heat at high temperature has the ability to do work, and the quality or the work potential depends on its temperature. Thus, when we add heat to a process we are equivalently adding a certain amount of work to the process that we could access if the process is designed for reversibility.

$$\Delta G_{process} = \Delta H_{process}\left(1 - \frac{T_o}{T_{Carnot}}\right) \qquad (17.6)$$

In order for a process to be reversible, the temperature, T_{Carnot}, of the heat that we add to the process must satisfy the above equation (Patel et al., 2005) as $\Delta G_{process}$ (minimum work load) and $\Delta H_{process}$ (minimum heat load) are fixed.

17.5.1 SUPPLYING WORK BY STAGING PROCESSES

Consider staging the CTL process, i.e., producing synthesis gas (CO and H_2) and then converting the syngas into liquid. Work is added to the process since in this case the process itself becomes a Carnot engine, as shown in Figure 17.3.

The two-stage process has Gibbs energies and enthalpies that allow much of the work to come in and out of the process as heat, i.e., via Carnot engines.

The mass balances for the two stages are as follows:

Stage 1: Gasification: $3/2C + 2H_2O = CO + 2H_2 + 1/2CO_2$

Stage 2: Synthesis: $CO + 2H_2 = -CH_2- + H_2O$

The enthalpy change and Gibbs energy of these two stages are given in Table 17.3.

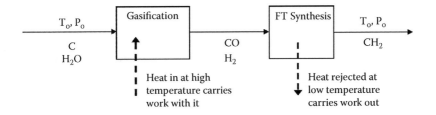

FIGURE 17.3 Carnot engine representation of the CTL process.

TABLE 17.3
Enthalpy Change and Gibbs Energy for a Two-Stage CTL Process

	$\Delta H_{process}$ kJ/mol	$\Delta G_{process}$ kJ/mol	T_{Carnot} [K]
Gasification	264	140	635
Synthesis	−207	−99	571

Both the enthalpy change and Gibbs energy change for stage 1 (gasification stage) are positive, which means that both heat and work are required for stage 1. What is very important, however, is that in the first stage the ΔG is less than the ΔH, implying that the heat and work requirements can potentially be matched according to Equation 17.6.

In terms of stage 2, both the enthalpy change and Gibbs energy are negative, implying that both heat and work need to be removed from the process. Again, the absolute ΔG is less than the ΔH, implying that the heat and work requirements can potentially be matched.

The Carnot temperatures of the first stage of the processes (gasification stage) correspond to a high-temperature process, whereas the second stage Carnot temperatures (synthesis stage) are lower.

Running each of the processes at their reversible temperature has certain advantages. A schematic diagram of workflows when running at the reversible temperature is shown in Figure 17.4. It must be remembered that the overall process is endothermic and requires work to be input. The work and heat input into the reformer are greater than those required by the overall process in all three cases. Thus, work and heat must be rejected in the second stage. When the second stage is run at its reversible temperature, then the heat that is rejected by the second stage carries the correct amount of work with it.

Let us first consider the implications of not operating the gasification stage at the Carnot temperature, $T_{Carnot,1}$. Usually the gasification stage is run at temperatures (T > 600°C) much higher than T_{Carnot}

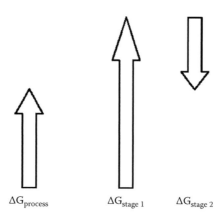

$\Delta G_{process}$ $\Delta G_{stage\ 1}$ $\Delta G_{stage\ 2}$

FIGURE 17.4 Schematic representation of the workflows in the overall process and stages 1 and 2. The length of the arrow is proportional to the quantity of work, and the direction indicates where the work is added or rejected.

If the gasification stage operated above its Carnot temperature (i.e., $T_1 > T_{Carnot}$; see Figure 17.5), this would imply that the heat added to the gasification stage carried more than the required work with it. This excess work would need to be rejected from the reformers, and if the excess work was not recovered, it would lead to irreversibility of the process and, hence, losses and higher carbon dioxide emissions.

Similarly, if the synthesis stage operates below its Carnot temperature (i.e., $T_2 < T_{Carnot,2}$), this means that insufficient work is rejected from the synthesis stage. Unless the excess work is recovered, it will be lost as irreversibilities, which is again not a desired outcome, as this again leads to higher carbon dioxide emissions.

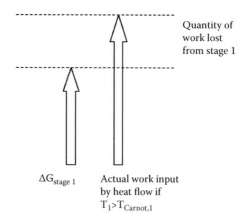

Quantity of work lost from stage 1

$\Delta G_{stage\ 1}$ Actual work input by heat flow if $T_1 > T_{Carnot,1}$

FIGURE 17.5 Schematic representation of the workflows for stage 1. The length of the arrow is proportional to the quantity of work, and the direction indicates where the work is added or rejected.

TABLE 17.4

Lost Work Associated with Two-Stage CTL Processes

	$\Delta H_{process}$ kJ/mol	$\Delta G_{process}$ kJ/mol	Actual Work Associated with Heat kJ/mol	Lost Work kJ/mol
Gasification (1,500 K)	264	140	212	72
Synthesis (500 K)	–207	–99	–83	15
Work to compress synthesis loop to 30 bar				25
				112 (Total)

The synthesis stage usually operates at lower temperatures (200–260°C, depending on the catalyst used) than T_{Carnot}.

Table 17.4 shows the actual work carried by the two stages when they operate at temperatures other than the Carnot temperature, as well as the work lost due to these operating temperatures. Usually the synthesis section is operated at high pressure (30 bar), which will also contribute to the lost work. In total, the lost work amounts to 112 kJ/mol.

17.6 REDUCING THE LOST WORK BY USE OF A NOVEL CHEMISTRY

It is apparent from the above discussion that to reduce the lost work in a CTL process, one would require that ΔG_{stage1} for the first stage is not too much larger than $\Delta G_{process}$, and one would also want to operate near the Carnot temperature of the processes.

A novel chemistry that satisfies these criteria is as follows and has been patented (Hildebrandt et al., 2007):

$$\text{Stage 1: } 3/2C + 3H_2O = 3/2CO_2 + 3H_2$$

$$\text{Stage 2: } CO_2 + 3H_2 = -CH_2- + 2H_2O$$

The thermodynamic variables as well as the lost work associated with this process are given in Table 17.5.

It is clear from Table 17.5 that in terms of the gasification stage, both enthalpy and Gibbs energy values are significantly smaller than those of the conventional CTL route. The lost work associated with this first stage is also comparatively smaller. In terms of the synthesis section, we note that it operates close to its Carnot temperature ($T_{Carnot} = 480$ K), and thus the lost work from this process is reduced significantly. Overall, the lost work amounts to 19 kJ/mol, as compared to the 112 kJ/mol for the conventional route.

TABLE 17.5

Lost Work Associated with the CO$_2$ Route CTL Process

	$\Delta H_{process}$ kJ/mol	$\Delta G_{process}$ kJ/mol	Actual Work Associated with Heat kJ/mol	Lost Work kJ/mol
Gasification	135	94	108	14
Synthesis	−209	−79	−85	5
				19 (Total)

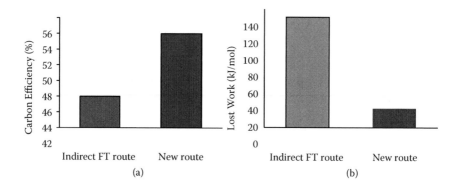

FIGURE 17.6   Comparison of two CTL routes in terms of (a) carbon efficiency and (b) lost work.

Figure 17.6 shows a comparison of the two CTL processes in terms of carbon efficiency and lost work. It is clear the newly developed process has the advantage of being more efficient, producing less carbon dioxide, and also less lost work.

17.7   CONCLUSION

The Fischer-Tropsch process has recently received renewed attention due to the increasing demand and rising costs of fuels. Fischer-Tropsch processes are either coal based or methane based. Coal-based processes face serious issues in terms of the increasing regulations regarding limiting CO$_2$ emissions worldwide. There is a firm need for innovative and novel solutions for dealing with CO$_2$ emissions from the CTL process.

The CTL process was assessed using a novel systematic approach of calculating targets for the process. It was found that current processes produce lower quantities of carbon dioxide per ton of product than the target value. An interesting question is whether such processes could be designed more efficiently in order to reduce the work lost from the process such that the process operates closer to the reversibility, and thus reduce the carbon dioxide emissions from the process. Ways of improving the reversibility of the Fischer-Tropsch process by adjusting the chemistry of the FT section were investigated. A novel chemistry was developed that reduces the lost work associated with the CTL process as well as increases the carbon efficiency of the process. In this way, carbon emissions may be reduced from CTL processes.

REFERENCES

Biegler, L.T., Grossmann, I. E., and Westerberg, A. W. 1997. *Systematic methods of chemical process design.* Englewood Cliffs, NJ: Prentice Hall.

Denbigh, K. G. 1956. The second-law efficiency of chemical processes. *Chem. Eng. Sci.* 6:1–9.

De Nevers, N., and Seader, J. D. 1980. Lost work: A measure of thermodynamic efficiency. *Energy* 5:757–69.

Hildebrandt, D., Glasser, D., Patel, B., and Hausberger, B. 2007. Patent WO/2007/122498.

Patel, B., Hildebrandt, D., Glasser, D., and Hausberger, B. 2005. Thermodynamic analysis of processes. 1. Implications of work integration. *Ind. Eng. Chem. Res.* 44:3529–37.

Patel, B., Hildebrandt, D., Glasser, D., and Hausberger, B. 2007a. Coal-to-liquid: An environmental friend or foe? Proceedings of the 24th International Pittsburgh Coal Conference, Johannesburg, South Africa.

Patel, B., Hildebrandt, D., Glasser, D., and Hausberger, B. 2007b. The synthesis and integration of chemical processes from a mass, energy and entropy perspective. *Ind. Eng. Chem. Res.* 46:8756–66.

18 Refining Fischer-Tropsch Syncrude

Perspectives on Lessons from the Past

Arno de Klerk

CONTENTS

An overview of commercial Fischer-Tropsch refineries to date is provided, namely, German technology (1930–1940s), U.S. Hydrocol (1940–1950s), Sasol 1 (1950s), Sasol 2 and 3 (1980s), Mossgas (1990s), Shell Middle Distillate Synthesis (1990s), and Sasol-Chevron technology (2000s). Changes in design philosophy are tracked, trends are identified, and with the benefit of hindsight, some general points of learning are highlighted. Some of the most important findings are:

1. Fischer-Tropsch syncrude is best refined to transportation fuels with chemicals co-production.
2. Oxygenates and olefins are key compound classes in Fischer-Tropsch refining.
3. HTFT syncrude is more efficient for the production of on-specification transportation fuels than LTFT syncrude.
4. Refinery designs taking cognizance of the properties of the syncrude are more efficient than refinery designs imposing a crude oil design approach on syncrude.

18.1 INTRODUCTION

The primary product from Fischer-Tropsch synthesis is a complex multiphase mixture of hydrocarbons, oxygenates, and water. The composition of this mixture is dependent on the Fischer-Tropsch technology and considerable variation in carbon number distribution, as well as the relative abundance of different compound classes is possible. The primary Fischer-Tropsch product has to be refined to produce final products, and in this respect, it is comparable to crude oil. The primary product from Fischer-Tropsch synthesis can therefore be seen as a synthetic crude oil (syncrude). There are nevertheless significant differences between crude oil and Fischer-Tropsch syncrude, thus requiring a different refining approach.[1]

The development of commercial refineries for processing various Fischer-Tropsch syncrudes is instructive in highlighting the main refining issues associated with the refining of Fischer-Tropsch syncrude.[2] Few refining technologies have been specifically developed for the conversion of Fischer-Tropsch syncrude. This does not imply that technologies developed for crude oil processing cannot be used, but the technology selection for Fischer-Tropsch syncrude is different.[3] In most instances, crude oil refining technologies have to be adapted to make them compatible with Fischer-Tropsch syncrude, mainly due to its high oxygenate and olefin content.[4–10]

The aim of this work is to make use of the historical development of Fischer-Tropsch refining to provide some perspective on the decisions made in Fischer-Tropsch refinery design. Specific aspects that will be highlighted are the design philosophy, response of refinery designs to changes in fuel specifications, co-production of chemicals, and impact of the Fischer-Tropsch technology selection on the refinery design.

18.2 CLASSIFICATION OF FISCHER-TROPSCH SYNCRUDES

Despite the limited number of commercial Fischer-Tropsch facilities that have been constructed to date, a wide variety of Fischer-Tropsch technologies have been employed (Table 18.1).

Fischer-Tropsch syncrude, just like crude oil, does not refer to a single feed mixture. The syncrude composition depends on many variables, and for industrially optimized technologies, the two main parameters that affect syncrude

TABLE 18.1
Commercially Applied Fischer-Tropsch Technologies

Technology	Catalyst	Reactor Type	First Commercial Facility	
			Year	Facility
German normal pressure	Precipitated Co	Fixed bed	1935	Ruhrchemie, Holten, Germany
German medium pressure	Precipitated Co	Fixed bed	1939	Hoesch, Germany
Hydrocol	Fused Fe	Fixed fluidized bed	1951	Hydrocol, Brownsville, TX
Arge	Precipitated Fe	Fixed bed	1955	Sasol 1, Sasolburg, South Africa
Kellogg Synthol	Fused Fe	Circulating fluidized bed	1955	Sasol 1, Sasolburg, South Africa
Sasol Synthol	Fused Fe	Circulating fluidized bed	1980	Sasol 2, Secunda, South Africa
Shell Middle Distillate Synthesis	Supported Co	Fixed bed	1993	Shell, Bintulu, Malaysia
Fe-LTFT Sasol Slurry Bed Process	Precipitated Fe	Slurry bubble column	1993	Sasol 1, Sasolburg, South Africa
Sasol Advanced Synthol	Fused Fe	Fixed fluidized bed	1995	Synfuels, Secunda, South Africa
Co-LTFT Sasol Slurry Bed Process	Supported Co	Slurry bubble column	2007	Oryx, Ras Laffan, Qatar

composition are Fischer-Tropsch catalyst type and operating temperature. From a refinery design perspective, the main commercial Fischer-Tropsch syncrude types are:

1. Cobalt-based low-temperature Fischer-Tropsch (Co-LTFT)
2. Iron-based high-temperature Fischer-Tropsch (Fe-HTFT)
3. Iron-based low-temperature Fischer-Tropsch (Fe-LTFT)

Within each syncrude type some variation is introduced by the operating conditions of Fischer-Tropsch synthesis, such as pressure and H_2:CO ratio, as well as by the Fischer-Tropsch reactor type. These variations cannot be ignored, and ultimately they have an impact on the refinery design. During the subsequent discussion it will become apparent that the selection of the Fischer-Tropsch technology influences not only the refinery design, but also the efficiency with which different products can be produced.

TABLE 18.2

Composition of the C_3 and Heavier Gas and Oil Fractions from the Syncrude Produced by the German Normal-Pressure and Medium-Pressure Co-LTFT Processes

Product Description	Boiling) Range (°C)	Normal-Pressure Syncrude Total (mass%)	Olefins (mass%)	Medium-Pressure Syncrude Total (mass%)	Olefins (mass%)
Liquid petroleum gas	C_3–C_4	14	43	10	40
Naphtha (Benzin)	C_5–180	47	37	26	24
Light distillate (Kogasin I)	180–230	17	18	24	10
Heavy distillate (Kogasin II)	230–320	11	8	13	—
Soft paraffin wax	320–460	8	—	17	—
Medium/hard paraffin wax	>460	3	—	10	—

18.3 GERMAN TECHNOLOGY (1930–1940s)

The first commercial Fischer-Tropsch facility was commissioned in 1935, and by the end of the Second World War a total of fourteen plants had been constructed. Of these, nine were in Germany, one in France, three in Japan, and one in China. Both German normal-pressure and medium-pressure processes (Table 18.1) were employed. The cobalt-based low-temperature Fischer-Tropsch (Co-LTFT) syncrude produced in these two processes differed slightly (Table 18.2), with the product from the medium-pressure process being heavier and less olefinic.[11] In addition to the hydrocarbon product, the syncrude also contained oxygenates, mostly alcohols and carboxylic acids.

The German Fischer-Tropsch facilities were not all the same and also varied in size from 11 kt/a (Wintershall) to 164 kt/a (Brabag) and by the type of Fischer-Tropsch technology employed.[12] The main design features found in most of the German Fischer-Tropsch refinery designs are illustrated by a generic refinery design (Figure 18.1) that does not represent any specific refinery. It will also be noted that syncrude from Fischer-Tropsch synthesis is not obtained as a single product, but as multiple fractions from stepwise condensation after synthesis. This is an important difference between Fischer-Tropsch syncrude and crude oil.

The C_3–C_4 light olefins were converted into liquid products by either H_3PO_4, or $AlCl_3$-catalyzed oligomerization processes (Figure 18.1). The products from H_3PO_4 oligomerization could be used for motor gasoline, but were more often employed as feed for hydroformylation.[13] The products from $AlCl_3$ oligomerization, where the LPG was typically co-processed with olefins derived from

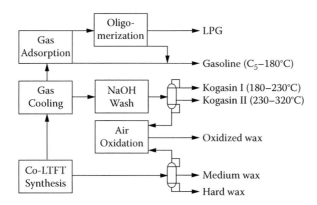

FIGURE 18.1 Generic German Fischer-Tropsch refinery.

thermal cracking, were used for synthetic lubricants.[14] Another processing route (not shown in Figure 18.1) was olefin absorption in H_2SO_4 followed by hydrolysis to produce alcohols, which was accompanied by the formation of some olefin oligomerization.[15] As a matter of historical interest, it is worthwhile to note that the production of motor gasoline by Fischer-Tropsch synthesis was not favored in Germany at that time, with synthetic motor gasoline production by direct coal liquefaction being the preferred route.[15]

Some naphtha is inevitably produced by Co-LTFT synthesis. The straight-run naphtha fraction was recovered from the gaseous Fisher-Tropsch product by carbon adsorption in the normal-pressure process and by condensation under pressure in the medium-pressure process. The straight-run motor gasoline had a low octane number (Table 18.3).[12,16] Further refining or blending of the Co-LTFT naphtha was consequently required to meet German motor gasoline specifications, which required a motor octane number (MON) of at least 72 at that time.[12] Thermal cracking, hot clay treating (acidic isomerization), and cracking and oligomerization have been considered for further refining,[16,17] but additional refining units were added in only some of the German Fischer-Tropsch refineries. In general, the octane number was improved by blending with alcohols or aromatics and a mild NaOH wash was required as a polishing step to remove dissolved carboxylic acids from the motor gasoline (not shown in Figure 18.1).[15]

TABLE 18.3
MON of Various Co-LTFT Straight-Run Naphtha Fractions

| Co-LTFT Straight-Run Naphtha | Motor Octane Number | | | |
	C_5–110°C	C_5–140°C	C_5–150°C	C_5–200°C
German normal-pressure process	67	62	57	43
German medium-pressure process	—	—	38	25

TABLE 18.4

Selected Diesel Fuel Properties from Different German Co-LTFT-Derived Straight-Run Distillate Fractions and the German Sonder Diesel Kraftstoff (SDK) Specifications of the 1940s

Fuel Property	German Co-LTFT Diesel Fuel		German SDK Specifications
	155–250°C	195–310°C	
Cetane number	75–78	80	45 min.
Density at 15°C (kg·m^{-3})	743–749	768	810–865
Flash point (°C)	27–49	78	55 min.
Cloud point (°C)	—	0	−10 max.
Pour point (°C)	< −37	−1	−30 max.

The straight-run distillate (Kogasin I and II) was obtained by condensation, NaOH washing, and atmospheric distillation. Mainly two types of diesel fuel were prepared from the Co-LTFT distillate: a light diesel (mixture of heavy naphtha and Kogasin I) and a heavy diesel (mixture of Kogasin I and II). The heavy diesel fuel also contained 2% oxygenates: alcohols, carbonyls, and carboxylic acids, with lesser amounts of esters and phenolic compounds.[18] As expected from Co-LTFT synthesis, the diesel fuels had a high cetane number and low density (Table 18.4). This fell short of the German diesel fuel specifications of that time. As Freerks[12] points out, there is little mention of diesel fuel production made in many of the reports on the German Fischer-Tropsch-based processes. In fact, the diesel fuel from Co-LTFT was not considered to be a good diesel fuel.[15] The fuel consumption of Fischer-Tropsch-derived diesel fuel was 5% higher than a 47 cetane number petroleum cut, and it had a 25% higher exhaust gas temperature. The Co-LTFT-derived diesel fuels were typically blended with distillate fractions from crude oil or direct coal liquefaction for use. The blend typically contained around 40 to 45% Co-LTFT material.

Much of the distillate range material was employed in chemicals production, which was considered the logical future use of the Fischer-Tropsch process by F. Martin, the director of Ruhrchemie during that time.[15] Kogasin II was converted into lubricating oils by thermal cracking and oligomerization.[19] It was also used as feed material for the production of detergents by sulfochlorination.

Fractionation of the heavier products by steam stripping resulted in different wax grades being produced. The soft wax and atmospheric bottoms (Gatsch) were typically air oxidized to produce oxygenated waxes. The medium, hard, and oxidized waxes all ended up as chemical products.

Some salient points from the design of German Fischer-Tropsch refineries that can be noted are:

1. The refineries were designed for the production of both transportation fuels and chemicals. The emphasis was on chemicals production, partly due to the poor quality of Co-LTFT-derived fuels that could not meet the German fuel specifications of that time. The Co-LTFT-derived fuel fractions required blending or further refining before it could be used as transportation fuels. Nevertheless, from a fuel refining perspective, the carbon number distribution (Table 18.2) was very favorable. In general, the German normal-pressure process with its lighter product (lower α-value) and higher olefin content was better suited for fuels production than the medium-pressure process. About 75% of the normal-pressure Co-LTFT syncrude was naphtha and distillate, with an even higher liquid fuel yield (>80%) possible by conversion of the C_3–C_4 olefins.

2. Olefins in the C_3–320°C range had significant synthetic value, and additional olefins were produced by thermal cracking in some facilities. Acid-catalyzed and thermal olefin oligomerization were important technologies for the upgrading of Fischer-Tropsch products.

3. The refinery design had to make provision for dealing with the oxygenates present in the syncrude. The beneficial use of alcohols has been noted. Carboxylic acids were neutralized and the resulting soaps recovered. However, the Fischer-Tropsch aqueous product was not refined.

18.4 U.S. TECHNOLOGY (1940–1950s)

After the Second World War a gas-to-liquids facility that employed an iron-based high-temperature Fischer-Tropsch (Fe-HTFT) process was constructed at Brownsville, Texas. The technology was developed by Hydrocarbon Research, Inc.,[20] and the commercial facility was operated by the Carthage Hydrocol Company. The Hydrocol plant was in commercial operation during the period 1951–1957, and it was shut down mainly for economic reasons (the oil price was around US$2 per barrel at that time).

The product from Hydrocol synthesis consisted mainly of motor gasoline range products and light gases (Table 18.5).[21] The Fe-HTFT syncrude had a high olefin content and was rich in linear hydrocarbons, especially linear α-olefins. Some aromatics were also co-produced on account of the high operating temperature and the C_8–C_{11} oil fraction contained around 6% aromatics.[22] In addition to the hydrocarbon products, the syncrude also contained oxygenates, namely, alcohols, aldehydes, ketones, and carboxylic acids.[23] Most of the short-chain oxygenates dissolved in the aqueous product from Fe-HTFT synthesis, with the longer-chain oxygenates being present mainly in the oil fraction.

The design aim of the Hydrocol refinery was to produce a better than 80% yield of motor gasoline from the Fe-HTFT syncrude at a quality that would be acceptable for the market (MON of 80 after tetraethyl lead addition). The refinery design (Figure 18.2) addressed the issues specific to the Hydrocol Fe-HTFT syncrude:

TABLE 18.5

Composition of the C_3 and Heavier Gas and Oil Fractions from the Syncrude Produced by the Hydrocol Fe-HTFT Process

Product Description	Boiling Range (°C)	Hydrocol Syncrude	
		Total (mass%)	Olefins (mass%)
Liquid petroleum gas	C_3–C_4	32	82
Naphtha	C_5–204	56	85–90
Distillate	>204	8	75–85
Residue		4	—

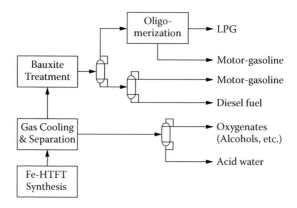

FIGURE 18.2   U.S. Hydrocol Fischer-Tropsch refinery.

1. Unwanted oxygenates were removed from the organic product and specifically the short-chain carboxylic acids that were known to cause problems.[24]
2. The octane number of the straight-run Fischer-Tropsch naphtha was increased.
3. Gaseous C_3–C_4 olefins, which constituted one-third of the syncrude, were oligomerized to liquid products, mainly in the naphtha boiling range, to boost motor gasoline production.
4. Alcohols and other valuable oxygenates dissolved in the aqueous product from FT synthesis were recovered to be sold as chemicals.[23]

The production of motor gasoline was further increased by blending a natural gas condensates (not shown in Figure 18.2), which were recovered from the natural gas feed, with the synthetic motor gasoline.[20]

The complete oil fraction from FT synthesis was treated over bauxite, a natural silica-alumina, at a temperature around 400°C. This bauxite treatment step was a commercial process, called the Perco process, which was used as a sulfur removal step in oil refineries. The acid-catalyzed conversion of the syncrude over bauxite reduced the amount of oxygenates in the syncrude and improved the octane number of the gasoline fraction (Table 18.6).[21,25]

After bauxite treatment the product was fractionated to produce C_3–C_4 and naphtha (C_5–204°C) fractions. The C_3–C_4 olefin-rich gas was oligomerized over a solid phosphoric acid (SPA) catalyst to produce an unhydrogenated "polymer" gasoline with a research octane number (RON) of 95 and MON of 82.[21] The bauxite-treated FT motor gasoline (RON of 87, MON of 76) was mixed with the polymer gasoline and some natural gas condensates (and crude-oil-derived naphtha) to produce the final motor gasoline product. In this respect it is noteworthy that the Fe-HTFT-derived material was the high-octane-blend stock.

The bauxite-treated distillate was not further refined, but it has been shown that this olefinic diesel could be hydrotreated to produce a diesel fuel with high cetane number (Table 18.7).[15,26] Although the cetane number of Fe-HTFT distil-

TABLE 18.6
RON and MON of the Hydrocol Fe-HTFT Syncrude before and after Upgrading over Bauxite, and with Different Levels of Tetraethyl Lead (TEL) Addition

Description	RON	MON
Straight-run Fe-HTFT naphtha	68	62
Straight-run Fe-HTFT naphtha + 1 ml TEL/gal	79	70
Straight-run Fe-HTFT naphtha + 3 ml TEL/gal	84	74
Bauxite-treated Fe-HTFT naphtha	87	76
Bauxite-treated Fe-HTFT naphtha + 1 ml TEL/gal	93	80
Bauxite-treated Fe-HTFT naphtha + 3 ml TEL/gal	94	82

TABLE 18.7
Selected Properties of Hydrocol Fe-HTFT Distillate before and after Hydrogenation

Fuel Property	Fe-HTFT Distillate	
	Straight Run	Hydrogenated
Cetane number	45–50	71
Density at 15°C (kg·m⁻³)	806	806
Pour point (°C)	−9 to −15	−1
Olefin content (g Br/100 g)	47	2
Distillation range (°C)		
T10	204	227
T50	232	260
T90	304	327

late is lower than that derived from Co-LTFT material (Table 18.4), the density is much higher, making blending to produce a final diesel fuel easier.

Some of the oxygenates in the Fischer-Tropsch aqueous product could be recovered by distillation to improve the overall carbon efficiency of the Fischer-Tropsch refinery. However, it was pointed out that complete separation of the short-chain oxygenates was difficult due to the formation of azeotropic mixtures.[23]

Some important features of the Hydrocol refinery design that can be noted are:

1. The stated aim of the Hydrocol refinery was to produce motor gasoline, but oxygenate chemicals were co-produced from the aqueous product. The carbon number distribution of the Fe-HTFT syncrude was ideally suited for motor gasoline production, with more than 50% of the straight-run material boiling in the naphtha range (Table 18.5). An even higher motor gasoline yield was possible by upgrading the olefinic C_3–C_4 material.
2. The selection of SPA oligomerization for upgrading the olefinic C_3–C_4 material was an excellent technology choice. SPA produces olefin oligomers with a narrow carbon number distribution and high naphtha selectivity.[27] This technology enabled the refinery to achieve a motor gasoline yield of close to 80%, the stated design objective of the refinery.
3. The potential processing problems associated with oxygenates were recognized by the refinery design. Although bauxite treatment was not able to remove all of the oxygenates, it significantly reduced the oxygenate content. All the alcohols and esters were converted to hydrocarbons, and the combined conversion of carboxylic acids and carbonyls was around 90%.[24] Furthermore, placing the bauxite unit before the primary distillation unit reduced thermal decomposition in the reboiler and allowed better carbon number separation (oxygenates and hydrocarbons that are co-boiling typically have a difference of two to four carbon numbers). This in turn allowed better separation during atmospheric distillation between the distillate and residue, thereby limiting the amount of distillate that is lost to residue due to the upper temperature limit placed on the reboiler by thermal cracking. It also simplified the routing of short-chain olefins (C_3–C_4) produced by oxygenate conversion over bauxite to the olefin oligomerization unit. Another subtle benefit of incomplete oxygenate conversion over bauxite was the retention of some polar compounds to provide boundary layer lubricity to the fuels.

18.5 SASOL 1 TECHNOLOGY (1950s)

The design of the Sasol 1 facility in Sasolburg, South Africa, was the result of a compromise between the experience that accompanied the Fe-LTFT processes from the Arbeitsgemeinschaft Ruhrchemie-Lurgi (Arge) in Germany and the economy of the Fe-HTFT process from Kellogg in the United States.[28] The Sasol

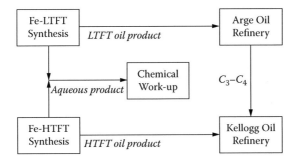

FIGURE 18.3 Integration of Sasol 1 refineries for processing HTFT and LTFT syncrudes.

1 facility was constructed using one-third German technology and two-thirds American technology. The two processes were integrated and employed a single gas loop. However, the refining of the products was only partially integrated (Figure 18.3). The Fischer-Tropsch aqueous products from Fe-LTFT and Fe-HTFT synthesis were combined and refined together, and the C_3–C_4 material from Fe-LTFT synthesis was refined in the Fe-HTFT refinery.

The Arge Fe-LTFT syncrude (Table 18.8)[29] was much heavier than the syncrude of the two German Co-LTFT processes (Table 18.2). The Arge Fe-LTFT syncrude exemplified a high α-value Fischer-Tropsch product with a significant linear paraffinic wax fraction. The syncrude (Table 18.8) from the Kellogg Fe-HTFT synthesis was very similar in carbon number distribution to that of Hydrocol Fe-HTFT synthesis (Table 18.5).

The separate stepwise condensation of the products from Fe-LTFT and Fe-HTFT synthesis produces streams of different carbon number distributions that serve as feeds to the oil refinery (Figure 18.4).[30] It is consequently not necessary to employ an atmospheric distillation unit as the first step in the refinery. The stepwise condensation products from Fe-LTFT are reactor wax (liquid at LTFT conditions), hot condensate (>100°C), cold condensate (produced by condensation with the aqueous product and then phase separated), and tail gas (typically C_4 and lighter). The stepwise condensation products from Fe-HTFT are decanted oil (liquid at 145°C; 1.6 MPa), light oil (produced by condensation with the aqueous product and then phase separated), and tail gas.

When comparing the Sasol 1 Arge Fe-LTFT oil refinery (Figure 18.4) with a generic German Co-LTFT oil refinery (Figure 18.1), the similarities may not be immediately apparent. Yet, the design principles were similar, and the Sasol 1 design included some of the thinking that has already been highlighted during the discussion of the German and Hydrocol technologies, for example: bauxite treatment (acidic isomerization) of the naphtha fraction to improve motor gasoline properties, thermal cracking to convert slack wax, and hydrogenation of some of the product fractions to improve their properties.

TABLE 18.8

Composition (Mass%) of the Sasol 1 Syncrudes from Arge Fe-LTFT and Kellogg Fe-HTFT Syntheses

Product Description	Syncrude Composition		C_3 and Heavier Oil Composition	
	Arge Fe-LTFT	Kellogg Fe-HTFT	Arge Fe-LTFT	Kellogg Fe-HTFT
Methane	5	10		
Ethylene	0.2	4		
Ethane	2.4	6		
Propylene	2	12	2	17
Propane	2.8	2	3	3
Butenes	3	8	3	11
Butanes	2.2	1	3	1
C_5–C_{12} naphtha	22.5	39	26	53
C_{13}–C_{21} distillate	21	6	24	8
C_{22}–C_{30} medium wax/heavy oil	17	3	19	4
C_{31} and heavier wax/residue	18	2	20	3
Aqueous phase products				
Nonacid oxygenates	3.5	6		
Carboxylic acids	0.4	1		

In the LTFT oil refinery the wax from Fe-LTFT synthesis was separated and hydrogenated to produce waxes with different congealing points as final products.[31] Material in the wax fraction with a boiling range of 320 to 370°C was thermally cracked, combined with the syncrude, and fractionated to produce fuel products. The cold condensate (heavy naphtha and distillate range material) was subject to a NaOH and water wash before being separated into final products. This removed some of the oxygenates, including most of the carboxylic acids. The light naphtha fraction was Bauxite treated (acidic isomerization) to produce motor gasoline. The C_3–C_4 material was transferred to the HTFT oil refinery, where it was oligomerized.

In the HTFT oil refinery the C_3–C_4 materials from both Arge Fe-LTFT and Kellogg Fe-HTFT syntheses were oligomerized over copper-pyrophosphate to produce an olefinic motor gasoline. The oligomerization technology is very similar to SPA-based technology, and over a wide range of operating temperatures the unhydrogenated motor gasoline has a RON of 96–98 and MON of 81–83.[32]

In the HTFT oil refinery the light oil and <345°C fraction of the decanted oil (obtained by passing the decanted oil through a vacuum flash drum) were clay treated. Clay treatment is similar to Bauxite treatment and is used to increase the octane number of the naphtha by acidic isomerization and to reduce the oxygenate content of the oil. Processing the LTFT and HTFT in separate (and different)

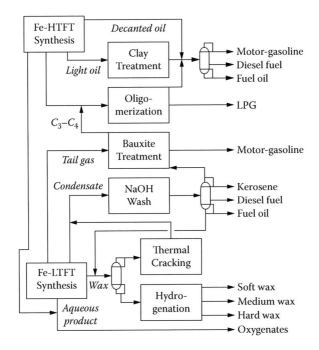

FIGURE 18.4 Sasol 1 oil refineries, original design (1950s).

acidic isomerization units was technically not necessary. The olefin oligomers and clay-treated oil were combined before distillation into different fuel fractions. The refined motor gasoline from Fe-HTFT synthesis typically had a RON of 86–90,[33] which was adjusted to meet final product specification by tetraethyl lead addition. The refined diesel fuel met the cetane number specification of 45 of that time. The low density of the diesel fuel was not an issue, since a specific density range was not prescribed by the South African diesel fuel specifications.

The refinery design included the recovery of nonacid oxygenates in the Fischer-Tropsch aqueous product that are lighter boiling than water.[30] The oxygenate chemicals recovered from the aqueous product included methanol (mainly from Fe-LTFT), ethanol (from Fe-HTFT, Fe-LTFT, and acetaldehyde hydrogenation), as well as mixed heavier alcohol and ketone streams. The carboxylic acids were not recovered and were processed with the wastewater.

In addition to the Fischer-Tropsch-derived material, coal-derived liquids were also recovered from low-temperature coal gasification (not shown in Figures 18.3 and 18.4). These products were processed separately to produce chemicals, such as phenols, cresols, and ammonia, as well as an aromatic motor gasoline blending stock.[34] The latter was mixed with the Fischer-Tropsch-derived motor gasoline.

The following important aspects from the design of the combined Fe-LTFT and Fe-HTFT refineries can be highlighted:

1. Neither the LTFT nor the HTFT refineries required an atmospheric distillation unit to prefractionate the syncrude before refining. The only separation step that was necessary was to recover the distillate from heavy fractions, namely, LTFT wax and HTFT decanted oil. The stepwise condensation of the syncrude after Fischer-Tropsch synthesis was sufficient prefractionation before refining. Atmospheric distillation was performed after refining, with many of the advantages noted for the Hydrocol design being equally valid for the Sasol 1 design.

2. The refinery catered for both fuels and chemicals production. Chemicals were mainly produced in the LTFT oil refinery and from the Fischer-Tropsch aqueous product. Transportation fuels were produced mainly in the HTFT oil refinery, although some fuels were also produced from LTFT syncrude. The ease of refining and the quality of the fuels from HTFT syncrude are better than those from LTFT, which is better suited to chemicals (rather than fuels) production.

3. Much attention has been devoted to the design of oxygenate refining. Beneficial recovery of the oxygenates from the Fischer-Tropsch aqueous product improved the overall carbon efficiency of the process. In the oil refineries, acidic isomerization was employed to improve motor gasoline quality by olefin isomerization and to reduce oxygenates. In addition to this, NaOH and water wash steps were included to reduce the oxygenate content of the fuels and specifically remove corrosive short-chain carboxylic acids. Furthermore, by converting most of the oxygenates before atmospheric distillation, thermal cracking in the reboiler was reduced and better carbon number separation of products could be achieved.

4. Olefin oligomerization was a key refining technology for converting normally gaseous olefins into liquid products and thereby substantially increasing the liquid yield. The selection of a copper-pyrophosphate-catalyzed process, rather than a SPA-catalyzed process, can be understood in historical context. Kellogg, supplier of the Fe-HTFT technology, also marketed a process based on this type of catalyst. Nevertheless, copper-pyrophosphate and solid phosphoric acid catalysts have similar conversion characteristics and produced a good-quality olefinic motor gasoline.

18.5.1 EVOLUTION OF THE SASOL 1 FACILITY

The Sasol 1 facility is still in commercial operation. Over a period of more than 50 years, many changes in fuel specifications and opportunities in chemicals markets developed. In order to remain profitable, the operation of the Sasol 1 site changed considerably over this period. The three most important changes that took place were:

1. Change in Fischer-Tropsch synthesis. In the 1990s the Kellogg Fe-HTFT synthesis section was decommissioned and additional Fe-LTFT synthesis capacity was added with the introduction of a slurry bed reactor.[35] This modified the syncrude feed to the refinery to Fe-LTFT only. This was accompanied by a significant change in the product slate being produced.
2. Emphasis on chemicals production. When the Sasol 1 facility was constructed in the 1950s, it had a nameplate capacity of 6,750 barrels per day oil equivalent (>200 kt/a). This was not an unusual size for a facility in the 1950s, but is small compared to present-day standards. Although the refinery was designed to produce fuels and chemicals, many changes took place after a decision in 1962 to expand production in the direction of chemicals. Some of the new units in and around Sasol 1 included ammonia synthesis, wax oxidation, naphtha cracking, and polyethylene production. In 1971 a new inland crude oil refinery was built next to Sasol 1, which opened new possibilities for fuels blending. However, with the low crude oil price in the 1990s and phasing out of tetraethyl lead, it became increasingly undesirable to produce fuels at the Sasol 1 site. The Sasol 1 refinery was therefore converted to an Fe-LTFT chemicals-only production facility that is based mainly on hydrogenation and distillation.
3. Conversion from coal to natural gas. Sasol 1 was designed as a coal-to-liquids facility. A natural gas pipeline was constructed and commissioned in 2004. This allowed the Sasol 1 facility to be converted to a gas-to-liquids plant. Although it implied that the associated coal tar refinery would become redundant, the decision was made by Sasol to keep the coal-to-chemicals units at Sasol 1 in operation by supplying coal pyrolysis products from its larger CTL facility in Secunda.

18.6   SASOL 2 AND 3 TECHNOLOGIES (1980s)

The Sasol 2 and 3 facilities (presently known as Sasol Synfuels) were constructed in Secunda, South Africa, in response to the 1973 oil crisis. Syncrude was produced by Fe-HTFT synthesis in improved Sasol Synthol circulating fluidized bed reactors[36]. The syncrude composition was similar to the Kellogg Fe-HTFT syncrude of Sasol 1 (Table 18.8). There was limited integration between the two facilities, which had a combined nameplate capacity of 120,000 barrels per day oil equivalent.

The refineries were almost exact copies of each other and were originally designed to produce only transportation fuels. The specific design objectives were:[37]

1. Convert normally gaseous C_3 and C_4 olefins to liquid range products.
2. Remove contaminants from the Fischer-Tropsch-derived oil to protect downstream catalysts and yield suitable products.
3. Upgrade the quality of the motor gasoline to meet octane specifications.

Although these primary design objectives were similar in many respects to those of the Hydrocol refinery, the design of the Sasol 2 and 3 refineries[37,38] was very different (Figure 18.5). The design approach followed for Sasol 2 and 3 was that of a typical third-generation (topping-reforming-cracking) crude oil refinery of that time.[39] This was a radical departure from the design of earlier Fischer-Tropsch refineries, and it took almost no cognizance of the difference in properties between Fischer-Tropsch syncrude and crude oil. The only refinery unit in the original design that catered specifically for Fischer-Tropsch syncrude was an acidic isomerization unit. However, instead of keeping with proven alumina-rich bauxite and acidic clays, a much more acidic rare-earth-exchanged Y-zeolite was employed. This catalyst was too acidic for the reactive Fe-HTFT syncrude and resulted in excessive cracking, thereby undermining the performance and purpose of this unit.

The transportation fuels produced and marketed (Table 18.9)[40] met the South African fuel specifications of that time and included some coal-derived liquids (not shown in Figure 18.5). Although the refinery originally produced no jet fuel, it was demonstrated that the hydrogenated kerosene range oligomers from olefin oligomerization over a solid phosphoric acid catalyst met the requirements for jet fuel.[38] (Semisynthetic jet fuel was approved in 1999 and fully synthetic jet fuel was approved in 2008; DEFSTAN 91-91/Issue 6).

In the original Sasol 2 and 3 refinery design (Figure 18.5), the light olefins were oligomerized over a solid phosphoric acid catalyst to produce liquid products. The product was fractionated into a naphtha and light distillate fraction, the latter being hydrogenated and used as diesel fuel. Some of the olefinic naphtha was also hydrogenated to reduce the olefin content of the motor gasoline, but resulted in a considerable drop in octane value (Table 18.10).[37] It was only later realized

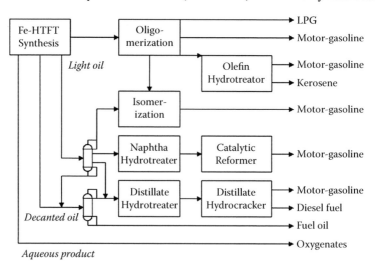

FIGURE 18.5   Sasol 2 and 3 oil refineries, original design (1980s).

TABLE 18.9

Selected Properties of the Motor Gasoline and Diesel Fuel Refined from Fe-HTFT Syncrude and Coal Liquids at Sasol 2, Which Were Marketed as Final Products in South Africa in the 1980s

Fuel Property	Fe-HTFT Sasol 2 Products	
	Motor Gasoline	Diesel Fuel
Liquid product yield (%)	55–65	35–45
Density (kg·m^{-3})	720	800
RON (+0.3 g/l Pb)	93	—
MON (+0.3 g/l Pb)	85	—
Cetane number	—	47
Distillation (°C)		
T50	90	219
FBP	200	378
Composition (%)		
Paraffins	49	96–98
Olefins	24	0
Aromatics	27	2–4
Sulfur	0.0001	0.0002

that a much better quality hydrogenated motor gasoline could be obtained from a butene-only oligomerization over solid phosphoric acid.[41]

Acidic isomerization of the C_5–C_6 naphtha and some heavy alcohols from the aqueous product refinery (not shown in Figure 18.5) produced a reasonable-quality olefinic motor gasoline (Table 18.10). The octane value varied depending on the carbon number distribution of the feed, which could result in a product with an octane number up to ten units higher.

The bulk of the naphtha was hydrotreated and catalytically reformed over a chlorided Pt/Al$_2$O$_3$-based catalyst to produce an aromatic motor gasoline. However, the hydrotreated Fischer-Tropsch naphtha is a poor feed for standard catalytic reforming on account of its high linear hydrocarbon content (>75%).[37] In order to limit liquid yield loss, typical operation resulted in a reformate with quite low octane value (Table 18.10). Higher octane reformate could be produced, but at the expense of significant liquid yield loss.

A light diesel fuel was produced by distillate hydrotreating of the straight-run Fe-HTFT material, while the heavier fraction was hydrocracked over a dewaxing catalyst, which produced a heavy diesel (Table 18.10). Some diesel fuel was also produced by C_3–C_4 olefin oligomerization over solid phosphoric acid by recycling the naphtha thus produced. It has previously been pointed out that solid phosphoric acid is not well suited for distillate production,[42] and the hydrogenated

TABLE 18.10

Selective Fuel Properties from Key Units in the Original Sasol 2 and 3 Fischer-Tropsch Oil Refineries

Sasol 2 and 3 Refinery Products	Selected Fuel Properties			
	Density (kg·m⁻³)	RON	MON	Cetane Number
Motor Gasoline				
Olefinic SPA oligomers	745	95–96	81–82	—
Hydrogenated SPA oligomers	730	64	70	—
C_5–C_6 isomerisate	690	85	75	—
Reformate (severity: 84% C_5+ yield)	770	87	80	—
Hydrocracker naphtha	720	77	71	—
Diesel Fuel				
Hydrogenated SPA oligomers	765	—	—	34
Hydrogenated straight-run distillate	810	—	—	55
Residue hydrodewaxing/cracking	860	—	—	67

distillate had good cold flow properties, but a low density and low cetane number (Table 18.10).

Oxygenates were recovered from the Fischer-Tropsch aqueous product, employing a separation strategy similar to that in the Sasol 1 refinery. The main difference was in volume, and this made further separation of the different alcohols and carbonyl compounds worthwhile. Some of the ethanol served as a blending component in motor gasoline, with the final blend containing around 10% ethanol.[38] Most of the alcohols and carbonyl compounds were sold as chemicals. In addition to the oxygenates, the C_2 hydrocarbons were also recovered and sold.

Aspects of the original Sasol 2 and 3 Fe-HTFT refinery design that are important to highlight are:

1. The design intent was mainly to produce transportation fuels, yet some chemicals were co-produced, rather than being converted into fuels. For example, the refinery design of the aqueous product refinery could have been simplified,[43] if the design intent had been to produce only fuels.

2. By approaching the refinery design from a crude oil perspective, the advantage of preseparation by stepwise condensation after HTFT synthesis was reduced. The refinery design included primary separation steps typically found in crude oil refineries, namely, an atmospheric distillation unit (ADU) that is followed by a vacuum distillation unit (VDU). Despite the design intent, the operation of these units, out of necessity, had to be different. The reboiler temperature of the ADU was

constrained by thermal cracking of the reactive Fe-HTFT syncrude to around 320°C, while little material was available for VDU (Table 18.8). In practice, this resulted in poor separation with considerable carbon number overlap.

3. Oxygenate refining was limited to chemicals recovery from the Fischer-Tropsch aqueous product and acidic isomerization of the C_5–C_6 naphtha. The naphtha and distillate range oxygenates were removed by hydrodeoxygenation (HDO) in hydrotreaters, before further refining.

4. Olefin oligomerization over solid phosphoric acid was a key technology in the refinery, and most of the C_3–C_4 gaseous olefins were converted to liquid products.

5. When comparing the refined motor gasoline and diesel fuel produced in the Hydrocol Fe-HTFT refinery (Tables 18.6 and 18.7) with those obtained from the original Sasol 2 and 3 Fe-HTFT refinery (Table 18.9), there is surprisingly little difference in quality. The crude oil refining approach followed for Sasol 2 and 3 resulted in a more complex and costly refinery design to achieve essentially the same product quality as the Hydrocol refinery.

18.6.1 EVOLUTION OF THE SASOL 2 AND 3 FACILITIES

The Sasol 2 and 3 facilities have over time become integrated to such an extent that they are presently collectively called Sasol Synfuels. Over a period of around 30 years, changes in fuel specifications and the exploitation of the chemicals potential of Fe-HTFT resulted in numerous changes to the refinery (Figure 18.6). Instead of moving away from the crude-oil-based design, growth occurred using the same crude oil refining paradigm that was employed in the original design. However, treating Fischer-Tropsch syncrude as a crude oil feedstock turned out to be costly, and significant capital investments were required to keep up with the changing demands placed on product slate and specifications.

The evolution of Sasol Synfuels was not limited to the refinery only, and the following important changes took place:

1. Change occurred in high-temperature Fischer-Tropsch reactor technology. The circulating fluidized bed Sasol Synthol reactors were replaced by fixed fluidized bed Sasol Advanced Synthol (SAS) reactors.[44] This did not meaningfully affect the Fe-HTFT syncrude composition, but it reduced the operating cost of HTFT synthesis.

2. Production of chemicals became increasingly important. The recovery of oxygenates from the Fischer-Tropsch aqueous product was expanded to include niche chemicals, such as 1-propanol.[45] Ethylene and propylene extraction was increased and even supplemented by the addition of a high-temperature catalytic cracker.[46] Linear α-olefin extraction units for the recovery of 1-pentene, 1-hexene, and 1-octene were added to the refinery,[45,47] and a new facility for the extraction of 1-heptene and its

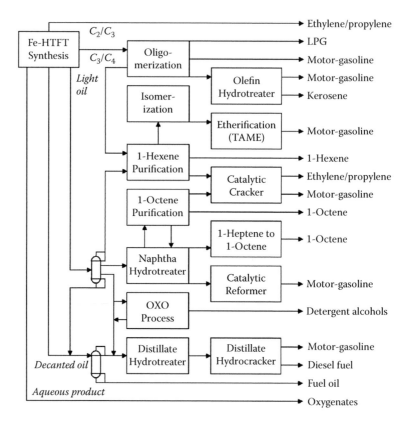

FIGURE 18.6   Sasol Synfuels current design (2008).

subsequent conversion to 1-octene[48] was commissioned in 2008. Longer-chain linear α-olefins are also extracted and converted to detergent range alcohols by hydroformylation.[45,47] All of these units exploited the unique chemical extraction possibilities of HTFT syncrude. In addition, chemicals are also recovered from coal liquids obtained during low-temperature coal gasification, such as ammonia, phenol, and cresols. The air separation units are also a source of chemicals, with products such as noble gases and liquid nitrogen being co-produced.

3. Refinery changes followed changes in fuel specifications. The original refinery design relied on the addition of tetraethyl lead (TEL) to meet the octane requirements of the motor gasoline. Phasing out of TEL in motor gasoline started in 1989 in South Africa, which caused an octane shortfall, with little high-octane refinery streams (Table 18.10) to offset the reduction in TEL. The extraction of light linear α-olefins helped, but also resulted in the isomerization unit being shut down. In order to bridge the octane shortfall, two new units were added to produce tertiary

amyl methyl ether (TAME): a pentene skeletal isomerization unit and an etherification unit. A further gain in octane number of the motor gasoline was achieved by optimizing the operation of the refinery[49] and the use of methylcyclopentadienyl manganese tricarbonyl (MMT) as TEL replacement. Increasing demand for 95 RON/85 MON motor gasoline and anticipated future changes in the fuel specifications prompted Sasol to preinvest in a high-temperature catalytic cracker.[46] The design intent was to convert the lowest-quality fuel components into chemicals, namely, ethylene and propylene, rather than upgrading the quality of the motor gasoline per se. (At the time of writing, this unit has not yet achieved full production capacity.) Yet, despite these improvements, some refining challenges remain on the road to Euro-4-type specifications, such as the olefin and benzene content of the motor gasoline. No changes have been made to the original refining strategy for diesel fuel, although diesel fuel also presents future refining challenges that are mainly related to increased volume demand and Euro-4-type diesel fuel specifications.

18.7   MOSSGAS TECHNOLOGY (1990s)

The South African government initiated the Mossgas project in the mid-1980s to investigate the conversion of gas and associated natural gas liquids into transportation fuel. This eventually led to the construction of the Mossgas gas-to-liquids plant (presently known as PetroSA) in Mossel Bay, South Africa. It was designed as a 33,000 barrels per day oil equivalent facility, with two thirds of the production being derived from Fischer-Tropsch synthesis and the remainder from associated gas liquids. This facility reached full commercial production in 1993 and was aimed at the production of transportation fuel only.[50]

Sasol Synthol Fe-HTFT circulating fluidized bed reactor technology was employed. The Fe-HTFT syncrude composition was consequently similar to that produced by Fe-HTFT synthesis in Sasol 1, 2, and 3 (Table 18.8), but the refinery feed was very different, since it contained natural gas liquids (mostly paraffins in the C_5–400°C boiling range). Co-processing of natural gas liquids made the design better suited to an oil refining approach, yet the Mossgas refinery design was cognizant of the nature of the Fischer-Tropsch syncrude. This resulted in a modern refinery (Figure 18.7), with the design emphasis on high-quality paraffinic motor gasoline and diesel fuel production (Table 18.11).[51]

The Fe-HTFT syncrude is fractionated in an atmospheric distillation unit to produce mainly naphtha and distillate, with a small amount of residue that is used as fuel oil (not shown in Figure 18.7). No vacuum distillation unit has been included in the design, since it would be superfluous with the limited residue production. The natural gas liquids are fractionated separately.

Most of the light olefins are oligomerized in the Conversion of Olefins to Distillate (COD) process[52] that was specifically developed for converting Fischer-Tropsch material to liquid products. Although the original design intent was to convert an oxygenate containing HTFT naphtha, co-production of carboxylic

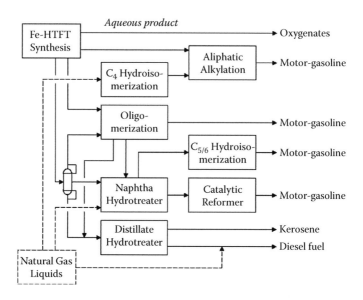

FIGURE 18.7   Mossgas HTFT refinery, original design (1990s).

TABLE 18.11
Final Products from the Mossgas HTFT Refinery

Final Product	Production Volume	
	(m³/day)	(bpd)
Propane	85	530
Liquid petrolum gas	265	1,665
Motor gasoline (unleaded 95 RON/85 MON)	2,760	17,360
Kerosene (illuminating paraffin)	600	3,775
Diesel fuel[a]	1,400	8,800
Fuel oil	100	630
Total production	5,210	32,760

[a] The production of 70,000 t/a low aromatic diesel (since 2003) from diesel fuel is not shown separately.

acids caused some postprocessing problems.[53] The product from oligomerization is mainly distillate (70%). The hydrogenated distillate has a high cetane number (>51) and low density.[54] Part of the naphtha is employed as an olefinic motor gasoline blending component, despite its low octane value (RON = 81–85, MON = 74–75), and part is further refined with the HTFT naphtha.

The co-refining synergy of natural gas liquids and Fe-HTFT was exploited for alkylate production. The natural gas liquids serve as a source of butane that can be hydroisomerized to yield isobutane that is alkylated (HF process) to produce a high-octane alkylate for motor gasoline. In the absence of natural gas liquids, this would not have been a preferred route for alkylate production.[55]

The C_5 and heavier natural gas liquids are combined with the heavy naphtha and part of the COD naphtha to serve as feed to a naphtha hydrotreater (NHT). This part of the refinery employs a typical crude oil design philosophy to produce isomerate and reformate for motor gasoline. Nevertheless, some cognizance of HTFT properties has been taken, and hydrotreating the combined naphtha fraction before further fractionation has a number of advantages. The oxygenates were converted to hydrocarbons before fractionation, thereby avoiding a broadening of the carbon number distribution. By hydrotreating the combined naphtha, duplication of feed pretreatment for downstream processing was avoided. (Both the hydroisomerization technology and the catalytic reforming technology that were selected employed chlorided catalyst systems that are sensitive to water/oxygenates.)

Diesel production involved a straightforward design. The olefinic distillate from olefin oligomerization was combined with the straight-run HTFT distillate and hydrotreated. The hydrotreated Fischer-Tropsch-derived distillate was blended with the distillate fraction from the natural gas liquids to produce diesel fuel. In 2003 another hydrotreater (noble metal catalyst) was added to the refinery to convert part of the hydrotreated HTFT distillate into low aromatic distillate to serve a niche market.[56]

Although chemicals co-production would be possible, it is, at the time of writing, limited to alcohol recovery from the aqueous product (not shown in Figure 18.7). In the Mossgas context this may or may not be viewed as chemicals production, since the alcohols can be used in fuels. The ethanol can be employed as blending component for motor gasoline, and the beneficial use of the heavier alcohols for diesel fuel has been demonstrated.[54]

Important aspects of the Mossgas refinery design that can be highlighted are:

1. The design intent was to produce transportation fuels, and the design did not specifically make provision for chemicals co-production. It is in principle possible to extract chemicals from the HTFT syncrude, such as the alcohols that are being recovered from the Fischer-Tropsch aqueous product. Extraction of linear α-olefins may also be considered, which has indeed been investigated,[57] and many other opportunities exist. However, it should be noted that the Mossgas facility is much smaller than the Sasol Synfuels facility, and recovery of valuable products in HTFT syncrude may not have economy of scale.

2. The refinery design included the co-refining of natural gas liquids, not only as a blending stock, but also as an integral part of the feed to the refinery. This created synergy and allowed refining pathways that would otherwise be less efficient.

3. The refinery design of the Mossgas facility displays some crude oil refinery design principles, but not without taking cognizance of HTFT syncrude properties. It produces on-specification transportation fuels for the South African market, with the motor gasoline being of Euro-4-type

quality, but diesel fuel having a lower density than that required for Euro-4-type diesel fuel. Yet, the Mossgas HTFT refinery (Figure 18.7) has a lower complexity than a generic fourth-generation crude oil refinery (topping-reforming-cracking-visbreaking-alkylation-isomerization),[39] and it has a much lower degree of complexity than the Sasol Synfuels HTFT refinery (Figure 18.6). The Mossgas HTFT refinery design demonstrates that it is easier to refine HTFT syncrude to on-specification transportation fuels than crude oil. It also demonstrates the value of catering for the difference in properties between Fischer-Tropsch syncrude and crude oil in the refinery design, rather than imposing a crude oil refinery design on a syncrude feed.

4. Olefin oligomerization is a key refining technology, and in the Mossgas refinery the COD process is used not only to convert normally gaseous olefins to liquid products, but also to refine oxygenates to fuel.

5. Apart from the oxygenate refining performed by the COD process, the oxygenates in the Fischer-Tropsch aqueous product are refined to alcohols. These alcohols are recovered and may be blended with the transportation fuels or sold as chemicals.

18.8   SHELL TECHNOLOGY (1990s)

The development of the Shell Middle Distillate Synthesis (SMDS) process began in 1983, when a pilot plant was constructed at the Shell Research and Technology Centre in Amsterdam. This eventually culminated in the design and construction of the Shell gas-to-liquids facility in Bintulu, Malaysia, which was completed and commissioned in 1993.[58]

In many respects the SMDS process (Figure 18.8) precipitated a change in the Fischer-Tropsch community with respect to the preferred catalyst for Fischer-Tropsch synthesis and the approach to product workup. It is therefore instructive to understand why Shell moved away from iron-based Fischer-Tropsch catalysts (and as a consequence also high-temperature synthesis) and opted for a Co-LTFT process with an uncomplicated refinery design that does not produce

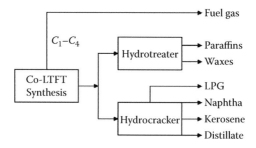

FIGURE 18.8   Shell Middle Distillate Synthesis process, Bintulu, Malaysia.

on-specification transportation fuels. The reasoning behind Shell's decision has been stated as follows:[59]

1. Remote and relatively small gas fields cannot justify the high investment cost associated with liquefied natural gas (LNG) production or a gas pipeline system. Conversion of the natural gas from such gas fields to liquids by a gas-to-liquids facility allows these gas fields to be exploited.
2. Synthetic fuels may attract local government subsidies or sell at a premium because they offset imports and improve the country's balance of payments. Furthermore, strategic considerations related to energy security may play an important role in promoting investment in synthetic fuels production.
3. The synthetic fuels that can be produced by low-temperature Fischer-Tropsch synthesis inherently have a high quality (being sulfur- and aromatics-free) and can therefore be used as quality improvers with conventional components.
4. In terms of global long-term demand, the demand for distillate may be more than the demand for motor gasoline.

The Co-LTFT syncrude produced by the SMDS process is comparable to Arge Fe-LTFT syncrude (Table 18.8), but it contains fewer olefins and oxygenates. It is also comparable to the German medium-pressure Co-LTFT syncrude (Table 18.2), but it is heavier. The refinery design is uncomplicated (Figure 18.8) and has only two conversion units: a hydrotreater and a hydrocracker. The hydrotreater is used to produce paraffins and waxes as final products for the chemicals market. Alternatively, the hydrocracker can be used to produce mainly kerosene and distillate range products (Table 18.12).[60] The hydrocracking catalyst has been developed specifically for LTFT waxes and the unit is operated at mild conditions (300–350°C, 3–5 MPa).[61] Although the composition of the hydrocracking catalyst has not been disclosed, the catalyst is likely to be a noble metal on mildly acidic support, for example, Pt or Pd on alumina.[58]

TABLE 18.12
Typical Product Properties Obtained from the Co-LTFT-Based SMDS Process

Property	Naphtha (43–166°C)	Kerosene (155–191°C)	Gas Oil (184–357°C)
Hydrocracker yield (mass%)			
Kerosene mode	25	50	25
Gas oil mode	15	25	60
Density at 15°C (kg·m^{-3})	690	738	776
Aromatics (%)	0	<0.1	<0.05
Cetane index	—	58	76

According to the original design objectives, the refinery has not been designed to produce on-specification transportation fuels. The distillate fraction (Table 18.12) is therefore only a blending component. The LTFT-derived distillate has to be mixed with crude-oil-derived distillate in order to produce an on-specification diesel fuel. This was a designed outcome and much thought went into the refinery design for the SMDS process. The compact design is in line with the potentially remote location of the facility, and the development of a noble metal hydrocracking catalyst keeps supply logistics simple (no sulfiding agent required) and keeps the product sulfur-free (no sulfur introduced by the use of a sulfiding agent).

The Fischer-Tropsch aqueous product was not further refined and was treated as a wastewater stream. This was in line with the simplicity of the refinery design, poor economy of scale for oxygenate recovery, and inherently low water-soluble oxygenate production from Co-LTFT synthesis.

Important points about the SMDS process as applied at Bintulu are:

1. The selection of a fixed bed Co-LTFT process supported the objective to apply the SMDS process for beneficiation of remote gas fields. The Co-LTFT catalyst has a useful lifetime of 5 years and the robustness of fixed bed reactor technology has been proven. For example, the fixed bed Arge Fe-LTFT process has now been in operation for more than 50 years at Sasol 1.
2. The emphasis on distillate production and more specifically the production of a diesel fuel blending component was a deliberate design decision. The design nevertheless included the flexibility to also produce chemicals, such as paraffins and paraffin waxes.
3. Short-chain olefins are not refined and the gaseous LTFT products are employed as fuel gas. Production of this fraction is limited by Co-LTFT synthesis, and with the product being less olefinic than iron-based Fischer-Tropsch syncrude, less benefit would be derived from the inclusion of an olefin oligomerization unit. Furthermore, adding complexity would go against the design objectives of the SMDS process.
4. No specific provision has been made for dealing with oxygenates.

18.9 SASOL-CHEVRON TECHNOLOGY (2000s)

The Sasol Slurry Bed Process (SSBP) was originally developed as an Fe-LTFT process that was commercialized at Sasol 1 in 1993.[35] The reactor technology was proven with a precipitated iron Fischer-Tropsch catalyst. It offered the advantage of on-line catalyst addition and removal, which enabled operation at constant product selectivity (Fe-LTFT catalyst selectivity changes with time on stream)[62] by keeping the average age of the catalyst in the reactor constant. In the Oryx gas-to-liquids (GTL) design for Ras Laffan, Qatar, the SSBP was adapted to make use of a newly developed Co-LTFT catalyst. Among the reasons cited for this change were higher activity and longer useful catalyst lifetime.[45]

Commissioning of the Oryx GTL facility started in 2006 and production of the first GTL products was announced in February 2007.[63] However, overproduction of a fine sediment as a result of catalyst attrition reduced throughput,[64] and at the time of this writing sustained full-scale production had not yet been achieved. The syncrude resembles other LTFT syncrudes (Tables 18.2 and 18.8), and the refinery consists of a single conversion unit, namely, a hydrocracker (Figure 18.9).

Superficially the Oryx GTL refinery design has much in common with the SMDS process, but there are important differences. There is no separate hydrotreater, which limits production of chemicals, such as waxes. The hydrocracker employs the Chevron Isocracking technology, which is based on a sulfided supported base-metal catalyst that was designed for crude oil conversion. The operating conditions of the hydrocracker are also more severe (>350°C, 7 MPa) than those required by the SMDS process (300–350°C, 3–5 MPa). Only intermediate products are produced (Table 18.13),[5] with the naphtha slated as cracker feed and the distillate as blending component for diesel fuel.

Some important aspects of the Oryx GTL design that can be highlighted are:

1. The selection of Co-LTFT synthesis, the associated refinery design, and the product slate for Oryx GTL all mimicked the SMDS process. Likewise, no provision has been made for the upgrading of short-chain olefins or oxygenates.
2. Unlike the SMDS process, the refining technology selection did not specifically cater to the properties of LTFT syncrude.[10]

18.10 DISCUSSION

18.10.1 REFINING OBJECTIVES

Most commercial Fischer-Tropsch refinery designs (Figures 18.1 to 18.9) included the co-production of chemicals with transportation fuels. The chemicals potential of Fischer-Tropsch syncrude has been pointed out repeatedly.[38,47,65,66] This is a natural consequence of the properties of Fischer-Tropsch syncrude, that is, richness in linear hydrocarbons, olefins (especially linear α-olefins), and oxygenates. Furthermore, it is sulfur-free and nitrogen-free, which enables access to synthetic routes sensitive to such compounds.

Chemicals co-production is attractive from an economic point of view, since chemicals generally have a higher value than fuels. It has been shown in the past

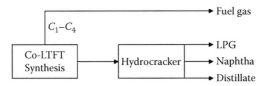

FIGURE 18.9 Oryx GTL process.

TABLE 18.13

Typical Product Properties Obtained from the Co-LTFT-Based Oryx GTL Facility

Property	Naphtha (51–131°C)	Distillate (151–334°C)
Hydrocracker yield (vol%)	25	75
Density at 15°C (kg·m⁻³)	685–687	769–777
Aromatics (%)	0.3	0.5
Cetane index	39	72

that for crude oil refineries the combined production of fuels and chemicals provides the best return on investment (better than fuels only or chemicals only).[67] Some molecules that are difficult to refine to fuels have value as chemicals and vice versa. A refinery design that takes cognizance of the molecular composition of the feed and selects an appropriate refining strategy for each fraction is bound to be more efficient. This is also true for the refining of Fischer-Tropsch syncrude.

It can therefore be said that Fischer-Tropsch syncrude is best refined to transportation fuels with chemicals co-production.

18.10.2 Selection of Fischer-Tropsch Syncrude

The selection of crude oil type is critical to the design and economic success of a crude oil refinery. Different crude oil types require a different refining strategy, and depending on the crude oil type, it may be easier or more difficult to achieve a specific product slate. The product slate is often determined by regional markets, unless the products are specifically earmarked for export. This also holds true for Fischer-Tropsch syncrude.

HTFT syncrude is easier to refine to on-specification transportation fuels than LTFT syncrude. This is partly due to its olefinic nature, giving it considerable synthetic ability, and partly due to the large proportion of material already in the fuels boiling range (C_5–360°C). Historically fuels refining from HTFT syncrude focused mainly on motor gasoline production and only to a lesser extent on diesel fuel production. Jet fuel production became possible only recently (2008) with the international qualification of fully synthetic jet fuel.

HTFT syncrude also has significant chemicals potential, especially when olefins and oxygenates are targeted.

LTFT syncrude is more difficult to refine to on-specification transportation fuels, but has become almost synonymous with distillate production from Fischer-Tropsch-based GTL conversion. In this application the SMDS process has been the trailblazer. However, there are two potential misconceptions that should be pointed out. First, Fischer-Tropsch distillate produced from LTFT syncrude is not a final on-specification diesel fuel, but a blending material. Second, LTFT syncrude is not inherently superior to HTFT syncrude for distillate production. LTFT syncrude may contain more straight-run distillate than HTFT syncrude, but most of the distillate produced during LTFT refining is produced by wax hydrocracking. Distillate production from HTFT syncrude occurs by the opposite process, namely, oligomerization of the olefinic C_3–C_{10} HTFT syncrude fraction. It is consequently possible to refine either LTFT or HTFT syncrude to distillate. Conversely, it must be pointed out that LTFT syncrude can also be refined to on-specification motor gasoline and fully synthetic jet fuel, albeit less efficiently than HTFT syncrude. It is therefore wrong to label LTFT syncrude as suitable for distillate production only.

LTFT syncrude has significant chemicals potential in the field of linear paraffins and waxes, with some potential for olefin and oxygenate chemicals too. It is also well suited for the production of lubrication oils by catalytic dewaxing.

Another important consideration is the raw material selection and synthesis gas production strategy. For example, in a GTL facility, some natural gas condensates may be available for co-refining with the Fischer-Tropsch syncrude, while in a coal-to-liquids (CTL) facility there may be coal liquids to be co-refined. Depending on the raw material, the liquid products that may be co-produced may complement or detract from the syncrude properties. Even if co-refining is not considered, blending opportunities with non-Fischer-Tropsch liquids may influence the refinery design, which is no longer dependent on just the syncrude type. This has been illustrated by the Hydrocol (Figure 18.2) and Mossgas (Figure 18.7) refineries for GTL applications and by the Sasol 1, 2, and 3 (Figures 18.3 to 18.5) refineries for CTL applications.

18.10.3 GAS-TO-LIQUIDS TRENDS

Commercial Fischer-Tropsch production has seen growth since the 1990s only in terms of gas-to-liquids facilities. In terms of design, Mossgas (1993) is the odd one out, since it is the only GTL facility that targeted on-specification fuels as final products. A new trend was started with Shell's Bintulu facility (1993), which employs the SMDS process. Since then the SMDS process (Figure 18.8) served almost as a blueprint for new GTL facilities that have been constructed or are under construction. All of these facilities are based on the combination of Co-LTFT and hydrocracking, albeit using different technologies. Escravos GTL, under construction in Nigeria, is very similar in design to Oryx GTL (Figure 18.9).[68] Pearl GTL, under construction in Qatar, is based on the SMDS process, but with the addition of a catalytic dewaxing unit for lubricating oil production.[69]

It is of interest to analyze the paradigm shift caused by the SMDS process in context of the original Shell design objectives (refer to Section 18.8) and the current situation.

The design objectives for the SMDS process were developed to exploit small or remote gas fields where the cost of liquefied natural gas production or a gas pipeline system is not warranted. In many of the present GTL applications this is not the case, with the expansion of LNG and GTL infrastructure in Qatar occurring in parallel. It is consequently possible to exploit economy of scale, and although robustness and simplicity are definitely virtues, there is no reason to limit the refinery design to the production of liquid intermediates. The inclusion of lubricating oil production in the Pearl GTL design is therefore not surprising.

The prediction that was made by Shell that the long-term demand for distillate will be more than the demand for motor gasoline was correct. Since the 1990s there has been a gradual shift from motor gasoline to diesel fuel, especially in Europe. The focus on distillate production is consequently justified, and securing a market should in principle not be a problem.[70] However, LTFT-derived distillate has lost some of its advantage as a quality improver for conventional crude-oil-derived distillate. The lowering of sulfur specifications (currently 50 $\mu g \cdot g^{-1}$ in Europe, going down to 10 $\mu g \cdot g^{-1}$) resulted in crude oil refiners having to hydrogenate their distillate more deeply. As a result, more aromatics are saturated and that in turn caused an increase in the cetane number and a decrease in the density of crude-oil-derived distillates. LTFT-derived distillate still has a significant blending advantage to reduce the density, but in low-sulfur markets there is little margin to use LTFT distillate to blend off-specification crude-oil-derived material to meet fuel specifications. For example, blending a 20 $\mu g \cdot g^{-1}$ sulfur crude oil distillate to meet a 10 $\mu g \cdot g^{-1}$ sulfur specification would require a 50:50 mixture with LTFT distillate, and the blended diesel fuel would fail a 820 to 845 $kg \cdot m^{-3}$ density specification if the crude oil distillate has a density of less than 860 $kg \cdot m^{-3}$.

It will be interesting to see whether the GTL trend started by the SMDS process will continue. The inclusion of lubricating oil production as part of the Pearl GTL design indicates that it may not. In a related field there also seems to be movement. In oil sands processing, where the product traditionally had been a synthetic crude oil that was sold for refining elsewhere, there seems to be increasing interest in adding value to the product.[71]

18.11 CONCLUSIONS

The designs of various commercial Fischer-Tropsch refineries have been explored to gain insight into the collective wisdom of many Fischer-Tropsch syncrude refiners. These are the lessons from the past. Some of these insights are bounded by historical context and should be interpreted as such. Nevertheless, there are some points of learning valid for Fischer-Tropsch refining in general:

1. Fischer-Tropsch syncrude is best refined to transportation fuels with chemicals co-production. However, it is possible to refine syncrude to just fuels or just chemicals.
2. Different refining strategies are possible, and different refinery designs are required for the refining of HTFT and LTFT syncrudes.
3. Selecting HTFT or LTFT syncrude to match the desired product slate is important for achieving an efficient design. HTFT syncrude is more

efficient for the production of on-specification transportation fuels than LTFT syncrude.

4. Refinery designs taking cognizance of the properties of the syncrude are more efficient than refinery designs imposing a crude oil design approach on syncrude.

5. Oxygenates present in Fischer-Tropsch syncrude require special refining and create opportunities for chemicals co-production.

6. Light olefin upgrading is an integral part of HTFT refining and in general becomes increasingly important as the α-value of the Fischer-Tropsch catalyst decreases. Olefins in general also create opportunities for chemicals co-production.

7. Olefin oligomerization is a key Fischer-Tropsch refining technology.

REFERENCES

1. De Klerk, A. 2007. Environmentally friendly refining: Fischer-Tropsch versus crude oil. *Green Chem.* 9:560–65.
2. De Klerk, A. 2008. Refining of Fischer-Tropsch syncrude: Lessons from the past. *Prepr. Pap.-Am. Chem. Soc. Div. Petrol. Chem.* 53:105–9.
3. De Klerk, A. 2008. Fischer-Tropsch refining: Technology selection to match molecules. *Green Chem.* 10:1249–1279.
4. De Klerk, A. 2003. Deactivation behaviour of Zn/ZSM-5 with a Fischer-Tropsch derived feedstock. In *Catalysis in application*, ed. S. D. Jackson, J. S. J. Hargreaves, and D. Lennon, 24–31. Cambridge: Royal Society of Chemistry.
5. Dancuart, L. P., De Haan, R., and De Klerk, A. 2004. Processing of primary Fischer-Tropsch products. *Stud. Surf. Sci. Catal.* 152:482–532.
6. Cowley, M. 2006. Skeletal isomerization of Fischer-Tropsch-derived pentenes: The effect of oxygenates. *Energy Fuels* 20:1771–76.
7. De Klerk, A. 2007. Effect of oxygenates on the oligomerization of Fischer-Tropsch olefins over amorphous silica-alumina. *Energy Fuels* 21:625–32.
8. Leckel, D. O. 2007. Selectivity effect of oxygenates in hydrocracking of Fischer-Tropsch waxes. *Energy Fuels* 21:662–67.
9. Mashapa, T. N., and De Klerk, A. 2007. Solid phosphoric acid catalysed conversion of oxygenate containing Fischer-Tropsch naphtha. *Appl. Catal. A* 332:200–8.
10. De Klerk, A. 2008. Hydroprocessing peculiarities of Fischer-Tropsch syncrude. *Catal. Today* 130:439–45.
11. Asinger, F. 1968. *Paraffins chemistry and technology*. Oxford: Pergamon.
12. Freerks, R. 2003. Early efforts to upgrade Fischer-Tropsch reaction products into fuels, lubricants and useful materials. Paper presented at the AIChE Spring National Meeting, New Orleans, 86d.
13. Asinger, F. 1968. *Mono-olefins chemistry and technology*. Oxford: Pergamon.
14. Horne, W. A. 1950. Review of German synthetic lubricants. *Ind. Eng. Chem.* 42:2428–36.
15. Weil, B. H., and Lane, J. C. 1949. *The technology of the Fischer-Tropsch process*. London: Constable.
16. Snodgrass, C. S., and Perrin, M. 1938. The production of Fischer-Tropsch coal spirit and its improvement by cracking. *J. Inst. Petrol. Technol.* 24:289–301.
17. Egloff, G., Nelson, E. S., and Morrell, J. C. 1937. Motor fuel from oil cracking production by the catalytic water gas reaction. *Ind. Eng. Chem.* 29:555–59.

18. Ward, C. C., Schwartz, F. G., and Adams, N. G. 1951. Composition of Fischer-Tropsch diesel fuel. *Ind. Eng. Chem.* 43:1117–19.
19. Seger, F. M., Doherty, H. G., and Sachanen, A. N. 1950. Noncatalytic polymerization of olefins to lubricating oils. *Ind. Eng. Chem.* 42:2446–52.
20. Keith, P. C. 1946. Gasoline from natural gas. *Oil Gas J.* 45:102–12.
21. Bruner, F. H. 1949. Synthetic gasoline from natural gas. Composition and quality. *Ind. Eng. Chem.* 41:2511–15.
22. Clark, A., Andrews, A., and Fleming, H. W. 1949. Composition of a synthetic gasoline. *Ind. Eng. Chem.* 41:1527–32.
23. Elliot, T. Q., Goddin, C. S., and Pace, B. S. 1949. Chemicals from hydrocarbon synthesis. *Chem. Eng. Progress* 45:532–36.
24. Schlesinger, M. D., and Benson, H. E. 1955. Upgrading Fischer-Tropsch products. *Ind. Eng. Chem.* 47:2104–8.
25. Helmers, C. J., Clark, A., and Alden, R. C. 1948. Catalytic treatment of synthetic gasoline. *Oil Gas J.* 47:86–92.
26. Tilton, J. A., Smith, W. M., and Hockberger, W. G. 1948. Production of high cetane number diesel fuels by hydrogenation. *Ind. Eng. Chem.* 40:1269–73.
27. McMahon, J. F., Bednars, C., and Solomon, E. 1963. Polymerization of olefins as a refinery process. In *Advances in petroleum chemistry and refining*, ed. K. A. Kobe and J. J. McKetta, 285–321. Vol. 7. New York: Wiley.
28. Meintjes, J. 1975. *Sasol 1950–1975*. Cape Town: Tafelberg.
29. Hoogendoorn, J. C. 1975. New applications of the Fischer-Tropsch process. In *Institute of Gas Technology's 2nd Clean Fuels and Coal Symposium*, Chicago, pp. 343–58.
30. Hoogendoorn, J. C., and Salomon, J. M. 1957. Sasol: World's largest oil-from-coal plant. III. *British Chemical Engineering*, July, pp. 368–73.
31. Le Roux, J. H., and Oranje, S., Eds. 1984. *Fischer-Tropsch waxes*. Sasolburg: Sasol.
32. Steffens, J. H., Zimmerman, M. U., and Laituri, M. J. 1949. Correlation of operating variables in catalytic polymerization. *Chem. Eng. Progress* 45:269–78.
33. Garrett, L. W., Jr. 1960. Gasoline from coal via the Synthol process. *Chem. Eng. Progress* 56:39–43.
34. Hoogendoorn, J. C., and Salomon, J. M. 1957. Sasol: World's largest oil-from-coal plant. IV. *British Chemical Engineering*, August, pp. 418–19.
35. Espinoza, R. L., Steynberg, A. P., Jager, B., and Vosloo, A. C. 1999. Low temperature Fischer-Tropsch synthesis from a Sasol perspective. *Appl. Catal. A* 186:13–26.
36. Holtkamp, W. C. A., Kelly, F. T., and Shingles, T. 1977. Circulating fluid bed catalytic reactor for the Fischer-Tropsch synthesis at Sasol II. *ChemSA*, March, pp. 44–45.
37. Swart, J. S., Czajkowski, G. J., and Conser, R. E. 1981. Sasol upgrades Synfuels with refining technology. *Oil Gas J.* 79(35):62–66.
38. Dry, M. E. 1988. The Sasol route to chemicals and fuels. In *Methane conversion*, ed. D. M. Bibby, C. D. Chang, R. F. Howe, and S. Yurchak, 447–56. Amsterdam: Elsevier.
39. Edern, Y. 2001. Introduction. In *Petroleum refining: Conversion Processes*, ed. P. Leprince, 1–9. Vol. 3. Paris: Editions Technip.
40. Hoogendoorn, J. C. 1982. Producing automotive fuels from coal in South Africa. *Hydrocarbon Process.* 61:34E–Q.
41. De Klerk, A., Engelbrecht, D. J., and Boikanyo, H. 2004. Oligomerization of Fischer-Tropsch olefins: Effect of feed and operating conditions on hydrogenated motor-gasoline quality. *Ind. Eng. Chem. Res.* 43:7449–55.
42. De Klerk, A. 2006. Distillate production by oligomerization of Fischer-Tropsch olefins over solid phosphoric acid. *Energy Fuels* 20:439–45.
43. Nel, R. J. J., and De Klerk, A. 2007. Fischer-Tropsch aqueous phase refining by catalytic alcohol dehydration. *Ind. Eng. Chem. Res.* 46:3558–65.

44. Steynberg, A. P., Espinoza, R. L., Jager, B., and Vosloo, A. C. 1999. High temperature Fischer-Tropsch synthesis in commercial practice. *Appl. Catal. A* 186:41–54.
45. Collings, J. 2002. *Mind over matter. The Sasol story: A half-century of technological innovation.* Johannesburg: Sasol.
46. Eng, K., Heidenreich, S., Swart, S., and Möller, F. 2005. Clean fuels and petrochemicals at SASOL via SUPERFLEX™. Paper presented at the 18th World Petroleum Congress, Johannesburg, cd122.
47. Redman, A. 2005. Production of olefins and oxygenated compounds from Fischer-Tropsch. Paper presented at the 18th World Petroleum Congress, Johannesburg, cd179.
48. McGurk, K. 2003. From 1-heptene to 1-octene: A new production route. Paper presented at the South African Chemical Engineering Congress, Sun City, P082.
49. Van Wyk, A. M., Moola, M. A., De Bruyn, C. J., Venter, E., and Collier, L. 2007. Molecule management for clean fuels at Sasol. *Chemical Technology* (South Africa), April, pp. 4–10.
50. Terblanche, K. 1997. The Mossgas challenge. *Hydrocarbon Engineering*, March, pp. 2–4.
51. Steyn, C. 2001. The role of Mossgas in Southern Africa. Paper presented at the 2nd Sub-Saharan Africa Catalysis Symposium, Swakopmund, Namibia.
52. Köhler, E., Schmidt, F., Wernicke, H. J., De Pontes, M., and Roberts, H. L. 1995. Converting olefins to diesel—The COD process. *Hydrocarbon Technology International*, Summer, pp. 37–40.
53. De Klerk, A. 2007. Properties of synthetic fuels from H-ZSM-5 oligomerization of Fischer-Tropsch type feed material. *Energy Fuels* 21:3084–89.
54. Knottenbelt, C. 2002. Mossgas "gas-to-liquids" diesel fuels—An environmentally friendly option. *Catal. Today* 71:437–45.
55. De Klerk, A., and De Vaal, P. L. 2008. Alkylate technology selection for Fischer-Tropsch syncrude refining. *Ind. Eng. Chem. Res.* 47:6870–77.
56. Mabena, N. 2005. Operating the worlds largest GTL facility (natural gas-to-liquids). Paper presented at the 18th World Petroleum Congress, Johannesburg, cd187.
57. Minnie, O. R., Petersen, F. W., and Samadi, F. R. 2003. Effect of 1-hexene extraction on the COD process conversion of olefins to distillate. Paper presented at the South African Chemical Engineering Congress, Sun City, P083.
58. Smith, R., and Asaro, M. 2005. *Fuels of the future. Technology intelligence for gas to liquids strategies.* Menlo Park: SRI.
59. Sie, S. T., Senden, M. M. G., and Van Wechem, H. M. W. 1991. Conversion of natural gas to transportation fuels via the Shell Middle Distillate Synthesis process (SMDS). *Catal. Today* 8:371–94.
60. Schrauwen, F. J. M. 2004. Shell Middle Distillate Synthesis (SMDS) process. In *Handbook of petroleum refining processes*, ed. R. A. Meyers, 15.25–40. New York: McGraw-Hill.
61. Eilers, J., Posthuma, S. A., and Sie, S. T. 1990. The Shell Middle Distillate Synthesis process (SMDS). *Catal. Lett.* 7:253–69.
62. Janse van Vuuren, M. J., Huyser, J., Kupi, G., and Grobler, T. 2008. Understanding Fe-LTFT selectivity changes with catalyst age. *Prepr. Pap.-Am. Chem. Soc. Div. Petrol. Chem.* 53:129–30.
63. Anon. 2007. Oryx plant produces GTL products for first time. *Oil Gas J.* 105:10.
64. Anon. 2008. GTL: Oryx breakthrough and oil-price surge lift industry spirits. *Petrol. Econ.* 75:36–37.
65. Steynberg, A. P., Nel, W. U., and Desmet, M. A. 2004. Large scale production of high value hydrocarbons using Fischer-Tropsch technology. *Stud. Surf. Sci. Catal.* 147:37–42.

66. De Klerk, A., Dancuart, L. P., and Leckel, D. O. 2005. Chemicals refining from Fischer-Tropsch synthesis. Paper presented at the 18th World Petroleum Congress, Johannesburg, cd185.

67. Chadwick, J. L. 1977. *Economics of chemical refineries.* Process Economic Program Report 107. Menlo Park: SRI.

68. Fraser, K. 2005. Escravos GTL: Delivering multiple benefits to Nigeria. In *Fundamentals of gas to liquids, 2005*, 15–16. London: Petroleum Economist.

69. Fabricius, N. 2005. Pearl GTL: Managing the challenges of scaling up. In *Fundamentals of gas to liquids, 2005*, 12–14. London: Petroleum Economist.

70. Corke, M. 2005. Securing a market. In *Fundamentals of gas to liquids, 2005*, 31–33. London: Petroleum Economist.

71. Fairbridge, C., Yang, H., Chen, J., Ng, S. H., and Rahimi, P. 2008. Oil sands processing. *Prepr. Pap.-Am. Chem. Soc. Div. Petrol. Chem.* 53:135.

19 Low-Temperature Water-Gas Shift

Assessing Formates as Potential Intermediates over Pt/ZrO$_2$ and Na-Doped Pt/ZrO$_2$ Catalysts Employing the SSITKA-DRIFTS Technique

Gary Jacobs, Burtron H. Davis,
John M. Pigos, and Christopher J. Brooks

CONTENTS

In this contribution, the steady-state isotopic transient kinetic analysis–diffuse reflectance Fourier transform spectroscopy (SSITKA-DRIFTS) method provides further support to the conclusion that not only are infrared active formates likely intermediates in the water-gas shift (WGS) reaction, in agreement with the mechanism proposed by Shido and Iwasawa for Rh/ceria, but designing catalysts based on formate C–H bond weakening can lead to significantly higher catalytic activity. This is in agreement with their proposal that the rate-limiting step involves cleaving the formate C–H bond. *In situ* DRIFTS experiments demonstrate that doping Pt/zirconia water-gas shift catalysts with alkali cations such as Na significantly weakens the formate C–H bond, such that the C–H stretching band position moves to lower wavenumbers, from 2,880 cm^{-1} for the Pt/ZrO$_2$ catalyst to the range of 2,804 to 2,845 cm^{-1} for the Na-doped Pt/ZrO$_2$ catalyst. Relative to undoped Pt/ZrO$_2$, the formate coverage during steady-state water-gas shift was very low at 225°C, since the formates were reacting too rapidly to accurately assess. However, by lowering the temperature to 185°C, the formate decomposition rate was slowed such that the coverage increased enough to monitor the formate reactive exchange from the ^{12}C to the ^{13}C label during ^{12}CO to ^{13}CO switching. In all tests, the formate C–H band reactive exchange rate was virtually the same as the product CO$_2$ exchange rate. Even at 185°C, the reactive exchange time of formate for the alkali-doped catalyst was shorter than that of the undoped Pt/zirconia catalyst at the higher temperature condition.

19.1 INTRODUCTION

Many researchers agree that a synergism in water-gas shift catalysis occurs when a metal such as Pt or Au is brought into contact with a partially reducible oxide carrier such as zirconia,[1–5] ceria,[6–22] titania,[23–31] thoria,[32–34] or α-Fe$_2$O$_3$.[35–43] To describe this behavior, many favor a support-mediated redox mechanism[7,9–12] involving reoxidation of the partially reduced oxide by H$_2$O to produce H$_2$. Others argue that H$_2$O is dissociated at the vacancy sites to yield active OH groups. In that case, those authors argue that an associative mechanism, most likely involving formate[1,3,4,6,14,15,20–22,24,27,29–33,38] or carbonate[5,16,38] or carboxyl[17] intermediates provides a better explanation. Most recently, in a review of the Au/ceria system by Burch,[44] the above mechanisms were compared and contrasted extensively, and he proposed that each mechanism may be valid under certain conditions. In this way, he argues, they may be subsets of a universal water-gas shift mechanism that involves a support-mediated redox process (i.e., support changing oxidation state during the course of the mechanism) at high temperature, and one of a number of associative mechanisms at low temperature, that likely depend on the feed conditions.[44]

For low-temperature shift conditions making use of a high H$_2$O/CO ratio, and including H$_2$ in the feed, researchers from the University of Kentucky Center for Applied Energy Research (CAER) have argued in favor of a formate-based mechanism over a number of Pt/partially reducible oxide catalysts, including ceria,[4,14,15] zirconia,[3,4] and thoria.[32,33] Though proposed long before, the formate mechanism gained momentum in 1981, when Grenoble and coworkers[45] reported an important influence of the support on the water-gas shift activity of various metal catalysts. For example, when Pt was supported on alumina instead of silica, the rate increased an order of magnitude. To describe these differences in activity, the authors used a bifunctional model, involving the chemisorption of water on alumina and CO on the metal. This was followed by the association of the CO with the activated water to form an adsorbed formic acid–like species via the

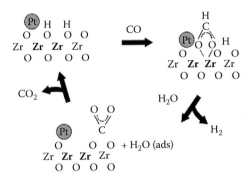

SCHEME 19.1 Surface formate mechanism over Pt/ZrO$_2$.

formate mechanism, with subsequent decomposition through dehydrogenation on the metal sites. In the early 1990s, Shido and Iwasawa[6,46] published detailed *in situ* DRIFTS studies of the water-gas shift reaction over ceria[46] and Rh/ceria,[6] which identified formates as the main intermediates in the WGS catalytic mechanism. In that mechanism, the metal assists in facilitating the formation of defect sites in the partially reducible oxide. H$_2$O dissociates into these vacancies to generate type II bridging OH groups, and these in turn react rapidly with CO to form formate species. In the presence of co-adsorbed H$_2$O, formate decomposes in the forward direction to H$_2$ and adsorbed CO$_2$ (as a carbonate), which further decomposes to yield the gas phase product, CO$_2$. Using temperature-programmed desorption–infrared (TPD-IR), a normal kinetic isotope effect was observed during steam-assisted forward formate decomposition. Based on this, the authors linked the rate-limiting step of the mechanism to C–H bond breaking of the formate species. An identical normal kinetic isotope effect was also observed with H$_2$O–D$_2$O switching during WGS reaction testing.

In 2005, Chenu and coworkers[3] reported that 1% Pt/ZrO$_2$ behaved through an analogous formate mechanism compared to metal/CeO$_2$ catalysts for low-temperature water-gas shift, as shown in Scheme 19.1. Characterization by TPR indicated that a partial reduction of zirconia, confined to surface defects, could be accomplished resulting in formation of type II bridging OH groups (i.e., vacancy + dissociated H$_2$O). CO adsorption was used to probe the active OH groups through the generation of formate species, as observed in DRIFTS. As with metal/CeO$_2$ catalysts before,[14,15] in the absence of H$_2$O, formates were found to be stable at low temperature, such that their band intensities in DRIFTS gave a good qualitative indication of the number of the active OH groups probed by the CO molecule. Formate band intensities observed over 1% Pt/ZrO$_2$ were lower than those over 1% Pt/ceria, suggesting a lower concentration of the defect-associated active OH groups. During steady-state water-gas shift tests with high H$_2$O/CO ratios, formate bands were reaction rate limited, implicating the species as a potentially important intermediate, in line with what was previously observed over metal/CeO$_2$ catalysts.[14,15] The concentration of type II OH groups over 1% Pt/ZrO$_2$ was suggested to be lower than the concentration observed for 1% Pt/CeO_2, and interestingly, water-gas shift rates followed an identical trend: Pt/ceria > Pt/m-zirconia > Pt/t-zirconia. More recently, yttrium-stabilized ZrO_2 tested in combination with metals by Panagiotopoulou and Kondarides[26,28] was found to exhibit higher activity over the metals supported on less reducible oxides, such as SiO_2 and Al_2O_3. However, the activity was not found to be as high as metal/CeO_2, in agreement with our previous findings, or metal/TiO_2 catalysts.

There have been recent reports of the use of promoters for Pt/ZrO_2 catalysts for low-temperature water-gas shift, and some that have aimed to increase water-gas shift activity by accelerating the formate turnover rate. For example, Iida and Igarashi[29,30] have reported enhanced activity by promoting Pt/ZrO_2 with Re to form a bimetallic catalyst. Infrared spectroscopic characterization of the carbonyls on Pt-Re/ZrO_2 indicated a correlation with the carbonyls on Pt/ZrO_2 and Re/ZrO_2 in an additive manner. Because Re/ZrO_2 alone did not exhibit significant WGS activity, the enhancement was argued to be the result of a synergistic effect of Re with Pt. The authors observed by infrared spectroscopy under steady-state WGS conditions that the formate species for Pt-Re/ZrO_2 were more reactive (i.e., lower coverage) than those observed over Pt/ZrO_2 alone. In addition, a band they assigned to Re-O upon the adsorption of steam to the reduced catalyst disappeared upon exposure to CO, leading the authors to conclude that Re imparted a redox cycle during active OH group formation on Re. This was suggested to accelerate the turnover rate of formate species, with Pt-CO providing sites for associating with Re-OH groups in order to form the bidentate formate species.

In 2005, Brooks et al.[47] from Honda Research Institute USA, Inc. (HRI) in conjunction with Symyx Technologies, Inc., reported findings from over 250,000 experiments conducted by combinatorial catalysis. Catalyst libraries were synthesized on 4-inch wafers in 16 × 16 arrays, and screened using a Symyx high-throughput scanning mass spectrometer. Among the promising compositions, an important improvement in catalyst activity was observed when Pt/ZrO_2 was doped with the alkalis Li, Na, or K, with Na yielding the greatest enhancement. In 2006, Pigos et al.[48,49] reported results of DRIFTS spectroscopy suggesting that formate species were more reactive on the Na-promoted catalysts relative to Pt/ZrO_2 alone. The conclusion was based on three separate tests specifically designed to probe formate stability during (1) steady-state CO adsorption and water-gas shift experiments at 225°C, (2) transient formate decomposition studies at 130°C in the presence of steam, and (3) formate decomposition tests under dry conditions using hydrogen-deuterium exchange at 225°C. In those studies, the formate C–H bands were significantly shifted to lower wavenumbers, suggesting an important electronic weakening of the formate C–H bond by the presence of the alkali on the catalyst surface.

While the above results appear to strongly favor a surface formate associative mechanism for low-temperature water-gas shift over the Pt/ZrO_2 catalysts, methods to provide direct support for the mechanism have remained elusive.

One method that is gaining momentum is *in operando* spectroscopy, which makes use of a combination of isotopic tracers with *in situ* infrared spectroscopy, such as SSITKA-DRIFTS, and is often used in combination with mass spectrometry. Recently, Tibiletti et al.[5] employed the technique and reported that the timescale for the exchange rate of the reaction product CO_2 was much shorter than that of the infrared active surface formates. Therefore, they ruled out infrared active formates as the main reaction pathway. They alternatively suggested that a carbonyl species or an infrared invisible complex (not excluding some type of formate) could be an intermediate, or that a redox mechanism[5] could be occurring. In the latter case, the *redox* term implies that the support oxide itself undergoes changes in oxidation state during the catalytic cycle, with CO spillover from Pt to the oxide. This results in reduction of the oxide surface and CO_2 formation, followed by reoxidation of the support by H_2O, with the generation of H_2. Whichever view holds, they conclude that contact between the Pt and ZrO_2 is necessary for the catalysis, indicating an important synergism relying on both metal and oxide sites during the catalytic cycle.

In previous independent and joint studies between HRI and CAER, an electronic weakening of the formate C–H bond due to the presence of the alkali promoter was deemed to be beneficial to accelerating the catalytic water-gas shift cycle.[47–49] To further explore the promoting effect of alkali addition to Pt/ZrO_2 and place the earlier hypothesis on a firmer footing, this study aims to apply the SSITKA-DRIFTS technique. Previously employed at the CAER, the approach was used to show that formate species over Pt/ceria catalysts[50] did reactively exchange at a timescale virtually identical to that of the CO_2 product. Most recently, CAER researchers have extended that viewpoint to include not only ceria, but zirconia, ceria-zirconia mixed oxide, and thoria as well.[51] Therefore, the SSIKA-DRIFTS method was applied to Na-doped Pt/ZrO_2 catalyst to determine (1) whether formates and CO_2 reactively exchange on a similar timescale and (2) whether formates/CO_2 reactively exchange faster than undoped Pt/ZrO_2.

19.2 EXPERIMENTAL

19.2.1 Catalyst Preparation

Catalyst samples were prepared by the incipient wetness impregnation method. Commercial high surface area zirconium support Gobain NorPro [Brunauer-Emmett-Teller (BET) surface area of 142 m²/g] was impregnated with aqueous platinum with and without sodium salt–containing solution. The precursor for Pt was tetraammineplatinum (II) hydroxide (9.09% Pt w/w), while the precursor for Na was sodium hydroxide (3.0 N). The impregnated catalysts were dried in an oven at 110°C for 24 h and then calcined at 300°C for 3 h in a furnace. Sequential impregnations were carried out by drying and calcining after each metal addition, with platinum added first and sodium added second.

19.2.2 BET SURFACE AREA

BET surface area measurements were carried out using a Micromeritics TriStar 3000 gas adsorption analyzer. Approximately 0.35 g of sample was weighed out and loaded into a 3/8-inch sample tube. Nitrogen was used as the adsorption gas and sample analysis was performed at the boiling temperature of liquid nitrogen.

19.2.3 SSITKA-DRIFTS METHOD

The infrared spectrometer was a Nicolet Nexus 870, equipped with a deuterated triglycine sulfate and thermoelectricity cooled (DTGS-TEC) detector. A chamber fitted with ZnSe windows capable of high temperature and high pressure served as the reactor for CO adsorption and water-gas-shift experiments. The gas lines leading to and from the reactor were heat traced and insulated with ceramic fiber wrap. Scans were taken at a resolution of 4 to give a data spacing of 1.928 cm^{-1}. For steady-state measurements, 256 scans were taken to improve the signal-to-noise ratio, and multiple scans were made to ensure reproducibility of the steady-state condition. During SSITKA, thirty-two scans were required, and the catalyst amount used was ~50 mg.

Feed gases were controlled by Brooks 5850 series E mass flow controllers. Iron carbonyl traps consisting of lead oxide on alumina (Calsicat) were placed on the CO gas line. All gas lines were filtered with Supelco O_2/moisture traps. Catalysts were activated with H_2:N_2 (100 cm^3/min:130 cm^3/min) at 300°C, purged in N_2 (130 cm^3/min), and cooled to a temperature of interest in N_2 to obtain background scans. For CO adsorption measurements, CO:N_2 (3.75 cm^3/min:130 cm^3/min) was flowed at 300°C and then cooled to the temperature of interest. Following CO adsorption experiments, steady-state water-gas shift measurements were carried out at either 225 or 185°C, using CO:H_2O:N_2 (3.75 cm^3/min:62.5 cm^3/min:67.5 cm^3/min) or CO:H_2O:H_2 (same ratio).

19.2.3.1 System Dynamics

In order to obtain enough scans to achieve an adequate signal-to-noise ratio by DRIFTS, it was necessary to take thirty-two scans per data point, which required approximately 1.1 min per point. Therefore, in order to have enough time to monitor the switching experiment, a linear input with time was utilized, as shown in Scheme 19.2. The time to achieve 50% ^{13}C exchange in a particular species actually included time delays in addition to the true transient kinetic half-life. First, there was a lag time from the point of the mass flow controller to the reactor itself, and this section included heat-traced tubing, a tubular steam generator, and the DRIFTS cell (with a small volume). This section was essentially a plug flow regime resulting in a time lag. Second, we were providing a linear input when we switched to the ^{13}CO feed. As shown in Scheme 19.2, the small volume pressurized to 25 psig (minimum pressure required to maintain mass flow controller operation) was gradually replaced, as the feed rate of CO is very low (~3.75 ccm).

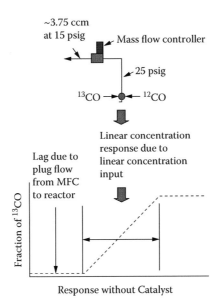

~3.75 ccm
at 15 psig

Mass flow controller

25 psig

$^{13}CO \rightarrow \quad \leftarrow {}^{12}CO$

Linear concentration
response due to
linear concentration
input

Lag due to
plug flow
from MFC
to reactor

Fraction of ^{13}CO

Response without Catalyst

SCHEME 19.2 Explanation of the time lag and linear concentration input.

This resulted in the linear concentration input that allowed us enough spectrometer scanning time such that we could improve the S/N ratio of the spectrum to the point where we could follow the signals by DRIFTS. Control experiments were made to determine the exchange time of the system in the absence of catalyst. To do so, a ^{12}CO to ^{13}CO switch in N_2 was carried out, and KBr was used as the blank. Results are reported in Figure 19.3. Removing the time lag, the time to achieve 50% replacement by ^{13}CO was ~2.2 min (Figure 19.3a) when carried out at 225° and ~2.7 min (Figure 19.3b) at 185°C.

To add transparency, we decided to use the intense and well-defined CO_2 bands from infrared spectroscopy to carry out the analysis, instead of using a mass spectrometer to independently analyze the product CO_2, as was carried out in an earlier study.[5] That is, our view is that the SSITKA-DRIFTS results are more meaningful if the signals for both species (formate and CO_2) are obtained from the same spectrum instead of utilizing two different instruments, especially considering that a separation of CO and H_2O is recommended such that both species are not present in the ionization chamber of the mass spectrometer. To use the CO_2 signal in DRIFTS, it was necessary to use a cover box and to purge ambient air from the system. This was accomplished by running a continuous nitrogen purge in the box, clearing out all traces of CO_2 gas due to atmospheric contamination.

During steady-state isotopic transient kinetic analysis, the ^{12}CO was switched to ^{13}CO and the carbon-containing adsorbed and gas phase species were monitored in the IR as they exchanged from the ^{12}C to the ^{13}C label. Particular attention was made to those species that exchanged on a timescale similar to that of the exchange of the product CO_2, as that species could be a likely intermediate to the water-gas

TABLE 19.1

BET Surface Area and Porosity Data for ZrO$_2$ and 2% Pt-Loaded Catalysts

Catalyst	BET SA (m^2/g)	Pore Volume (cm^3/g)	Pore Radius (nm)
ZrO$_2$	142.2	0.233	2.33
2% Pt/ZrO$_2$	136.3	0.226	2.33
2% Pt/2.5% Na/ZrO$_2$	118.1	0.203	2.33

shift mechanism. However, in addition, control experiments were carried out in the absence of steam to determine whether any of the species exchanged rapidly in the absence of the water-gas shift reaction. That is, we identified those species that were likely to undergo a purely nonreactive exchange process.

To further assess the stability of the formate and decouple the promoting impact of steam (which promotes both the rate and the formate decomposition selectivity to favor the forward the decomposition to CO_2 and H_2), the formate C–H bond was subjected to ^{12}C–^{13}C exchange under dry conditions. This resulted primarily in reverse formate decomposition back to CO and –OH. Following catalyst activation, background collection, and CO adsorption, the catalyst was cooled to 225°C in $^{12}CO:N_2$ (3.75 cm^3/min:130 cm^3/min). This was followed by switching the feed to $^{13}CO:N_2$. The formate ^{12}C to ^{13}C exchange rates were followed, with 128 scans per spectrum taken between intervals.

19.3   RESULTS AND DISCUSSION

BET results are reported in Table 19.1. A decrease in the BET surface areas was observed with addition of Pt and Na to zirconia; however, the surface areas remained well above 100 m^2/g. The activation of reduced defect sites over zirconia has been suggested[49] to occur via either (1) the formation of an oxygen vacancy defect or (2) the formation of a type II bridging OH group. The latter is interpreted to be the result of the dissociative adsorption of H_2O at the oxygen vacancy sites, or from the dissociation and spillover of H_2 from the metal to the oxide surface. TPR experiments[49] demonstrated that 2% Pt addition shifted reduction peaks for the zirconia surface to <200°C.

Following activation of zirconia at 300°C in H_2, and cooling to the temperature of interest and upon CO adsorption, formate bands are readily identified and include the C–H stretching bands of bidentate formate (Figure 19.2). The main band, υ(C–H), is situated at 2,880 cm^{-1}, with minor bands for δ(C–H) + υ_s(OCO) and 2δ(C–H) positioned at 2,966 and 2,745 cm^{-1}, respectively. In addition, OCO asymmetric and symmetric stretching bands were also identified, with υ_{as}(OCO) positioned at 1,568 cm^{-1}, and υ_s(OCO) located at 1,386 and 1,366 cm^{-1}. As Pt promoted the formation of reduced defect sites in the surface shell of zirconia, the number of type II bridging OH groups increased when the Pt loading was

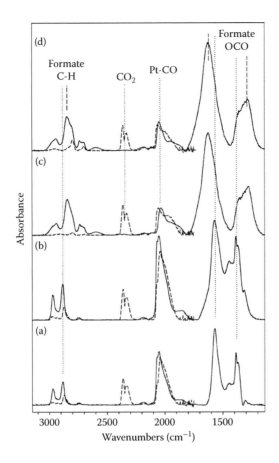

FIGURE 19.1 (Solid lines) CO adsorption and (dashed lines) steady-state WGS for (a) 1% Pt/ZrO$_2$ at 225°C, (b) 2% Pt/ZrO$_2$ at 225°C, (c) 1% Pt/2.5% Na/ZrO$_2$ at 225°C, and (d) 1% Pt/2.5% Na/ZrO$_2$ at 185°C.

increased from 1% to 2%, and this in turn caused the formate band intensities upon CO adsorption to increase (Figure 19.2). However, the band positions for formate did not change.

As shown in Figure 19.1, the addition of alkali metals to 1% (or 2%) Pt/ZrO$_2$ catalysts in the alkali to Pt molar ratio of ~21 for 1% Pt (~11 for 2% Pt) (i.e., on a wt% basis, this results in a value of 2.5% Na) produced a pronounced increase in both the intensity (×3 relative to 1% Pt/ZrO$_2$) and band positions of formate. In addition, the υ(C–H) band decreases from 2,880 cm^{-1} for 1% Pt/ZrO$_2$ alone to the range of 2,842–2,804 cm^{-1} for 1% Pt/2.5% Na/ZrO$_2$, while the difference between the OCO asymmetric υ_{as}(OCO) and symmetric bands υ_s(OCO) displays a pronounced increase when comparing 1% Pt/ZrO$_2$ (1,568 to 1,386 = 182 cm^{-1}) to 1% Pt/2.5% Na/ZrO$_2$ (1,614 to 1,310 = 304 cm^{-1}).

The addition of Na also impacted the Pt–CO linear υ(CO) to bridged υ(CO) band ratio, referred to as L/B.[52,53] The presence of the Na dopant favored an increase in the amount of bridge-bonded CO, where the ratio of 8.3 for no alkali doping decreased to 1.5 for the 2.5% Na-doped sample. Competing viewpoints (e.g., electronic, geometric) regarding the nature of the change in the L/B ratio are summarized in previous works.[1,49]

Over metal/CeO$_2$ catalysts, it has been observed that at high H$_2$O/CO ratios, the coverage of formate is regulated by the water-gas shift rate at low temperature.[15] This suggests not only that it is a likely intermediate, but that the coverage itself provides an important measurement of the water-gas shift rate. Moreover, based on the observed normal kinetic isotope effect,[6,54,55] the formate decomposition by dehydrogenation (which involves formate C–H bond breaking) was suggested to be the rate-limiting step over Pt/ceria under conditions relevant to low-temperature shift as applied to fuel processing for fuel cell applications. A similar normal kinetic isotope effect was found in the water-gas shift rate over Pt/ZrO$_2$[3]. Therefore, if the formation rate of the formate is sufficiently rapid, then the rate of the decomposition step should ultimately determine the surface coverage. It follows, then, that if the formate coverage is low during steady-state water-gas shift, the formate is suggested to be reacting faster. As shown in Figure 19.1, when one switches from CO adsorption to steady-state water-gas shift (while maintaining constant CO partial pressure using N$_2$ as the balancing gas), the formate coverage becomes significantly limited, and this suggests that the water-gas shift rate is regulating the coverage. Assuming that CO adsorption probes the total number of active sites of bridging OH groups available, then the coverage during steady-state water-gas shift can be defined as the intensity during water-gas shift divided by the intensity during CO adsorption. The results of the formate fractional coverage are reported in Table 19.2. Using the 1% Pt/ZrO$_2$ catalyst as a basis (Figure 19.1a), the coverage is ~0.36 at 225°C. Increasing the Pt loading is suggested to increase the rate at which formate decomposes at the Pt-ZrO$_2$ interface, and consequently

TABLE 19.2
Formate Fractional Coverage $\theta_{formate}$ at Steady-State Low-Temperature Shift Conditions at 225°C or 185°C

Catalyst Description	T (°C)	$\theta_{formate}$ (CO + H$_2$O + N$_2$)
1% Pt/ZrO$_2$	225	0.36
2% Pt/ZrO$_2$	225	0.29
1% Pt/2.5% Na/ZrO$_2$	225	0.05
1% Pt/2.5% Na/ZrO$_2$	185	0.16

Note: $\theta_{formate}$ was calculated by dividing the formate area during steady-state water-gas shift by the formate area from CO adsorption.

decrease the coverage to ~0.29 at 225°C (Figure 19.1b). However, doping the catalyst with Na weakens the formate C–H bond and makes it more reactive. Therefore, the coverage is profoundly impacted by the addition of the dopant, dropping the coverage to just 0.05 at 225°C (see Figure 19.1c). However, it is not possible to carry out a meaningful SSITKA experiment with such a low formate coverage. In order to increase the formate coverage, it was necessary to decrease the reaction rate by lowering the temperature. As shown in Figure 19.1d, the coverage increases to 0.16 when the temperature is lowered to 185°C, a measurable value for SSITKA testing. Note that the Pt carbonyl band is hardly impacted when switching to steady-state water-gas shift, suggesting that it may not be an important intermediate.

Figure 19.2 displays reference spectra for the ^{12}C- and ^{13}C-containing species either after CO adsorption or during steady-state water-gas shift. The difference between the formate C–H band position during CO adsorption and steady-state water-gas shift is important. Coverage effects result in a difference of 2,883 − 2,872 = 11 cm^{-1} (Figure 19.2a) for Pt/ZrO$_2$ (^{12}C basis) and 2,844 − 2,800 = 44 cm^{-1} (Figure 19.2c) for PtNa/ZrO$_2$ (^{12}C basis). The differences are quite pronounced and stress the importance of using the formate C–H bands obtained during steady-state water-gas shift testing (i.e., as opposed to those obtained from CO adsorption) as references for the SSITKA analysis. This is especially critical when considering that the differences between the formate bands carrying the ^{12}C and ^{13}C labels during steady-state water-gas shift are 2,872 − 2,854 = 18 cm^{-1} (Figure 19.2 a,b) in the case of Pt/ZrO$_2$ and 2,800 − 2,791 = 9 cm^{-1} (Figure 19.2 c,d) for PtNa/ZrO$_2$, during steady-state water-gas shift. Therefore, in the course of an isotopic transient study, the resulting formate band can be fitted using a linear combination of reference spectra of ^{12}C–H and ^{13}C–H recorded during steady-state water-gas shift.

Figure 19.3 shows the results of isotopic switching over a blank KBr sample, providing measurements of the system dynamics. The results of the control experiments carried out in the absence of H$_2$O are displayed in Figures 19.4 (Pt/ZrO$_2$) and 19.5 (PtNa/ZrO$_2$), respectively, after removal of the time lag from the data. Figure 19.4 demonstrates unequivocally that formates display remarkable stability in the absence of H$_2$O. This suggests that the rate-limiting step more likely involves formate decomposition instead of their formation. Removing the time lag, even after 40 min, only a small fraction (<15%) has undergone exchange to the ^{13}C label. Hereafter, all reported exchange times exclude the time lag. In contrast, Pt–CO undergoes rapid exchange to the ^{13}C label (time to achieve 50% ^{13}C incorporation, ~4.5 min). Since there is no water-gas shift reaction taking place in this condition, the result suggests that the rapid Pt–CO exchange during WGS may likely be due to a replacement exchange of Pt–^{12}CO by ^{13}CO, and not indicative of being involved in the WGS pathway. On the other hand, any exchange of formate during water-gas shift would suggest that it is likely a reactive process, since formate is otherwise quite stable in the low-temperature region—in contrast to what has been reported in a recent review.[56] Figure 19.5 shows that the formate for the PtNa/ZrO$_2$ catalyst is more reactive under the dry condition than Pt/ZrO$_2$,

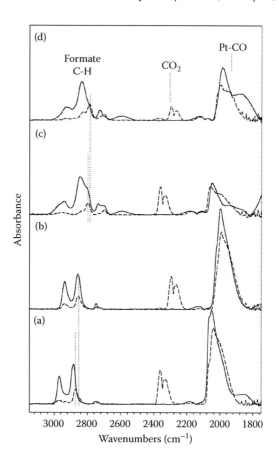

FIGURE 19.2 (Solid lines) CO adsorption and (dashed lines) steady-state WGS for (a) 2% Pt/ZrO$_2$ at 225°C using ^{12}CO, (b) 2% Pt/ZrO$_2$ at 225°C using ^{13}CO, (c) 2% Pt/2.5% Na/ZrO$_2$ at 185°C using ^{12}CO, and (d) 2% Pt/2.5% Na/ZrO$_2$ at 185°C using ^{13}CO.

and the time to achieve 50% ^{13}C incorporation is ~27.5 min. Thermal formate decomposition, which is the reaction back to CO and –OH, is suggested to involve C–H bond breaking of the formate, as in water-gas shift. Therefore, this decrease in formate stability (via C–H bond weakening from alkali doping) as observed in Figure 19.5, provides a further indication that the formate is more reactive under true water-gas shift conditions. In other words, the decomposition is faster in the forward direction and easier with the influence of sodium.

Figures 19.6 through 19.11 detail the isotopic exchange rates during water-gas shift for the formate, CO$_2$, and Pt–CO bands, in switching from the ^{12}C to ^{13}C label. In all cases, the reactive exchange rates of formate and CO$_2$ were virtually identical, implicating the formate species as the likely intermediate to the water-gas shift catalytic mechanism over Pt/ZrO$_2$ and PtNa/ZrO$_2$ catalysts. The DRIFTS spectra at the top of each figure show the switching of these species from

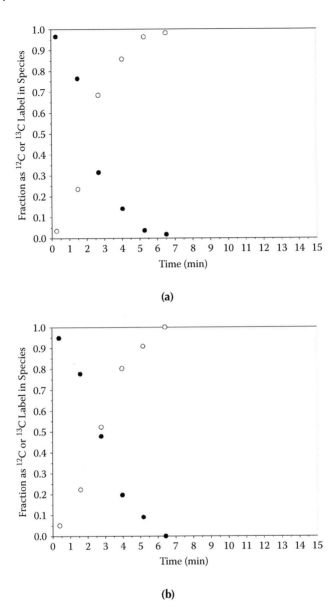

(a)

(b)

FIGURE 19.3 Isotopic switching control experiments at (a) 225°C and (b) 185°C to assess the exchange time of the DRIFTS cell. ^{12}CO to ^{13}CO switching was conducted in flowing N_2 using the same CO flow and partial pressure that were used in the WGS experiments.

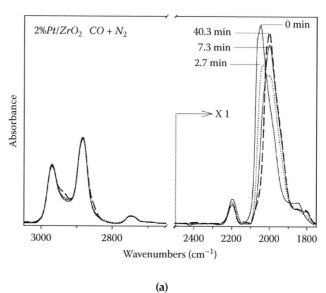

(a)

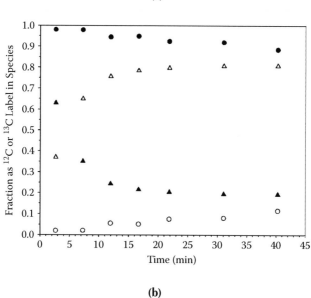

(b)

FIGURE 19.4 Isotopic tracer-DRIFTS experiment over 2% Pt/ZrO$_2$ at 225°C, including ((a) left) formate C–H stretching region and ((a) right) Pt–CO stretching regions. Results of the linear combination fitting (b) showing the fraction of (filled) ^{12}C label or (unfilled) ^{13}C label in (circles) formate C–H band or (triangles) Pt carbonyl bands. Feed contained CO + N$_2$. In the absence of H$_2$O, formates show remarkable stability. Pt–CO was found to undergo isotopic exchange readily.

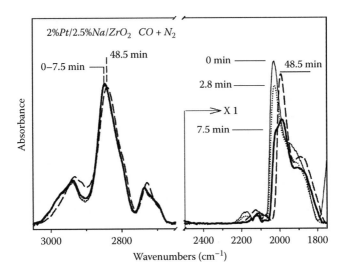

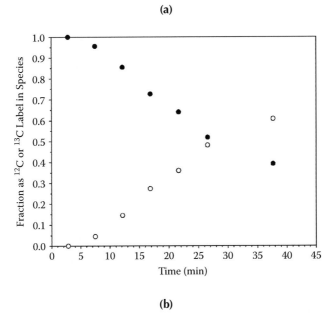

FIGURE 19.5 Isotopic tracer-DRIFTS experiment over 2% Pt/2.5% Na/ZrO$_2$ at 225°C, including ((a) left) formate C–H stretching region and ((a) right) Pt–CO stretching regions. Results of the linear combination fitting (b) showing the fraction of (filled) ^{12}C label or (unfilled) ^{13}C label in (circles) formate C–H band or (triangles) Pt carbonyl bands. Feed contained CO + N$_2$. In the absence of H$_2$O, formates also show stability, but are less stable than those found on the nonalkali 2% Pt/ZrO$_2$ catalyst (see Figure 19.4). Pt–CO ((a) right) was found to undergo isotopic exchange readily, but the bands were too complex to be amenable to the linear combination fitting technique.

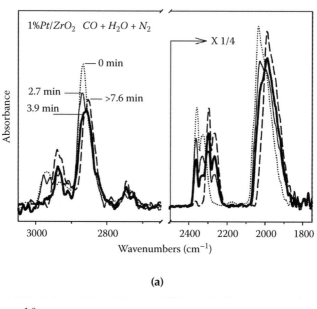

(a)

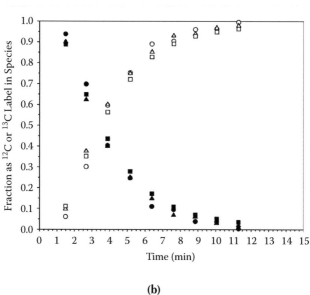

(b)

FIGURE 19.6 SSITKA-DRIFTS WGS experiment over 1% Pt/ZrO$_2$ at 225°C, including ((a) left) formate C–H stretching region and ((a) right) gas phase CO$_2$ and Pt–CO stretching regions. Results of the linear combination fitting (b) showing the fraction of (filled) ^{12}C label or (unfilled) ^{13}C label in (circles) formate C–H band, (triangles) gas phase CO$_2$ band, or (squares) Pt carbonyl bands. Feed contained CO + H$_2$O + N$_2$.

the ^{12}C to ^{13}C label, while the bottom of each figure provides the results of linear combination fitting with reference to each labeled species taken under steady-state water-gas shift.

Figures 19.6, 19.7, and 19.9 provide results of steady-state water-gas shift in the absence of co-fed hydrogen. In Figure 19.6, the reactive exchange rates of formate and CO_2 were faster for the 1% Pt/ZrO_2 catalyst at 225°C (time to achieve 50% ^{13}C incorporation, ~3.5 min) than for the 2% Pt/ZrO_2 (~5.7 min), as shown in Figure 19.7. Replacing N_2 with H_2 did not measurably impact the exchange time (Figure 19.8).

To preface the discussion of these figures, as mentioned previously, CO adsorption is a useful tool to probe the bridging OH group active sites over a number of metal/partially reducible oxide catalysts, including Pt/zirconia.[2–5] This is because, in the absence of H_2O, CO reacts with the bridging OH groups to yield pseudostable formate species, with band intensities that are strong and can be readily assessed. Interestingly, in previous investigations, it was found that over a series of activated Pt/ceria catalysts with different Pt loadings (e.g., 0.5%, 1%, 2.5% Pt) but the same surface area (~125 m²/g), the formate band intensities were virtually identical.[15] This indicated that the Pt facilitated nearly complete reduction of the ceria surface shell during formation of the defect-associated bridging OH groups. This assumption was corroborated when the extent of reduction was quantified by x-ray absorption near-edge spectroscopy (XANES),[15] which showed that ~20 to 25% of the ceria was in the Ce^{3+} oxidation state, consistent with virtually complete surface shell reduction.[57,58] Yet, in switching to steady-state water-gas shift conditions, the formate coverages exhibited a decreasing trend with increasing Pt loading, indicating that the formates were reacting faster, consistent with CO conversion rates measured from fixed bed catalytic testing. Most recently, the same series of catalysts was tested using the SSITKA-DRIFTS method, and the time intervals to achieve 50% ^{13}C incorporation for formate and CO_2 were very similar, and decreased as a function of increasing Pt loading. Again, this implies formates react faster with increasing Pt content.

This led to the inclusion of a formate surface diffusion step in the model from the oxide, where the formate is formed, to the metal-oxide interface, where the metal can abstract H from formate via C–H bond cleaving. The relationship between formate mobility and O-mobility, as recently emphasized in a review by Duprez,[19] was considered but not assessed in recent work.[51] Following from that line of reasoning, a three-zone surface model was envisioned to describe the bifunctional catalysts over metal/ceria.[51] At low temperature, and with very low Pt levels, diffusional lengths can become excessive (or formate mobility too slow), such that only formates directly in the vicinity of the Pt particles will be reactive. In this situation, designated case 1, an important problem arises in interpreting data obtained by the SSITKA-DRIFTS technique. That is, since CO_2 and its monodentate carbonate precursor stemming from formate decomposition would only evolve from the reaction of formates in the "rapid reaction zone" in the vicinity of the metal particles, and not those at excessive surface diffusional lengths, CO_2 and mondentate carbonate should display a faster overall fractional reactive

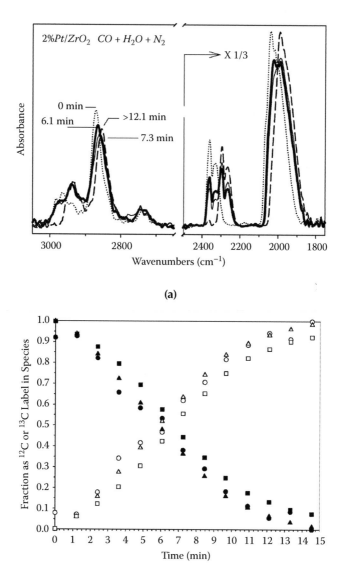

FIGURE 19.7   SSITKA-DRIFTS WGS experiment over 2% Pt/ZrO$_2$ at 225°C, including ((a) left) formate C–H stretching region and ((a) right) gas phase CO$_2$ and Pt–CO stretching regions. Results of the linear combination fitting (b) showing the fraction of (filled) ^{12}C label or (unfilled) ^{13}C label in (circles) formate C–H band, (triangles) gas phase CO$_2$ band, or (squares) Pt carbonyl bands. Feed contained CO + H$_2$O + N$_2$.

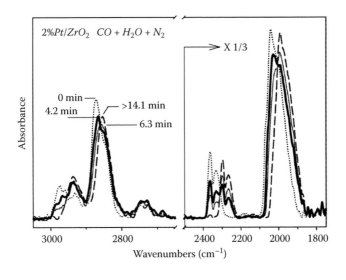

(a)

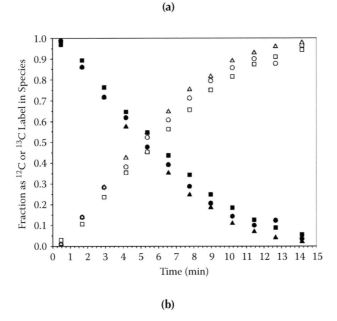

(b)

FIGURE 19.8 SSITKA-DRIFTS WGS experiment includes H_2 over 2% Pt/ZrO$_2$ at 225°C, including ((a) left) formate C–H stretching region and ((a) right) gas phase CO$_2$ and Pt–CO stretching regions. Results of the linear combination fitting (b) showing the fraction of (filled) ^{12}C label or (unfilled) ^{13}C label in (circles) formate C–H band, (triangles) gas phase CO$_2$ band, or (squares) Pt carbonyl bands. Feed contained CO + H$_2$O + H$_2$.

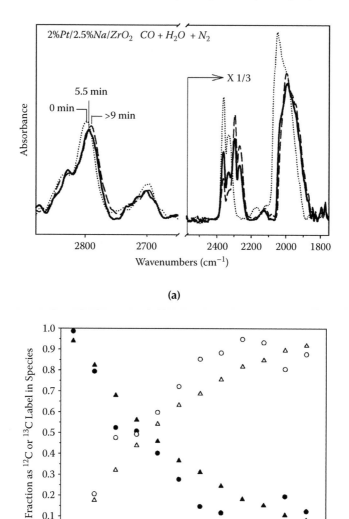

(a)

(b)

FIGURE 19.9 SSITKA-DRIFTS WGS experiment over 2% Pt/2.5% Na/ZrO$_2$ at 185°C, including ((a) left) formate C–H stretching region and ((a) right) gas phase CO$_2$ and Pt–CO stretching regions. Results of the linear combination fitting (b) showing the fraction of (filled) ^{12}C label or (unfilled) ^{13}C label in (circles) formate C–H band, (triangles) gas phase CO$_2$ band, or (squares) Pt carbonyl bands. Feed contained CO + H$_2$O + N$_2$.

exchange than that of formate. This is because formate will inevitably carry multiple components within its signal, including those from the rapidly reacting formates in the zone close to the metal particle, as well as those further from the metal particle at longer diffusional lengths, including a stranded intermediate zone at excessive diffusional lengths. Increasing temperature or number of Pt particles leads to case 2, which involves reducing the diffusional path lengths such that the stranded intermediate zone is effectively removed, although a surface diffusional zone will still remain in place. It is important to consider that formates rapidly re-forming in the rapid reaction zone may likely hinder diffusion of the formate in the surface diffusional zone to the metal particle. With further increases in temperature or number of Pt particles, the formates will all be located and reacting in the rapid reaction zone, leading to case 3. CAER researchers previously noted that a different problem can exist for SSITKA-DRIFTS data interpretation. If all the formates are reacting fast in the rapid reaction zone, their surface coverages during steady-state water-gas shift may be below detectable limits.

The case of Pt/ZrO_2, though in many ways similar to that of Pt/ceria, poses some significant differences. Using CO adsorption to probe the type II bridging OH group active sites, it was found in this case that an increase in the Pt loading led to increases in the formate band intensity upon CO adsorption, and this suggests that the number of defect-associated bridging OH groups on ZrO_2, unlike Pt/ceria, are confined to the area surrounding the metal particles. Analogous defect vacancy sites have been proposed to be important during CO_2 reforming of CH_4 over Pt/ZrO_2 catalysts.[59] For Pt/ceria, the surface of ceria was virtually completely reduced after activation over a wide range of Pt loadings (0.5 to 2.5% Pt). Our interpretation of Figure 19.1 is that there is likely a higher population of formate on 2% Pt/ZrO_2 relative to 1% Pt/ZrO_2. Therefore, longer times needed to achieve 50% ^{13}C incorporation in formate and CO_2 may be related to the longer diffusional path lengths for those species that are formed at defect sites farther away from the Pt particle. However, presumably those formates that are situated close to Pt react rapidly at the $Pt-ZrO_2$ interface. For the lower loaded 1% Pt/ZrO_2 catalyst, the number of defects is lower (and hence the formate population arising from CO + −OH is lower), and therefore the formates that are likely only to be present on the ZrO_2 defects are located close to the rim of the Pt metal particles. A comparison of the proposed models for the different cases of Pt/ceria and Pt/zirconia is provided in Figure 19.12.

Regarding the impact of alkali addition, the time to achieve 50% ^{13}C incorporation in formate on 2% Pt/2.5% Na/ZrO_2 was ~4.8 min; however, that measurement was at 185°C, which was even lower than the time required for 2% Pt/ZrO_2 at 225°C. This finding is a testimony to the much higher reactivity of formate species over the Na-doped Pt/ZrO_2 catalysts. In the presence of co-fed hydrogen, formate was still found to exchange at a rate close to the CO_2 product exchange rate, as shown in Figures 19.8 and 19.10. The reactive exchange rate of formate with the co-fed hydrogen stream was very similar to that in the case without H_2 co-feeding (i.e., using N_2 balancing gas).

Moving to a lower Pt loading of 1% for the Na-doped catalyst (Figure 19.11), unlike its Pt/ZrO_2 counterpart, resulted in formate bands that were very similar in intensity to those observed over the 2% $Pt/2.5\%$ Na/ZrO_2 catalyst. In that case, if the rate of formate decomposition is related to the metal-ZrO_2 interfacial area, the formates should take longer to diffuse and react over the lower loaded 1% Pt catalyst. As shown in Figure 19.11, this was indeed the case, as the time to achieve 50% ^{13}C incorporation in formate was ~7.6 min relative to ~4.8 min for the case of 2% $Pt/2.5\%$ Na/ZrO_2. A summary is provided in Table 19.3.

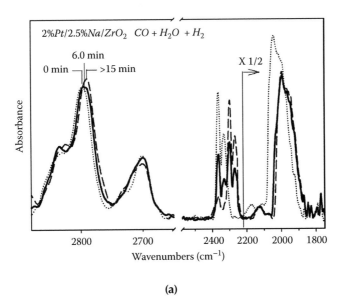

(a)

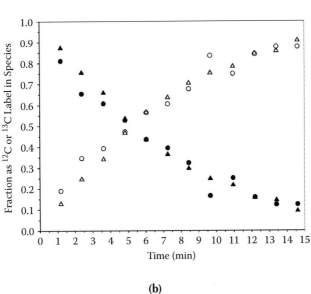

(b)

FIGURE 19.10 SSITKA-DRIFTS WGS experiment including H_2 over 2% Pt/2.5% Na/ ZrO$_2$ at 185°C, including ((a) left) formate C–H stretching region and ((a) right) gas phase CO_2 and Pt–CO stretching regions. Results of the linear combination fitting (b) showing the fraction of (filled) ^{12}C label or (unfilled) ^{13}C label in (circles) formate C–H band, (triangles) gas phase CO_2 band, or (squares) Pt carbonyl bands. Feed contained CO + H_2O + H_2.

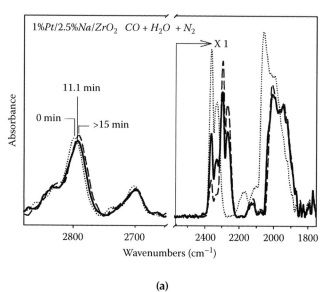

(a)

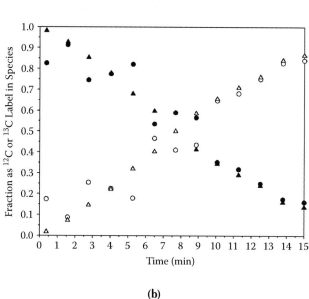

(b)

FIGURE 19.11 SSITKA-DRIFTS WGS experiment over 1% Pt/2.5% Na/ZrO$_2$ at 185°C, including ((a) left) formate C–H stretching region and ((a) right) gas phase CO$_2$ and Pt–CO stretching regions. Results of the linear combination fitting (b) showing the fraction of (filled) ^{12}C label or (unfilled) ^{13}C label in (circles) formate C–H band, (triangles) gas phase CO$_2$ band, or (squares) Pt carbonyl bands. Feed contained CO + H$_2$O + N$_2$.

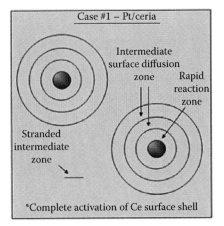

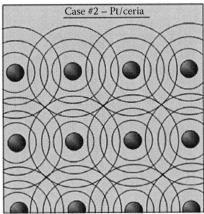

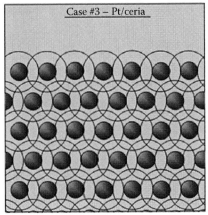

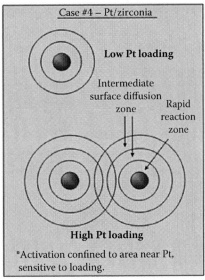

FIGURE 19.12 Considerations for the interpretation of SSITKA data. **Case 1:** Three formates can exist, including (a) rapid reaction zone (RRZ)—those reacting rapidly at the metal-oxide interface; (b) intermediate surface diffusion zone (SDZ)—those at path lengths sufficient to eventually diffuse to the metal and contribute to overall activity, and (c) stranded intermediate zone (SIZ)—intermediates are essentially locked onto surface due to excessive diffusional path lengths to the metal-oxide interface. **Case 2:** Metal particle population sufficient to overcome excessive surface diffusional restrictions. **Case 3:** All rapid reaction zone. **Case 4:** For Pt/zirconia, unlike Pt/ceria, the activated oxide is confined to the vicinity of the metal particle, and the surface diffusional zones are sensitive to metal loading.

TABLE 19.3
Exchange Times of Isotopic Transient

Catalyst Description	Condition	T (°C)	Time to Achieve 50% ^{13}C Incorporation in Formate/CO_2 (min)[a]
No catalyst	$CO + N_2$	225	2.2
No catalyst	$CO + N_2$	185	2.7
1% Pt/ZrO$_2$	$CO + H_2O + N_2$	225	3.5
2% Pt/ZrO$_2$	$CO + H_2O + N_2$	225	5.7
2% Pt/ZrO$_2$	$CO + H_2O + H_2$	225	5.4
2% Pt/ZrO$_2$ (formate)	$CO + N_2$	225	Hours
2% Pt/ZrO$_2$ (Pt–CO)	$CO + N_2$	225	4.5
2% Pt/2.5% Na/ZrO$_2$	$CO + H_2O + N_2$	185	4.8
2% Pt/2.5% Na/ZrO$_2$	$CO + H_2O + H_2$	185	5.0
2% Pt/2.5% Na/ZrO$_2$ (formate)	$CO + N_2$	225	27.5
1% Pt/2.5% Na/ZrO$_2$ (formate)	$CO + H_2O + N_2$	185	7.6

Note: All experiments refer to formate and CO_2 switching rates (~ equivalent), except for linear input response experiments (the time it takes to switch from ^{12}CO to ^{13}CO in the absence of catalyst, excluding lag time).

[a] Note that the values exclude the lag time. Also note that a linear input was used and the reported times are not true kinetic half-life values.

19.4 CONCLUSIONS

The SSITKA-DRIFTS method was employed to assess the rates of exchange of formate, gas phase CO_2, and Pt–CO during switching from a feed containing ^{12}CO to one containing ^{13}CO. Over Pt/ZrO$_2$ and PtNa/ZrO$_2$ catalysts containing 1% and 2% loadings of Pt, with and without co-fed H$_2$, the formate species were found to undergo reactive exchange along an almost identical trajectory with the CO_2 product. Pt–CO exchanged rapidly, but it also exchanged rapidly during control experiments with H$_2$O absent, whereas formate was quite stable in the dry switching experiment. The results suggest that formate is reacting under water-gas shift conditions, while Pt–CO is probably undergoing merely a replacement process. Addition of Na dopant significantly accelerated the formate exchange rate. The coverage during steady-state water-gas shift was very low at 225°C (due to the formate rapidly reacting), such that the SSITKA test temperature had to be lowered to 185°C to obtain enough formate coverage for the experiment. Even then, the time required to achieve 50% ^{13}C incorporation in formate (i.e., formate C–H band) and CO_2 at 185°C over 2% Pt/2.5% Na/ZrO$_2$ was still less than that of the 2% Pt/ZrO$_2$ catalyst recorded at a much higher temperature. The results provide further support to the conclusion that not only are infrared active formates likely intermediates in the water-gas shift reaction, but designing catalysts based on formate C–H bond weakening (e.g., by alkali doping) can lead to significantly higher catalytic activity.

ACKNOWLEDGMENTS

This work was supported by a grant from the Kentucky Office of Energy Policy and by the Commonwealth of Kentucky.

REFERENCES

1. Pigos, J.M., Brooks, C.J., Jacobs, G., and Davis, B.H. 2007. Low temperature water-gas shift: Characterization of Pt-based ZrO2 catalyst promoted with Na discovered by combinatorial methods. *Appl. Catal. A Gen.* 319:47–57.
2. Tabakova, T., Idakiev, V., Andreeva, D., and Mitov, I. 2000. Influence of the microscopic properties of the support on the catalytic activity of Au/ZnO, Au/ZrO$_2$, Au/Fe$_2$O$_3$, Au/Fe$_2$O$_3$–ZnO, Au/Fe$_2$O$_3$–ZrO$_2$ catalysts for the WGS reaction. *Appl. Catal. A Gen.* 202:91–97.
3. Chenu, E., Jacobs, G., Crawford, A.C., Keogh, R.A., Patterson, P.M., Sparks, D.E., and Davis, B.H. 2005. Water-gas shift: An examination of Pt promoted MgO and tetragonal and monoclinic ZrO$_2$ by in situ DRIFTS. *Appl. Catal. B Environ.* 59:45–56.
4. Ricote, S., Jacobs, G., Milling, M., Ji, Y., Patterson, P.M., and Davis, B.H. 2006. Low temperature water–gas shift: Characterization and testing of binary mixed oxides of ceria and zirconia promoted with Pt. *Appl. Catal. A Gen.* 303:35–47.
5. Tibiletti, D., Meunier, F.C., Goguet, A., Reid, D., Burch, R., Boaro, M., Vicario, M., and Trovarelli, A. 2006. An investigation of possible mechanisms for the water–gas shift reaction over a ZrO$_2$-supported Pt catalyst. *J. Catal.* 244:183–91.
6. Shido, T., and Iwasawa, Y. 1993. Reactant-promoted reaction mechanism for water-gas shift reaction on Rh-doped CeO$_2$. *J. Catal.* 141:71–81.
7. Padeste, C., Cant, N.W., and Trimm, D.L. 1993. The influence of water on the reduction and reoxidation of ceria. *Catal. Lett.* 18:305–16.
8. Barbier, Jr., J., and Duprez, D. 1993. Reactivity of steam in exhaust gas catalysis. I. Steam and oxygen/steam conversions of carbon monoxide and of propane over PtRh catalysts. *Appl. Catal. B Environ.* 3:61–83.
9. Bunluesin, T., Gorte, R.J., and Graham, G.W. 1998. Studies of the water-gas-shift reaction on ceria-supported Pt, Pd, and Rh: Implications for oxygen-storage properties. *Appl. Catal. B Environ.* 15:107–14.
10. Li, Y., Fu, Q., and Flytzani-Stephanopoulos, M. 2000. Low-temperature water-gas shift reaction over Cu- and Ni-loaded cerium oxide catalysts. *Appl. Catal. B Environ.* 27:179–91.
11. Hilaire, S., Wang, X., Luo, T., Gorte, R.J., and Wagner, J. 2001. A comparative study of water-gas-shift reaction over ceria supported metallic catalysts. *Appl. Catal. A Gen.* 215:271–78.
12. Fu, Q., Weber, A., and Flytzani-Stephanopoulos, M. 2001. Nanostructured Au-CeO$_2$ catalysts for low-temperature water-gas shift reaction. *Catal. Lett.* 77:87–95.

13. Andreeva, D., Idakiev, V., Tabakova, T., Ilieva, L., Falaras, P., Bourlinos, A., and Travlos, A. 2002. Low-temperature water-gas shift reaction over Au/CeO$_2$ catalysts. *Catal. Today* 72:51–57.

14. Jacobs, G., Williams, L., Graham, U., Sparks, D., Thomas, G., and Davis, B.H. 2003. Low temperature water–gas shift: In situ DRIFTS-reaction study of ceria surface area on the evolution of formates on Pt/CeO$_2$ fuel processing catalysts for fuel cell applications. *Appl. Catal. A Gen.* 252:107.

15. Jacobs, G., Graham, U.M., Chenu, E., Patterson, P.M., Dozier, A., and Davis, B.H. 2005. Low-temperature water–gas shift: Impact of Pt promoter loading on the partial reduction of ceria and consequences for catalyst design. *J. Catal.* 229:499–512.

16. Goguet, A., Shekhtman, S.O., Burch, R., Hardacre, C., Meunier, F.C., and Yablonsky, G.S. 2006. Pulse-response TAP studies of the reverse water–gas shift reaction over a Pt/CeO$_2$ catalyst. *J. Catal.* 237:102–10.

17. Germani, G., and Schuurman, Y. 2006. Water-gas shift reaction kinetics over μ-structured PtCeO$_2$Al$_2$O$_3$ catalysts. *AIChE J.* 52:1806–13.

18. Panagiotopoulou, P., and Kondarides, D.I. 2006. Effect of the nature of the support on the catalytic performance of noble metal catalysts for the water–gas shift reaction. *Catal. Today* 112:49–52.

19. Duprez, D. 2006. Study of surface reaction mechanisms by ^{16}O/^{18}O and H/D isotopic exchange. *Catal. Today* 112:17–22.

20. Meunier, F.C., Tibiletti, D., Goguet, A., Shekhtman, S., Hardacre, C., and Burch, R. 2007. On the complexity of the water-gas shift reaction mechanism over a Pt/CeO$_2$ catalyst: Effect of the temperature on the reactivity of formate surface species studied by operando DRIFT during isotopic transient at chemical steady-state. *Catal. Today* 126:143–47.

21. Leppelt, R., Schumacher, B., Plzak, V., Kinne, M., and Behm, R.J. 2006. Kinetics and mechanism of the low-temperature water–gas shift reaction on Au/CeO$_2$ catalysts in an idealized reaction atmosphere. *J. Catal.* 244:137–52.

22. Denkwitz, Y., Karpenko, A., Plzak, V., Leppelt, R., Schumacher, B., and Behm, R.J. 2007. Influence of CO$_2$ and H$_2$ on the low-temperature water–gas shift reaction on Au/CeO$_2$ catalysts in idealized and realistic reformate. *J. Catal.* 246:74–90.

23. Sakurai, H., Ueda, A., Kobayashi, T., and Haruta, M. 1997. Low-temperature water-gas shift reaction over gold deposited on TiO$_2$. *Chem. Commun.* 271.

24. Andreeva, D. Ch., Idakiev, V.D., Tabakova, T.T., and Giovanoli, R. 1998. Low-temperature water-gas shift reaction on Au/TiO$_2$, Au/α-Fe$_2$O$_3$ and Au/Co$_3$O$_4$ catalysts. *Bulg. Chem. Commun.* 30:59–68.

25. Boccuzzi, F., Chiorino, A., Manzoli, M., Andreeva, D., and Tabakova, T. 1999. FTIR study of the low-temperature water–gas shift reaction on Au/Fe$_2$O$_3$ and Au/TiO$_2$ catalysts. *J. Catal.* 188:176–85.

26. Panagiotopoulou, P., and Kondarides, D.I. 2004. Effect of morphological characteristics of TiO$_2$-supported noble metal catalysts on their activity for the water–gas shift reaction. *J. Catal.* 225:327–36.

27. Sato, Y., Terada, K., Hasegawa, S., Miyao, T., and Naito, S. 2005. Mechanistic study of water–gas-shift reaction over TiO$_2$ supported Pt–Re and Pd–Re catalysts. *Appl. Catal. A Gen.* 296:80–89.

28. Panagiotopoulou, P., Christodoulakis, A., Kondarides, D.I., and Boghosian, S. 2006. Particle size effects on the reducibility of titanium dioxide and its relation to the water–gas shift activity of Pt/TiO$_2$ catalysts. *J. Catal.* 240:114–25.

29. Iida, H., Kondo, K., and Igarashi, A. 2006. Effect of Pt precursors on catalytic activity of Pt/TiO$_2$ (rutile) for water gas shift reaction at low-temperature. *Catal. Commun.* 7:240–44.

30. Iida, H., and Igarashi, A. 2006. Characterization of a Pt/TiO$_2$ (rutile) catalyst for water gas shift reaction at low-temperature. *Appl. Catal. A Gen.* 298:152–60.

31. Sato, Y., Soma, Y., Miyao, T., and Naito, S. 2006. The water-gas-shift reaction over Ir/TiO$_2$ and Ir–Re/TiO$_2$ catalysts. *Appl. Catal. A Gen.* 304:78–85.

32. Jacobs, G., Crawford, A., Williams, L., Patterson, P.M., and Davis, B.H. 2004. Low temperature water–gas shift: Comparison of thoria and ceria catalysts. *Appl. Catal. A Gen.* 267:27–33.

33. Jacobs, G., Patterson, P.M., Graham, U.M., Crawford, A.C., Dozier, A., and Davis, B.H. 2005. Catalytic links among the water–gas shift, water-assisted formic acid decomposition, and methanol steam reforming reactions over Pt-promoted thoria. *J. Catal.* 235:79.

34. Tabakova, T., Idakiev, V., Tenchev, K., Boccuzzi, F., Manzoli, M., and Chiorino, A. 2006. Pure hydrogen production on a new gold–thoria catalyst for fuel cell applications. *Appl. Catal. B: Environ.* 63:94–103.

35. Basinska, A., and Domka, F. 1997. The influence of alkali metals on the activity of supported ruthenium catalysts for the water-gas shift reaction. *Catal. Lett.* 43:59–61.

36. Basinska, A., and Domka, F. 1999. The effect of lanthanides on the Ru/Fe$_2$O$_3$ catalysts for water-gas shift reaction. *Appl. Catal. A Gen.* 179:241–46.

37. Andreeva, D., Idakiev, V., Tabakova, T., and Andreev, A. 1996. Low-temperature water–gas shift reaction over Au/α-Fe$_2$O$_3$. *J. Catal.* 158:354–55.

38. Andreeva, D., Idakiev, V., Tabakova, T., Andreev, A., and Giovanoli, R. 1996. Low-temperature water-gas shift reaction on Au/α-Fe$_2$O$_3$ catalyst. *Appl. Catal. A Gen.* 134:275–83.

39. Andreeva, D., Tabakova, T., Idakiev, V., Christov, P., and Giovanoli, R. 1998. Au/α-Fe$_2$O$_3$ catalyst for water–gas shift reaction prepared by deposition–precipitation. *Appl. Catal. A Gen.* 169:9–14.

40. Venugopal, A., Aluha, J., Mogano, D., and Scurrell, M.S. 2003. The gold–ruthenium–iron oxide catalytic system for the low temperature water–gas-shift reaction: The examination of gold–ruthenium interactions. *Appl. Catal. A Gen.* 245:149–58.

41. Venugopal, A., and Scurrell, M.S. 2004. Low temperature reductive pretreatment of Au/Fe$_2$O$_3$ catalysts, TPR/TPO studies and behaviour in the water–gas shift reaction. *Appl. Catal. A Gen.* 258:241–49.

42. Hua, J., Wei, K., Zheng, Q., and Lin, X. 2004. Influence of calcination temperature on the structure and catalytic performance of Au/iron oxide catalysts for water–gas shift reaction. *Appl. Catal. A Gen.* 259:121–30.

43. Hua, J., Zheng, Q., Zheng, Y., Wei, K., and Lin, X. 2005. Influence of modifying additives on the catalytic activity and stability of Au/Fe$_2$O$_3$–MOx catalysts for the WGS reaction. *Catal. Lett.* 102:99–108.

44. Burch, R. 2006. Gold catalysts for pure hydrogen production in the water–gas shift reaction: Activity, structure and reaction mechanism. *PCCP* 8:5483–500.

45. Grenoble, D.C., Estadt, M.M., and Ollis, D.F. 1981. The chemistry and catalysis of the water gas shift reaction. 1. The kinetics over supported metal catalysts. *J. Catal.* 67:90–102.

46. Shido, T., and Iwasawa, Y. 1992. Regulation of reaction intermediate by reactant in the water-gas shift reaction on CeO$_2$, in relation to reactant-promoted mechanism. *J. Catal.* 136:493–503.

47. Brooks, C.J., Hagemeyer, A., Yaccato, K., Carhart, R., and Herrmann, M. 2005. Combinatorial methods for the discovery of novel catalysts for the WGS reaction. Paper presented at the 19th North American Catalyst Society Meeting, Philadelphia, May 22–27.

48. Pigos, J.M., Brooks, C.J., Jacobs, G., and Davis, B.H. 2006. Evidence of enhanced LTS water-gas shift rate with sodium promoted Pt-ZrO_2-based catalyst discovered by combinatorial methods. Paper presented at the AICHE Annual Meeting, San Fransisco, November 12–17.

49. Pigos, J.M., Brooks, C.J., Jacobs, G., and Davis, B.H. 2007. Low temperature water–gas shift: The effect of alkali doping on the CH bond of formate over Pt/ZrO_2 catalysts. *Appl. Catal. A Gen.* 328:14–26.

50. Jacobs, G., Crawford, A.C., and Davis, B.H. 2005. Water-gas shift: Steady state isotope switching study of the water-gas shift reaction over Pt/ceria using in-situ DRIFTS. *Catal. Lett.* 100:147–52.

51. Jacobs, G., and Davis, B.H. 2007. Low temperature water–gas shift: Applications of a modified SSITKA–DRIFTS method under conditions of H_2 co-feeding over metal/ceria and related oxides. *Appl. Catal. A Gen.* 333:192–201.

52. Mojet, B.L., Miller, J.T., and Koningsberger, D.C. 1999. The effect of CO adsorption at room temperature on the structure of supported Pt particles. *J. Phys. Chem. B* 103:2724–34.

53. Menacherry, P.V., and Haller, G.L. 1997. The effect of water on the infrared spectra of CO adsorbed on Pt/K L-zeolite. *Catal. Lett.* 44:135–44.

54. Jacobs, G., Khalid, S., Patterson, P.M., Sparks, D.E., and Davis, B.H. 2004. Water-gas shift catalysis: Kinetic isotope effect identifies surface formates in rate limiting step for Pt/ceria catalysts. *Appl. Catal. A Gen.* 268:255–66.

55. Jacobs, G., Patterson, P.M., Graham, U.M., Sparks, D.E., and Davis, B.H. 2004. Low temperature water-gas shift: Kinetic isotope effect observed for decomposition of surface formates for Pt/ceria catalysts. *Appl. Catal. A Gen.* 269:63–73.

56. Gorte, R.J., and Zhao, S. 2005. Studies of the water-gas-shift reaction with ceria-supported precious metals. *Catal. Today* 104:18–24.

57. El Fallah, J., Boujani, S., Dexpert, H., Kiennemann, A., Majerus, J., Touret, O., Villain, F., and Le Normand, F. 1994. Redox processes on pure ceria and on Rh/CeO_2 catalyst monitored by x-ray absorption (fast acquisition mode). *J. Phys. Chem.* 98:5522–33.

58. Laachir, A., Perrischon, V., Badri, A., Lamotte, J., Catherine, E., Lavalley, J.C., El Fallah, J., Hilaire, L., Le Normand, F., Quemere, E., Sauvion, G.N., and Touret, O. 1991. Reduction of CeO_2 by hydrogen. Magnetic susceptibility and Fourier-transform infrared, ultraviolet and x-ray photoelectron spectroscopy measurements. *J. Chem. Soc. Faraday Trans.* 87:1601–10.

59. Noronha, F.B., Shamsi, A., Taylor, C., Fendley, E.C., Stagg-Williams, S.M., and Resasco, D.E. 2003. Catalytic performance of Pt/ZrO_2 and Pt/Ce-ZrO_2 catalysts on CO_2 reforming of CH_4 coupled with steam reforming or under high pressare. *Catal. Lett.* 90:13–21.

Index